SPERM WHALES

THE UNIVERSITY OF CHICAGO PRESS   CHICAGO AND LONDON

Hal Whitehead

SPERM WHALES
Social Evolution in the Ocean

HAL WHITEHEAD is Killam Professor in the Department of Biology at Dalhousie University. He is the author of *Voyage to the Whales* (1989) and co-editor of *Cetacean Societies: Field Studies of Dolphins and Whales* (2000, with Janet Mann, Richard C. Connor, and Peter L. Tyack), the latter published by the University of Chicago Press.

The University of Chicago Press, Chicago 60637
The University of Chicago Press, Ltd., London
© 2003 by The University of Chicago
All rights reserved. Published 2003
Printed in the United States of America
12 11 10 09 08 07 06 05 04 03   1 2 3 4 5

Library of Congress Cataloging-in-Publication Data

Whitehead, Hal.
    Sperm whales : social evolution in the ocean / Hal Whitehead.
        p.   cm.
    Includes bibliographical references and index.
    ISBN 0-226-89517-3 (cloth : alk. paper)—ISBN 0-226-89518-1 (pbk. : alk. paper)
        1. Sperm whale. I. Title.

    QL737.C435 W465 2003
    599.5'47—dc21

                                                             2002155088

⊗ The paper used in this publication meets the minimum requirements of the American National Standard for Information Sciences—Permanence of Paper for Printed Library Materials, ANSI Z39.48-1992.

To Lindy Weilgart

Contents

Acknowledgments

The writing of this book is built upon the contributions of many.

Unfortunately, studying sperm whales is costly. A number of organizations and individuals provided the funds that made our studies possible. World Wildlife Fund Netherlands began it all by sending Jonathan Gordon and me off into the Indian Ocean aboard *Tulip* between 1981 and 1984. The Natural Sciences and Engineering Research Council of Canada has provided the base funding for my research continuously since 1986, as well as grants for a number of important pieces of equipment, graduate scholarships for several of the students who did so much of the work (Mary Dillon, Susan Dufault, Kenny Richard, Linda Weilgart), a postdoctoral fellowship to Linda Weilgart, and the vital University Research Fellowship that kickstarted my research program and academic career. The Dalhousie University Isaak Walton Killam Foundation has supported many of the graduate students (Jenny Christal, Mary Dillon, Nathalie Jaquet, Luke Rendell, Kenny Richard, Linda Weilgart) who worked on the sperm whale projects, as well as awarding me, early on, a Killam Postdoctoral Fellowship and, more recently, a Killam Professorship. The National Geographic Society funded our two most ambitious South Pacific projects, and the International Whaling Commission, the Green Island Foundation, Mary Clark, Frances Whitehead, and the Connecticut Cetacean Society provided vital help during times when funding from other sources was scarce.

Especially during the early stages of my efforts to study sperm whales, I needed encouragement and assistance from more experienced scientists in planning and funding the studies. Particularly crucial were Sidney Holt and Roger Payne, who largely set up the *Tulip* project and guided it to completion. Roger Payne and the late George Nichols encouraged us to start work off the Galápagos Islands. Our research in other parts of the South Pacific has been enabled by Michael Poole (French Polynesia), Steve Dawson and Elizabeth Slooten (New Zealand), and Anelio Aguayo and Angel Crovetto (Chile).

Many crew on *Elendil-Tulip* and later *Balaena* helped collect data on the whales. I thank them all, but am especially grateful to the graduate students who spent months at sea collecting data and years in the laboratory analyzing it, originated many of our hypotheses and techniques, and in whose theses many of the results that I discuss first appeared. In chronological order, they are Jonathan Gordon, Vassili Papastavrou, Tom Arnbom, Susan Waters, Linda Weilgart, Sean Smith, Benjamin Kahn, Kenny Richard, Mary Dillon, Nathalie Jaquet, Susan Dufault, Jenny Christal, Luke Rendell, and Amanda Coakes. My ideas about sperm whale societies developed during continuing vigorous discussions with these scientists and others, especially Richard Connor, Janet Mann, and Peter Tyack, my co-editors on *Cetacean Societies.*

Godfrey Merlen has been key to the success of our Galápagos studies by keeping our engine working, skippering our boats, and collecting data from his own vessel, *Ratty.* He has also been an inspiration through his commitment to the Galápagos Islands and their conservation. The Charles Darwin Research Station has helped us greatly through all the studies around the islands. The Galápagos National Park Service, the Armada of Ecuador, and the National Institute of Fisheries sanctioned the studies and allowed them to proceed. Our work off Chile in 2000 was carried out in conjunction with Anelio Aguayo of the Chilean Antarctic Institute.

I am extremely grateful to Eberhardt Gwinner, who made it possible for me to write this book in the lovely environment of the Research Center for Ornithology of the Max Planck Foundation in Seewiesen, Germany. While there, and also at Dalhousie University, Jana Bock very kindly volunteered many hours of assistance in preparing the book. Revision of the book was also carried out at Seewiesen, this time within the group of Wolfgang Wickler.

A number of people generously reviewed parts of the book: Luke Rendell, chapters 1 and 7; Andrea Ottensmeyer, chapter 6; John Heyning, section 1.2; John Steele, section 2.1; Nathalie Jaquet, section 2.2; Jay Barlow, section 4.2; Karen Evans, section 5.10; Stefan Leitner and Linda Weilgart, section 8.3; and Sean Smith, section 2.3 and table 2.2. Two anonymous reviewers of the book proposal were most encouraging and had a number of useful suggestions. Andrew Read and Bernd Würsig constructively reviewed the entire draft for the University of Chicago Press and provided many important comments.

Michel André, Jay Barlow, Robert Brownell, Paula Canton, Karen Evans, Alexandros Frantzis, Jonathan Gordon, Rodrigo Hucke-Gaete, David Janiger, Thierry Jauniaux, Benjamin Kahn, Peter Madsen, David Mellinger, Keith Mullin, Deborah Palka, Robert Pitman, and Christophe Richter kindly sent preprints, reprints, and unpublished results. John Hampton and Valerie Alain provided information on large pelagic fish, and Ian McLaren on zooplankton and seabirds. Godfrey Merlen allowed me to reproduce his account of an encounter between sperm whales and killer whales.

Sperm whales spend their lives beneath the ocean, where they are almost invisible to us. However, National Geographic photographer Flip Nicklin has penetrated this realm, and has kindly allowed me to reproduce some of his wonderfully revealing photographs. Robert Pitman offered the photograph of the whales in Marguerite formation (fig. 5.31). But much is hidden from even the most skillful photographer. Emese Kazar kindly offered to prepare illustrations, and did so with such skill, care, and artistry that they have become central elements of the book. The illustrations of squids (figs 2.12 and 2.13) were checked by Allison King, Sean Smith, and, especially, Clyde Roper, and that of the social introduction of the newborn sperm whale (fig. 5.26) is based upon a sketch by Linda Weilgart. Luke Rendell very kindly prepared the figures depicting the sounds of the sperm whales (figs 1.7, 5.1, 5.2, 5.3, 5.18, 5.19, 5.20, 7.1, 7.2, 7.3, and 7.4). Susan Dufault, characteristically thorough and efficient, organized, arranged, and often found the illustrations.

Throughout my work on this book, Christie Henry of the University of Chicago Press has been extremely supportive and helpful. Norma Roche edited the manuscript meticulously, greatly improving it.

The whales themselves have put up with our sometimes clumsy tracking with great tolerance.

Finally, the greatest thanks to Linda Weilgart for anchoring me and our family.

Conventions

In this book I have tried to maintain the following conventions:

- Mathematical derivations are restricted to footnotes and boxes.
- The results of statistical significance tests (type of test, P value) are provided only when this information has not been previously published, and the published sources are cited when it has.
- Illustrations by the artist Emese Kazar are used to portray important parts of the sperm whale's biology and behavior for which no clear photographs exist, but there is sufficient observational or other information to produce an informative picture.
- The scientific names of species are given only when the species are first mentioned in each chapter, except for species, such as *Dosidicus gigas,* without a well-accepted common name.
- Unless otherwise specified, photographs, data, and diagrams are from the H. Whitehead laboratory, Dalhousie University.
- "Immature" is used to refer to those members of social units that are not sexually mature females.

Symbols and Abbreviations

ATOC	Acoustic Tomography of Ocean Climate program of the Scripps Institute of Oceanography
CV	Coefficient of variation (standard deviation or standard error divided by mean)
dB	Logarithmic scale of relative sound level
DNA	Deoxyribonucleic acid
DDT	Dichlorodiphenyltrichloroethane
$g(0)$	Probability that an animal on the survey line is counted
G_{ST}	A measure of the genetic variability among demes within a population (one minus the ratio of the probability that two homologous genes from the same deme are different to the probability that two genes from different demes are different)
Ma	Million years ago
Mt	Million metric tons
mtDNA	Mitochondrial DNA
n	Sample size
P	Statistical significance (probability of obtaining the given value of the statistic, or a more extreme value, if no effect was present)
PCB	Polychlorinated biphenyl
Q	Quality of photograph depicting sperm whale flukes, ranging from $Q = 1$ (very poor) to $Q = 5$ (excellent)
r	Correlation coefficient
Rms	Root-mean-square, a form of average that emphasizes large values
SD	Standard deviation
SE	Standard error

Prologue:
Observations of Sperm Whale Societies

The following observations of the social lives of sperm whales are mostly extracts from "popular" articles or books, but nearly all were written by professional scientists who have published journal articles on the same, or related, observations. Exceptions are the excerpts from a book titled *The Natural History of the Sperm Whale* by Thomas Beale, who was a whaleship surgeon, and from an article by W. D. Boyer, a deck officer on a merchant ship.

Herding, and other particulars, of the sperm whale
> The female is much smaller than the male; her size, when generally considered, being not more than one-fifth that of the adult "large whale." The females are very remarkable for attachment to their young, which they may be frequently seen urging and assisting to escape from danger with the most unceasing care and fondness. They are also not less remarkable for their strong feeling of sociality or attachment to one another; and this is carried to so great an extent, as that one female of a herd being attacked and wounded, her faithful companions will remain around her to the last moment, or until they are wounded themselves. This act of remaining by a wounded companion is called by whalers "heaving-to," and whole "schools" have been destroyed by dextrous management, when several ships have been in company, wholly from these whales possessing this remarkable disposition. The attachment appears to be reciprocal on the part of the young whales, which have been seen about the ship for hours after their parents have been killed.
> —Beale 1839, 52–53

Call me gentle [a popular account from our first Galápagos sperm whale study in 1985]
> During the weeks that we spent with sperm whales, the subjects of our research showed themselves to be gentle animals. They are usually shy but occasionally curious in the presence of humans and their boats. They show no shyness, however, with each other, displaying very

sociable behavior. Although adjacent sperm whales probably separate [by a hundred meters or so] when feeding at depth, off the Galápagos they often appeared to form a line several [kilometers] long, with the whales swimming parallel to, and roughly abreast of, one another. These ranks swept through the deep ocean at a steady rate of [about 3 to 6 km/hr], for twenty-four hours or more. Individuals surfaced about every forty minutes to breathe, but the formations advanced relentlessly.

When foraging about [450 m] beneath the surface, each individual made the characteristic, regular (about once per second) click of the sperm whale, which is almost certainly a form of echolocation. The jumble of clicks of a group of hunting sperm whales, which together sound like radio static, foretells approaching death for many squid, the whales' preferred food. But for us on board the *Elendil,* the clicks were an important key to the whales' position. We listened regularly to a directional hydrophone (an underwater microphone) and adjusted the *Elendil's* course and speed depending on the direction and volume of the clicks. With our hydrophone we could tell the bearing of a clicking sperm whale from about [7 km] away, and thus were able to track groups of whales for days at a time.

Between forty-minute feeding bouts, the sperm whales remained at the surface breathing for about eight minutes. During these periods the whales seemed irresistibly drawn to one another; if two whales surfaced within [250 m], they usually sidled up to one another for companionship during their few minutes at the surface. The small calves, which did not dive deeply, were particularly active in joining the adults.

But it was during their social times, when [clusters] of five to forty whales congregated at the surface for an hour or more, that the significance of their communal relationships was most apparent. Although from the deck of our boat the whales resembled a raft of inanimate logs, when seen beneath the surface they were revealed as extraordinarily flexible, tactile, and tender animals. Snorkeling behind *Elendil,* we saw them turn gracefully to watch us, gently stroke one another with their small flippers, or nuzzle a smooth, bulbous brow against a wrinkled flank.

—Whitehead 1986

Social behavior and family life

The breeding system is another aspect of sperm whale biology that is poorly understood. We do know that large bulls travel down into warmer waters to join female groups to breed. It had once been assumed that these males were "harem masters" controlling a social group for long periods of time; some even thought they provided the "social glue" that held the groups together. In fact, their mating strategy seems to be quite different. They travel between groups searching for receptive females and staying with each group for only a few hours at a time.

It is a striking sight to see a large male in a mixed group, they are so much larger, and their heads are so much more prominent. When visiting female groups they announce their presence by producing extremely loud resonant clicks, known as "clangs." Most surprising to me was the behavior of the members of mixed groups during these visits. I had expected these huge males to be forcing their attentions on unwilling females; what I observed under water could not have been more different. The male was the focus of intense attention from all group members, who crowded in on him, rolling themselves along his huge body. They just seemed delighted that he was there. For his part the male was all calm serenity and gentleness. Even the calves were interested and on one occasion we saw a male gently carrying a calf in its mouth.

—Gordon 1998, 22–25

Observations of a sperm whale birth

Three crew members were on deck at 0848 hours, watching a solitary adult lying stationary, 25 meters from *Tulip*. The nearest other visible whale was 350 meters away. Suddenly the nearby whale began making unusual movements, flexing at the middle to show both fluke-tips and head, and then an arched back. It repeated this body contortion, then rolled on its side, belly towards our boat. . . . To our great surprise, a rush of blood and a dark object were expelled from the genital area of the whale, at water surface level. At 0855 hours, a calf had been born next to the boat! The wrinkled, 3 to 4 meter long calf was then seen bobbing next to the mother's head as more blood was expelled from the mother. The calf had a very flattened dorsal fin, and a pitifully thin peduncle [tailstock]. It swam ineffectually and awkward in a rocking fashion, throwing its head out of the water each time it blew.

At 0900 hours, the mother and calf were joined by another adult sperm whale who began jostling and pushing the calf. Soon, two more adults came, and further shoved the poor calf about. The calf was sometimes sandwiched between the adults; at other times it was seen sliding down their backs or over their heads as it was practically tossed out of the water.

—Weilgart 1985

Terror in Black and White

Nine sperm whales have gathered to form a "rosette," their heads pointing to the center, their bodies radiating out like the spokes of a wheel. . . . The reason for this defensive formation quickly becomes apparent: three or four adult killer whales are rapidly circling just outside the rosette.

Killer whales hunt in packs like wolves, and this group may have spent decades together honing the cooperative skills necessary to bring down large prey. This morning they seem intent on breaking up the rosette and isolating individuals. During one of their sorties, a sperm whale is pulled away from the rosette and immediately set on by four or five attackers. We can see several black-and-white shapes beneath the water; the group is charging the sperm whale from both sides. Twisting their bodies and violently shaking their heads like hungry sharks, the killer whales try to wrench off mouthfuls of what must be very tough flesh. The tempo of the attack picks up, as though the killer whales sense they are gaining the advantage. The sperm whale cannot survive this punishment for long.

Then, to our astonishment, two sperm whales leave the rosette formation and approach their isolated companion. One on each side, the two begin to herd the severely injured whale back to the rosette. For a time, the killer whales redirect their attack to the escorts, then retreat once again. We see this same heroic scenario several times: one or two members of the rosette invite attack on themselves in an effort to bring one of their own back into the formation.

—Pitman and Chivers 1999

A stranding of male sperm whales on Sable Island, Nova Scotia

The group stranding event of three males in January 1997 was observed by researchers working on the island at the time. When first found, at about 09:00, two whales had just stranded, and both were active, slapping flukes and rolling. A third whale was in the distance offshore and appeared to be swimming roughly parallel to the beach. Eventually this individual moved closer to the beach, appearing to head directly toward the first two, and soon after stranded within

50 m of the others (S. Iverson, Biology Department, Dalhousie University, Halifax, personal communication).
—Lucas and Hooker 2000

Large accumulation of sperm whales

On the morning of August 28, 1945, we were northbound off Aguja Cape, Perú approximately 6°S 82°W. Shortly before 9.00 A.M. individual groups from two to six sperm whales were seen dotting the visible surface of the ocean. They were all travelling south. A short while later the number of groups increased until the entire ocean, to all visible limits of the horizon, seemed spotted with them. The sum total was a school of gigantic proportions—all headed south. It took the vessel nearly an hour to travel through the main body of the school, and the ship was proceeding north at full speed. Several times a collision with a whale was narrowly averted, as they apparently held little fear of the ship and as often as not would stay on the surface rather than sound. This afforded close inspection and positive identification of the whales.

During the remainder of the morning small groups or single straggling whales were seen, the last of them being sighted shortly after noon. It is impossible to estimate the number of whales in the school, because the east and west limits could not be ascertained. However, approximately 400 to 600 whales were to be seen at one time from the centre of the school and it can safely be assumed that the entire school consisted of well over 1,000 whales.
—Boyer 1946

1 The Sperm Whale: An Animal of Extremes

1.1 Social Complexity in the Ocean

The open sea is an environment where technical knowledge can bring little benefit and thus complex societies—and high intelligence—are contraindicated (dolphins and whales provide, maybe, a remarkable and unexplained exception).
—Humphrey 1976

Nicholas Humphrey's essay "The social function of intellect," from which this quote is taken, sparked the "Machiavellian intelligence" hypothesis, which proposes that sophisticated animal intelligence evolved as selection favored those animals that effectively exploited social complexity (Byrne and Whiten 1988). This hypothesis is now widely accepted, but the "remarkable and unexplained exception" of the cetaceans (dolphins and whales) has received little attention in writings about the evolution of intelligence and social complexity, except from a few cetologists (e.g., Connor et al. 1998) and open-minded primatologists (e.g., Whiten and Ham 1992).

This situation is changing, and cetacean science is gradually adding to our knowledge of the fabric of nonhuman social systems and cognitive abilities. However, of the eighty-five or so extant species of whales and dolphins, four have received almost all of the attention (Mann 1999): the killer whale (*Orcinus orca*), bottlenose dolphin (*Tursiops* spp.), humpback whale (*Megaptera novaeangliae*), and sperm whale (*Physeter macrocephalus*). These are the species whose behavior has been studied in greatest detail. A variety of factors draw science toward particular organisms. For the sperm whale, these factors include the largest brain on Earth, its commercial significance, and its highly social lifestyle.

As the observations in the prologue illustrate, sperm whales interact with one another conspicuously, both during their daily routine and at the critical junctures of birth, predation, and death. Some of their social behavior is unusual, and perhaps unexpected, when considered from the perspective of terrestrial mammalian socioecology. In this book, I will describe the social life of the sperm whale as quantitatively and rigorously as current data allow and relate it to the sperm whale's ecological role in the deep and open ocean, an environment structured very differently from the habitats of our terrestrial model organisms (Steele 1991). The structure and scale (see Levin 1992) of the physical, biotic, and social environments of the sperm whale and the importance of culture will be recurring issues throughout the book. First, though, in this chapter, I will summarize some other elements of the biology of the sperm whale that bear on their ecology and behavior, and I will explain my approach to the study of sperm whale society and culture. The subsequent chapters will consider the sperm whale's habitat (chapter 2), how the whales move through it (chapter 3), behave (chapter 5), and organize their societies (chapter 6), and what we know of their culture (chapter 7). At the end of each of these chapters, I will summarize those results that seem to have significant bearing on social and cultural evolution.

In a range of characteristics as diverse as brain size, nasal complexity, sexual dimorphism, ecological success, social complexity, and diving ability, sperm whales have evolved unusual or extreme states. The evolutionary and ecological links between these extremes and the animals' social ecology will be explored toward the end of the book (chapter 8). In several places I will compare sperm whales with other marine and terrestrial animals, using the comparative method informally to examine the processes of social and cultural evolution in the ocean.

1.2   Evolutionary History of the Sperm Whale

The sperm whale has had a distinct, interesting, and controversial evolutionary history, but all agree that it is of the order Cetacea, the whales and dolphins. The cetaceans evolved from ungulate-like creatures that made their way back into the oceans perhaps 60 Ma (million years ago). About 25–35 Ma, the mysticetes, or baleen whales, separated from the odontocetes, or toothed whales (Berta and Sumich 1999, 59), and here the controversy begins.

Milinkovitch et al. (1993) used results from molecular studies to make the radical suggestion that sperm whales were more closely related to the baleen whales (Mysticeti) than to the toothed whales (Odontoceti). This conclusion seemed odd, as sperm whales, unlike baleen whales but like odontocetes, have substantial teeth and a single, rather than double, blowhole. More recent morphological and molecular analyses, especially those of Heyning (1997) and Nikaido et al. (2001), have come down strongly in favor of the traditional arrangement, with the sperm whale as an odontocete. Milinko-vitch et al.'s conclusion may have resulted, at least partially, from the problem of rooting the cetacean phylogenetic tree. The morphological evidence and some recent molecular analyses (e.g., Nikaido et al. 2001) strongly indicate that the split between the mysticetes and all other extant cetaceans is the old-est division in the tree. Other molecular data are less well founded and more equivocal, with some analyses agreeing with the evidence from morphology and others supporting Milinkovitch et al.'s primacy of the division of the sperms and mysticetes from all other odontocetes (Heyning 1997).

In any case, not too long after the odontocete-mysticete division, sperm whales started to go their own way. Species possessing characteristic sperm whale features—the family Physeteridae—are found from about 25 Ma on-ward (Berta and Sumich 1999, 73), and soon evolved some highly special-ized features (Mchedlidze 2002). The sperm whales had radiated into a num-ber of different species by about 15 Ma (Rice 1998, 82; Kazar 2002), but only three survive today: the sperm whale itself (*Physeter macrocephalus*)* and the pygmy sperm whale (*Kogia breviceps*) and dwarf sperm whale (*Kogia simus*), which at 2–4 m long are much smaller. The kogiids seem to have separated from the lineage that led to the sperm whale at least 8 Ma (Berta and Sumich 1999, 73), making the sperm whale the most phylogenetically distinct of all the seventy-five-odd species of living odontocetes.

1.3   Morphology: What Does the Sperm Whale Look Like?

A sperm whale is to my fancy the most uncomely shaped animal that I can think of.
—Ellsworth 1990, 49

The cetaceans include some lovely creatures. The finback whale (*Bal-aenoptera physalus*) and right whale dolphins (*Lissodelphis* spp.) are as well-proportioned, beautifully colored, and graceful as anything in the animal kingdom. But the right whale (*Eubalaena* spp.) is a fat, lumpy animal, and Baird's beaked whale (*Berardius bairdii*) seems a misshapen spindle. And then there is the sperm whale. From just about any angle, it looks very strange.

---

* There has been some dispute as to whether *Physeter macrocephalus* or *Physeter catodon* is the cor-rect scientific name for the sperm whale (Husson and Holthuis 1974; Schevill 1986, 1987; Holthuis 1987). Both originate from the work of Linnaeus (1758), who named four species of sperm whale. However the International Code for Zoological Nomenclature, the International Whaling Commis-sion, and most scientists currently working on sperm whales prefer *P. macrocephalus*. Rice (1989, 1998) reviews the dispute, and settles firmly for *P. macrocephalus*.

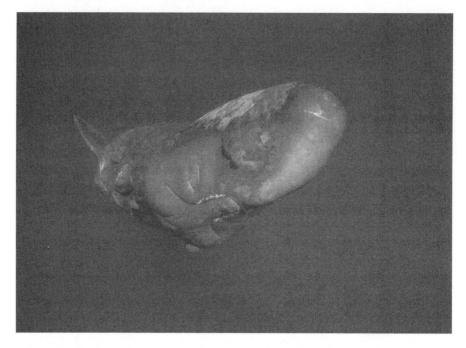

Figure 1.1 The head of an advancing sperm whale. (Photograph by F. Nicklin, courtesy of Minden Pictures.)

During its final moments, if in daylight and not too deep, a doomed cephalopod might see the sperm whale advancing formidably on a broad front from the gloom of the ocean (fig. 1.1). Coming directly toward it is the long and narrow, but powerful, lower jaw, outlined in white, which will soon open to display two rows of large conical teeth and the white lining of the folds of the mouth and tongue (the teeth in the upper jaw rarely emerge). Above, or perhaps below (the sperm whale may feed upside down), the jaw is the mass of the spermaceti organ, expanding from the narrow jaw to the "high and mighty god-like dignity inherent in the brow" (Melville 1851, 454). The front of the spermaceti organ is often scarred from encounters with past prey, other sperm whales, or other hazards of the ocean. Protruding on either side of the head are the eyes, and behind them the rest of the body fades into the murk of the ocean. Such is the vision of this devastating predator as seen by its prey.

Sperm whales usually see one another from a different perspective, in profile from the side (fig. 1.2). From this angle, the spermaceti organ and jaw are also prominent, particularly the spermaceti organ, which overhangs the jaw and extends the whole vertical profile of the animal, giving the head a boxlike appearance. The spermaceti organ is smooth and rounded, but ap-

Figure 1.2 The sperm whale in profile. (Photograph by Flip Nicklin, courtesy of Minden Pictures.)

pears to consist of two barrels stacked on top of each other, rather like a double-barreled shotgun. If seen from the left, the asymmetric S-shaped blowhole is apparent at the tip of the spermaceti organ, capping a remarkable nose. The nose is remarkable in both shape and size, for it does not end until a quarter to a third (in large males) of the body has passed. And then, at about the position of the eye, the smooth case of the spermaceti organ gives way to a deep dimpling that covers almost all of the rest of the body. At the transition between nose and trunk there is often a crease along the top of the head. This crease is particularly apparent in the largest males, whose spermaceti organs appear swollen. Behind the eye, and normally tucked in close to the body, is a small, paddle-shaped flipper. The following portion of the back is usually quite straight until we reach the dorsal fin, which is low and rounded, and is followed by a series of "knuckles" as the trunk narrows to the peduncle leading to the flukes. If the whale rotates slightly, the belly becomes visible. Just behind the jaw are a few ventral grooves (see fig. 1.1), which presumably add flexibility to the throat, allowing the whale to suck in its prey. Farther back may be large patches of white pigmentation, especially around the genital slits. There are three slits in females—the mammaries on either side of the urogenital opening—and a single long slit in males, from which the penis sometime protrudes, with the anus behind. Farther back, perhaps hidden in the murk of the ocean, is the profile of the large tail, or flukes, of the whale.

We humans viewing whales from above the surface usually see very little of this. From our perspective, the sperm whale is just a long, loglike back stretching between the offset bulge of the blowhole and the curve of the dorsal fin (fig. 1.3), perhaps indented toward its middle by the crease at the end of the spermaceti organ. But we do see, and may hear, the blow—a rather low, bushy blow for such a large whale, pointed forward and to the left (fig. 1.4). We also can get a good view of the dorsal fin, and may note a white, or whitish, callus on the top (fig. 1.5). This callus is a secondary sexual character that indicates a mature female, although not all mature females possess calluses and some immature males do (Kasuya and Ohsumi 1966; Clarke and Paliza 1994). Other parts of the sperm whale are seen only briefly from our above-water perspective (see fig. 5.4): the spermaceti organ during a spyhop, the flippers and belly during a roll, and most spectacularly and briefly, almost the entire body during a breach, or leap from the water. More frequently, during lobtails, sideflukes, or fluke-ups, only the flukes of the sperm whale emerge. When compared with those of other cetaceans, the tail of the

Figure 1.3 The sperm whale as seen from above the water surface.

Figure 1.4 The asymmetric blow of the sperm whale.

Figure 1.5 The dorsal fin of a sperm whale, showing a callus indicative of a mature female.

sperm whale is one of its least unusual external features, but the morphology of the triangular flukes is especially important for our research. On the trailing edge of the flukes is a pattern of marks and scars that can be reliably photographed as the whale raises its flukes to dive (see fig. 5.13) and used to identify individuals (see appendix, section A.3).

From snout to flukes a mature female stretches about 11 m, and she may weigh about 15 t. A mature male is very much larger, often surpassing 15 m and 45 t. Thus the sperm whale is the largest of the toothed whales, being surpassed among extant animals only by the largest baleen whales, and is the most sexually dimorphic of all cetaceans.

1.4   The Peculiar Anatomy of the Sperm Whale: The Spermaceti Organ

Beneath its thick skin, the sperm whale is a mixture of the standard cetacean anatomy and physiology (see Berta and Sumich 1999; Elsner 1999; Pabst et al. 1999 for summaries) and some unusual features specific to sperm whales (described by Berzin 1972 and Rice 1989). One of these features is the largest brain on Earth (see section 8.3), but by far the most prominent and unusual is the spermaceti organ. It takes up 25–33% of the animal's body, dominates the head, and affects the surrounding organs, including the skull, jaw, and nasal passages.

The spermaceti organ, or "case," as the whalers called it—the upper barrel of the shotgun—is long and has a shape somewhere between a barrel and a cone (fig. 1.6; see also fig. 8.1). It is covered by a muscular sheath and contains spongy tissue soaked in spermaceti oil.* Before, behind, and beneath it are other strange structures. At the wide posterior (rear) of the spermaceti organ, the skull forms a huge semicircular basin. The two are separated by an air-filled cushion, supplied by the right nasal passage. This passage also runs beneath the spermaceti organ and above another large mass of tissue, also saturated with oil, which the whalers called the "junk"—the lower barrel of the shotgun. At the narrow anterior (front) end of the spermaceti organ, the right nasal passage reaches another air sac. Air enters this distal air sac through a valvelike clapper system, the museau du singe (or "monkey's muzzle"). Meanwhile, the left nasal passage runs more directly to the blowhole, through which the whale breathes.

---

* It was a strangely mistaken view of the nature of the spermaceti oil—that it was semen—that gave the sperm whale its common name.

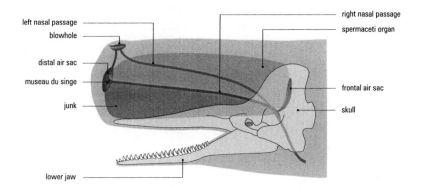

Figure 1.6 The spermaceti organ and surrounding structures. (Copyright Emese Kazár.)

As might be expected, the huge and strange spermaceti organ has attracted its share of scientific attention. Why should an animal carry around an enormous barrel of oil? Why is it partially abutted by air sacs? What is the junk? There have been a number of theories about the function of the spermaceti organ, ranging from the battering ram of Melville (1851, 443) and Carrier et al. (Carrier et al. 2002) to the buoyancy regulator of M. R. Clarke (1970, 1978). Most can be largely dismissed (see box 8.1). Instead, modern science has focused attention on the acoustic theory of Norris and Harvey (1972). It now seems that the spermaceti organ is involved in the production of echolocation clicks, but the exact mechanism by which clicks are produced is not quite clear.

The most informed speculation as to how the spermaceti organ functions is that of Cranford (1999) and Madsen (2002). Like Norris and Harvey (1972), they suggest that a pulse of sound is initially produced by air being forced through the museau du singe. The pulse then passes through the spermaceti organ until it is reflected off the air-filled cushion at its far end. The reflected pulse is partially redirected into the junk, from where it is broadcast into the ocean through a series of acoustic lenses present in the junk. However, part of the pulse makes an additional transit back and forth along the spermaceti organ, leading to a secondary pulse entering the junk, and then the environment, a little later. Some clicks contain three or more pulses (fig. 1.7). In fact, as several authors have pointed out, the inter-pulse interval is proportional to the size of the spermaceti organ and is thus a function of the length of the whale, giving rise to the possibility of acoustic measurement (Norris and Harvey 1972; Adler-Fenchel 1980; Gordon 1991a; Goold

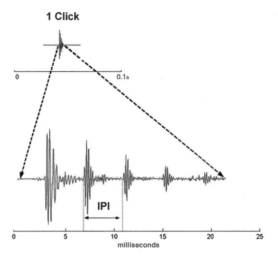

Figure 1.7 The multipulse structure of the click of the sperm whale, as shown by oscilloscope records: this is one click from a coda. The click has a duration of about 18 ms, but it is made up of five pulses with an inter-pulse interval (IPI) of 4 ms.

1996). Although the multiple-pulse structure of the sperm whale click is characteristic of the species, multiple pulses are generally heard only off the axis of this highly directionalized sound (Møhl 2001). This model of sperm whale sound production has recently been supported by an experiment in which sounds were played into the head of a recently dead whale and recorded (Møhl 2001): when the sound was made in front of the museau du singe or junk, a multipulse structure with an inter-pulse interval corresponding to the predictions of Gordon (1991a) was recorded, but when the sound was made above the head, no pulse structure was recorded.

Although the spermaceti organ is unique in size and appearance, other odontocetes make their echolocation clicks by a process homologous to that of the sperm whale (Cranford 1999). However, sperm whale clicks seem to be more powerful (Møhl et al. 2000), and to contain substantially more low-frequency (<1 kHz) energy, than those of other odontocetes (Cranford 1999). Furthermore, this energy is well directionalized (Møhl et al. 2000). It is difficult to directionalize low-frequency sound, but, because low-frequency sound attenuates little, if this can be done by using a large structure (Madsen 2002), then energy can be transmitted substantial distances, increasing the potential range of a signal. And so emerges a scenario for the evolution of this huge organ, discussed further in section 8.2.

Cranford (1999) draws attention to another feature of the spermaceti organ: it is relatively larger in males than in females. Given that mature males are so much larger anyway, this means that they probably have a much more powerful sound producer. Cranford (1999) considers the spermaceti organ to have been subject to "sexual selection on a grand scale," and some of what we know of sperm whale vocalizations supports this conclusion (see sections 5.2 and 6.11).

During its lifetime, a sperm whale may produce half a billion clicks (Madsen 2002). The sperm whale's clicks allow it to echolocate and communicate over ranges of hundreds of meters to tens of kilometers (see section 5.2). So, in its most fundamental terms, the spermaceti organ extends the sperm whale's sensory environment far out into the ocean.

1.5   The Lives and Populations of Sperm Whales

The subsequent chapters of this book describe and discuss the ecology, behavior, and social structure of the sperm whale. To put these in perspective for the reader, it may be useful to sketch the life of a sperm whale and what we know of their populations.

Following a gestation of about 15 months, a sperm whale is born into the vigorous social milieu of its mother's social unit. A newborn sperm whale weighs about a ton and is approximately 4.0 m long (Best et al. 1984). The sex ratio is about equal at birth, and, as is the case for other marine mammal species, births are almost always single. Twin sperm whale fetuses have been found very occasionally by whalers, but it is unclear whether twins survive even when their mother escapes the whalers (Best et al. 1984). Also following the universal cetacean pattern, the newborn is precocial, being able to swim after a fashion immediately following birth and reasonably effectively within a few hours (Weilgart and Whitehead 1986). The principal activities of infants seem to be breathing, keeping up with a mobile mother, and suckling. A sperm whale suckles until about 2 years of age, but this is highly variable, with animals as old as 13 years having been found with milk in their stomachs (Best et al. 1984). However, weaning is gradual, as animals begin to be found with solid food in their stomachs at about 1 year of age (Best et al. 1984).

As far as we can tell, young sperm whales stay within their mothers' social units. As these units are nomadic, the juveniles and their female elders may travel about 35,000 km during a year within a home range, measuring perhaps 1,500 km along any bearing (see section 3.4), in tropical or temperate waters. Both male and female juveniles take an active role in the social lives of their social units, but in some respects the two sexes begin to diverge. The males grow faster, and are about 9% longer than the females at age 10 (Best et al. 1984). But in their teens, with the onset of sexual maturity, the lives of the two sexes separate radically.

Females generally stay within their mother's social unit, first conceive at about age 9, and for the next decade or two give birth at an average rate of about one calf every 4–6 yr (Best et al. 1984; Rice 1989). They raise their calves communally within their social units (see section 6.9). During subsequent decades of life the birth rate drops, reaching about one calf every 15 yr for animals in their forties (Best et al. 1984). The drop in pregnancy rate seems greatest for those animals that have had the most pregnancies (Marsh and Kasuya 1986). Female sperm whales can live into their eighties, and probably sometimes reach a century, but we know little about reproduction at these older ages. Although evidence for the existence of postreproductive females is inconclusive, extrapolating from what we know about younger female sperms, as well as other mammals with similar life history characteristics, it seems likely that these older females rarely give birth (Marsh and Kasuya 1986). Instead, their energy probably goes toward assisting the other members of their social unit in ways we do not yet understand. But, it is clear that the female sperm whale is intensely social throughout her life.

The life of a male takes a very different turn when he leaves his mother's social unit. This may occur at any age between about 3 and 15. Males in their teens and twenties are found in loose "bachelor groups," often consisting of animals of about the same age. Within these groups there is little overt evidence for more sophisticated social structures, such as preferred companionships (but see section 6.10). As the males age, their groups become smaller and are found at higher latitudes, until the largest males, in their forties and older, at about 18 m long and 45 t (three times the mass of a mature female), may be seen singly near the ice edges in both hemispheres, thousands of kilometers from the warm-water haunts of the females. The males become gradually sexually mature in their teens, but do not seem to take much of an active role in breeding until their late twenties, when they make migrations from their cold-water feeding grounds to the warm-water habitat of the females. We know very little of the frequency, duration, or extent—latitudinally or longitudinally—of these migrations, although at least some males travel many thousands of kilometers (Kasuya and Miyashita 1988). The males cease growing between the ages of 35 and 60 (Rice 1989).

Of all the aspects of the sperm whale's life history, we know least about mortality. In the population models that the International Whaling Commission's Scientific Committee used to try to manage sperm whaling, annual mortalities of 0.066 for males, 0.055 for females, and 0.093 for infants were assumed (International Whaling Commission 1982). However, these figures, especially that for infants, were based on extremely shaky evidence, and we can probably get closer to the mark by looking at the mortality rates of other large mammals (see section 4.3). Although some agents of mortality have been identified (disease, stranding, predation, whaling, and collisions with vessels), we do not know their relative importance for sperm whales.

With its long life and low reproductive rate, the sperm whale is the epitome of a "*K*-selected" organism, a species in which populations are controlled mainly by competition among their members for resources (probably principally deep-water squid). Such species tend to have low rates of population increase. If we take the population parameters used by the International Whaling Commission at face value, the maximum rate of increase for a population with a stable age structure is just 0.9% per year. This figure is so abnormally low that it calls into question the International Whaling Commission's population parameters, which certainly do have problems (see section 4.3). However, sperm whale populations were, in pre-whaling times, probably fairly stable and large. With a worldwide population numbering a million or so (see section 4.4), the sperm whale had one of the largest biomasses of any mammal.

Although sperm whales can be found almost anywhere over deep water, they are particularly common in certain areas that the whalers called "grounds" (see section 2.2). The grounds of the females, which are usually, but not always, coincident with high oceanic productivity, are at tropical and temperate latitudes. Although there are seasonal peaks in reproductive activity, sperm whales are much less bound by seasonal migrations than are most large baleen whales.

The ecology, movements, and populations of sperm whales are the subjects of the next three chapters. I will consider these topics principally with the purpose of setting the scene for the evolution of social structure, which is considered in the later chapters.

1.6   Sperm Whales and Humans

Such a portentious and mysterious monster raised all my curiosity.
—Melville 1851, 98

Sperm whales were a major, and probably a dominant, element in the oceanic ecosystem for most of their recent evolutionary history, but this has changed as we humans have moved out from the land. It seems likely that indigenous people in Lamalera, Indonesia, were taking a few sperm whales in the seventeenth century (Barnes 1991), and there may have been similar small operations elsewhere, but during this era most interactions between sperms and humans were passive. As humans made increasing use of the deep ocean between the fourteenth and eighteenth centuries, the whales and oceanic voyagers noticed one another, and occasionally took evasive action, as their tracks crossed.

This relationship began to change in 1712, when Captain Hussey, cruising for right whales in the shallow waters off the southern coast of Nantucket Island, New England, was blown offshore, where he found sperm whales (Starbuck 1878, 20). He killed one and towed it ashore, and so began commercial whaling for sperm whales. With the invention of the clean-burning spermaceti oil candle and the onboard tryworks (rendering plant) in the mid-eighteenth century, sperm whaling became both profitable and efficient. New England whalers started to voyage widely after sperms, and they were followed by the British, the French, and others. By 1800, oil from sperm whales was lubricating the machines of the Industrial Revolution, and sperm whalers were moving out of the Atlantic and into the Pacific and Indian Oceans. During the first part of the nineteenth century sperm whaling continued to increase, as the vast Pacific whaling grounds were opened up and many thousands of animals were killed annually (Best 1983). The catch, still dominated by American vessels, began to level off between 1830 and 1850, then declined in the last half of the nineteenth century as sperm whale populations were reduced and other sources of oil, especially petroleum, became available.

From about 1750 to its later stages in the 1880s, the sperm whaling industry operated in the same basic manner. A stubby square-rigged sailing ship would sail through sperm whale grounds, with crew members searching visually for sperm whale blows from a crow's nest high up in the rigging. When sperms were sighted, slim whaleboats were launched and rowed (or, in later years, sailed) to their targets (fig. 1.8). A harpoon attached to a long line was thrown into the animal. The sperm whale then dragged the whaleboat until it tired and could be approached for a final lancing. The whalers towed the carcass to the ship, where the blubber and spermaceti organ were removed and their oil was rendered. There are many descriptions of this form of whaling (see box 1.1), which in many ways was a cross between a primitive and dangerous hunt and a sophisticated industry. It produced large amounts of valuable product through the complex collaborative efforts of skilled technicians (the harpooners, captains, and mates) and laborers (the

Figure 1.8 "The Stove Boat," a watercolor by an unknown whaleman (ca. 1840), portrays traditional open-boat sperm whaling of the late eighteenth and early nineteenth centuries. The humans were generally more dangerous to the whales than vice versa. (Whaling Museum, New Bedford, MA.)

crew) using specialized technology (the whaleship, whaleboats, tryworks, harpoons, etc.). In addition to being one of the most important economic endeavors of its time, eighteenth- and nineteenth-century sperm whaling had major effects on global exploration (many Pacific islands were made known to the Western world by sperm whalers), literature (see box 1.1), and the societies and ecologies of the places that the whalers visited (Ellis 2002). From the islands of the Atlantic, Pacific, and Indian Oceans, the whalers took young men for their crews and wildlife, such as Galápagos giant tortoises (*Geochelone elephantopus*), for food. They left behind diseases, non-native animals (especially rats), technology, and their genes.

Box 1.1 **The Sperm Whale in Print**

Few nonhuman animals have inspired as much writing as the sperm whale. It is not surprising that we should have written about an animal that was the basis of a vital industry, that we pursued "to the ends of the

Box 1.1 continued

Earth," that killed a fair number of us, and that is undeniably a strange beast to any member of the general public that happens upon it, as well as to the whalers and scientists who set out to hunt and study it. But, even considering these factors, the sperm whale has been a remarkable magnet for our pens, typewriters, and word processors. Much of this is due to one book, *Moby-Dick; or, The Whale,* published by Herman Melville in 1851, which has been the inspiration and subject for a large literature that followed it.

But *Moby-Dick* did not spring from a void. It was preceded by the memoirs of other whalers. For instance, Captain James Colnett published an account of his pioneering 1793–1795 voyage "to the South Atlantic and round Cape Horn into the Pacific Ocean, for the purpose of extending the spermaceti whale fisheries" (Colnett 1798), and Owen Chase described the sinking of the whaleship *Essex* by a large male sperm whale in 1820 (Chase 1963). Most significant, though, are the books on sperm whales and sperm whaling by two English whaleship surgeons of the 1830s, Thomas Beale and Frederick Bennett (Beale 1839; Bennett 1840). Beale and Bennett were good observers and, given the slow pace of nineteenth-century whaling, had plenty of time to observe. Even from today's perspective, their accounts of sperm whale biology are largely accurate.

Most of the sections on sperm whale biology in *Moby-Dick* were based on these books. However, Melville chose judiciously from the information presented by Beale and Bennett. Most significantly, incorporating Beale's (1839, 6) largely accurate description of sperm whales as "timid and inoffensive" would have made for a very different novel. Instead, Melville adopted Bennett's (1840) perspective on the sperm whale as having a "disposition to employ these weapons offensively, and in a manner at once so artful, bold, and mischievous, as to lead to its being regarded as the most dangerous . . . species of the whale tribe."

Moby-Dick was largely ignored in its day, and had no effect on subsequent descriptions of sperm whaling, such as Frank Bullen's colorful, but perhaps not very accurate, *Cruise of the Cachalot* (Bullen 1899). Recognition of Melville's masterpiece came in the early twentieth century as the type of whaling that he was describing drew to its close, and the flow of writing about it has continued until today.

There was little scientific literature on sperm whales between the time of Beale and when they started being killed by modern techniques

Box 1.1 continued

in the middle years of the twentieth century. One of the first fairly comprehensive papers was Harrison Matthews's account of the sperm whales killed off South Africa and in the Antarctic during the 1920s and 1930s (Matthews 1938). After the Second World War, modern whaling for sperm whales hit its stride, and after a lag of a few years, scientists working with the whaling industry began to publish papers and monographs stemming from operations all around the world. As might be expected of studies associated with the whaling industry and largely using dead animals, the foci of this research were population biology, life history, and to a lesser extent, diet, morphology, physiology, and anatomy. From about 1955 onward, many papers on these subjects were published by scientists working with the whaling industry in journals such as *Reports of the International Whaling Commission, Discovery Reports,* and *Scientific Reports of the Whales Research Institute of Tokyo.* In 1980, the International Whaling Commission published a special issue of its *Reports* devoted to sperm whale population biology (International Whaling Commission 1980b).

Fortunately, there have been several useful attempts to summarize this large, and sometimes inaccessible, literature. Alexander Berzin's book covers many areas of sperm whale biology in some depth (Berzin 1972); Dale Rice's chapter in the *Handbook of Marine Mammals,* although shorter, is more up-to-date and authoritative (Rice 1989). There are also useful reviews of sperm whale reproduction (Best et al. 1984) and diet (Kawakami 1980).

This scientific writing on sperm whales was very largely based upon whaling operations and carcasses, so it is not surprising that behavior is generally little mentioned. However, there are two important exceptions. D. K. Caldwell et al. (1966) made a very useful summary of what information was available on behavior, principally in the open-boat whaling literature of the nineteenth century. A chapter on social structure by Best (1979) makes good use of the twentieth-century whaling literature and is prescient about the results of studies of the social relationships of living animals that were soon to follow.

The first major studies of sperm whale behavior unconnected with the whaling industry were the studies on vocalizations published by William Schevill, William Watkins, and their colleagues (Worthington and Schevill 1957; Backus and Schevill 1966; Watkins and Schevill 1975, 1977a,b). These dry reports contrast powerfully with the contemporaneous writing of another U.S. scientist: John Lilly. Lilly's work

Box 1.1 continued

on the brains and vocalizations of captive dolphins was important and original, but it led him to speculate further and further (Samuels and Tyack 2000), until he reached the sperm whale. He believed that the abilities of the sperm whale brain "exceed those of the human regarding past and future used in computations of the current situation" (Lilly 1978, 7). Lilly may have been right, but he had no evidence for this, nor for some of his other radical pronouncements. He was so roundly condemned by most other cetologists that there was a backlash (Samuels and Tyack 2000). Scientific consideration of the intelligence of sperm whales, and other cetaceans, languished, while some scientists drew conclusions about cetacean intelligence that were almost the reverse of Lilly's, and almost equally unsupported (e.g., Wilson 1975, 474: "In intelligence the bottle-nosed dolphin probably lies somewhere between the dog and rhesus monkey"). The truth lay at sea.

In 1982, Jonathan Gordon and I started our study of the social behavior of living sperm whales in the Indian Ocean—the *Tulip* project. The first scientific papers resulting from this work were published in 1986 (Weilgart and Whitehead 1986; Whitehead and Gordon 1986), and the next year saw Jonathan's paper on the social behavior of Sri Lankan sperm whales (Gordon 1987b), as well as his remarkably wide-ranging thesis (Gordon 1987a). Also published in 1987 were the first papers from our studies off the Galápagos Islands, which began in 1985 (Arnbom 1987; Arnbom et al. 1987; Whitehead 1987; Whitehead and Arnbom 1987). These, and the subsequent publications from studies off the Galápagos and in other parts of the world, form a base for much of the material in chapters 5 and 6 of this book.

Other themes have appeared in modern scientific writing about the sperm whale. These themes include the function of the spermaceti organ (e.g., Cranford 1999), the effects of chemical pollution (e.g., Law et al. 1996), genetics (e.g., Richard et al. 1996a; Lyrholm et al. 1999), distributions (Jaquet 1996b), and the properties of sperm whale myoglobin (e.g., Garner et al. 2001).

Readers who want a broader perspective on the biology and conservation of cetaceans and other marine mammals have been well served in recent years. Available books include *Conservation of Whales and Dolphins* (Simmonds and Hutchinson 1996), *Marine Mammals: Evolutionary Biology* (Berta and Sumich 1999), *Biology of Marine Mammals* (Reynolds and Rommel 1999), *Conservation and Management of Marine Mammals* (Twiss and Reeves 1999), *Cetacean Societies* (Mann et al.

Box 1.1 continued

2000), the *Encyclopedia of Marine Mammals* (Perrin et al. 2002), *Marine Mammal Biology: An Evolutionary Approach* (Hoelzel 2002), and *Marine Mammals: Biology and Conservation* (Evans and Raga 2002). Except for the last book on this list, which I have not yet seen, all these books contain substantial reference to sperm whales and have been useful in preparing this book.

Returning specifically to sperm whales, both Jonathan Gordon's book for nonspecialist readers (Gordon 1998) and my personal account of the *Tulip* project (Whitehead 1989b) try to give a feel for what it is like to take part in the new research on living sperm whales from sailing boats. These experiences can perhaps be compared with the accounts of the whalers who sailed the same waters seeking the same species, but with a different goal, a century or two earlier. Other contemporary writers also return to themes of earlier centuries. Alison Baird retells *Moby-Dick* from the whale's perspective (Baird 1999), and Nathaniel Philbrick's bestseller *In the Heart of the Sea* reexamines the events surrounding the sinking of the whaleship *Essex* (Philbrick 2000).

The whaling industry naturally also affected sperm whales, but to what degree is debated (Best 1983). The sperm whale's social system means that the killing of a whale may mean more than just a reduction of the population size by one. When social systems are fragmented and cultural knowledge lost, the remaining animals may have lower birth and survival rates, as is found with similarly structured elephant populations (McComb et al. 2001) and for which there is some evidence in sperm whales (Whitehead et al. 1997).

Between 1880 and 1946, sperm whales received something of a reprieve (fig. 1.9). There was still a little open-boat pelagic whaling at that time; the famous whaler *Charles W. Morgan* made her final return to New Bedford in 1921 with 700 barrels of sperm whale oil (Church 1938, 20), and the last such voyage was in 1925 (Rice 1989). A few shore-based sperm whaling operations, offshoots of pelagic open-boat whaling, continued in the Azores, Madeira, and the West Indies, but these were fairly small-scale operations (Rice 1989).

The modern whaling industry, in which steam-powered catcher vessels replaced rowed or sailed whaleboats and large guns delivered heavy, and often explosive, harpoons, was started in the late nineteenth century, but initially its main targets were the large baleen whales—particularly blue (*Balaenop-*

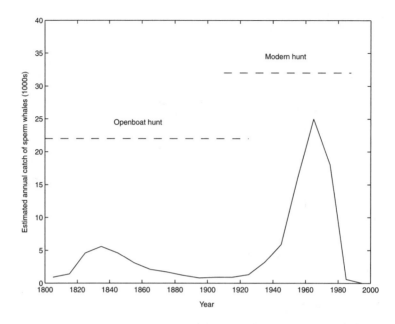

Figure 1.9 Sperm whale catches between 1800 and 1999. The open-boat hunt began in 1712, and catches from it may be underestimated here (Best 1983), as may be the modern catch record (e.g., Zemsky et al. 1995). (From Best 1983 and FAO Fisheries Department, Fishery Information, Data and Statistics Unit, FISHSTAT Plus: Universal software for fishery statistical time series, Version 2.3, 2000.)

tera musculus), fin, and humpback whales. With the development in 1924 of floating factory ships with stern slipways by which whale carcasses could be hauled on deck (Tønnessen and Johnsen 1982, 354), modern whaling became extremely efficient and mobile. Sperm whales were taken by these modern techniques, but this catch was largely an adjunct to the baleen whale hunt. Annual takes were usually on the order 1,000–2,000 until the end of the Second World War (see fig. 1.9).

But in 1946, after humans had laid down their arms, they took up their harpoons in earnest. The large baleen whales were soon down to relict populations, and sperm whales, generally viewed as unpalatable except in Japan and Indonesia, found new commercial applications in cosmetics, soap, and machine oil (Rice 1989). Catches rose through the 1950s and into the 1960s, with a record catch of 29,255 in 1964 (Rice 1989). During the late 1960s and 1970s, the management of whaling operations became more serious, and the catch fell as concern grew about the effects of modern whaling and our ability to manage it (see fig. 1.9). Sperm whale populations were given full protection by the International Whaling Commission in 1985, although Japan continued its operations in the western North Pacific until 1988.

The population consequences of modern whaling for sperm whales are even less well understood than those of open-boat whaling. A number of capable scientists and technicians collected data on the operations of the whaling industry and examined sperm whale carcasses. These data were fed into sophisticated mathematical and computer models by members of the International Whaling Commission's Scientific Committee, which included some of the world's best biometricians (see, for instance, International Whaling Commission 1980b). But all this effort largely failed to achieve its principal goals: estimating sperm whale population sizes and the effects of the whaling industry. Several interacting factors contributed to this failure, including misreporting of data, both on the gross scale of numbers of animals killed and on details such as animal lengths and pregnancies (Best 1989; Zemsky et al. 1995; Kasuya 1999), difficulties in defining discrete "stocks" to be managed (see section 4.1), lack of knowledge about sperm whale behavior, and technical problems with the assessment methods used (e.g., de la Mare and Cooke 1985; Cooke 1986b).

During the second half of the twentieth century, the relationship between sperm whales and humans evolved from that of a prey and its predator (or a resource and its harvester, as the whalers often liked to put it) into something very different. This shift began with the scientists who went out on whaling vessels or scouting boats to observe the whales, but principally examined carcasses on flensing platforms, where the animals were butchered. They compiled an impressive body of knowledge on many aspects of sperm whale biology (for summaries see Berzin 1972 and Rice 1989), but, with a few exceptions (e.g., Nishiwaki 1962; Ohsumi 1971; Best 1979), gained no knowledge of behavior or social structure from their brief and unnatural encounters with living animals.

One aspect of sperm whale biology, however, was receiving attention from scientists unconnected with the whaling industry: sounds. The first study of sperm whale vocalizations, which could be recorded through hydrophones, was published in 1957 (Worthington and Schevill 1957). This was the start of a series of important studies made by William Watkins, William Schevill, and their colleagues at Woods Hole Oceanographic Institution, who used arrays of hydrophones to document and interpret the sounds of sperm whales and other marine mammal species that they encountered at sea.

This work, like the first scientific accounts of sperm whale social behavior (e.g., Nishiwaki 1962), were "snapshot" studies, describing what was happening during recordings or observations made over short time windows. However, during the 1960s, with the work of Jane Goodall on the chimpanzees (*Pan troglodytes*) at Gombe, Tanzania (Goodall 1986), and other dedicated research efforts, the biological community began to realize the value of long-term studies of individually identifiable animals. Although the ocean is a more difficult environment for such work than even the African forests, a few farsighted cetologists recognized that if individuals could be followed, or at least repeatedly identified, over long periods, this would open a whole new field of insight into social relationships and social structure as well as populations, migrations, and ecology. The first scientists to seriously try such an approach were Roger and Katy Payne in their studies of right whales at Peninsula Valdes, Argentina. The Paynes and their colleagues identified and measured the whales photographically, watched them from cliffs, light aircraft, and boats, and listened to them through hydrophones (Payne 1983, 1995). By the late 1970s, long-term research on individually identifiable animals was blossoming in a variety of coastal locations. Humpback, right, and killer whales were being studied in some detail and several other species in a more preliminary fashion (see Samuels and Tyack 2000 for a review).

But not sperm whales. The prevailing opinion was that sperm whales lived too far from shore for long-term studies using small vessels to be feasible or economical, that they dived for too long to be visually trackable, and that there were too many of them for individual photographic identification to be practicable. As S. Ohsumi wrote in 1971, "Long time of observation will be taken to identify the same school clearly. . . . However, such an observation is almost impracticable technically and economically."

A corollary of such arguments was that a whaling industry was necessary if we were to understand sperm whale social structure. In 1981, World Wildlife Fund Netherlands decided to challenge this view. With money raised at the opening of a flower auction hall, they sent Jonathan Gordon and me into the newly created Indian Ocean Whale Sanctuary to study living sperm whales. In 1982, using the 10-meter ocean-going sloop *Elendil* (renamed *Tulip* in honor of the Dutch origins of the study), we started tracking and identifying sperm whales off the island of Sri Lanka (fig. 1.10). When the project ended in 1984, we had made a start on developing a suite of methods for long-term research on individually identified sperm whales (Whitehead and Gordon 1986; see the appendix for a summary of methods), and Jonathan and I were both fascinated with the challenge of studying this species. While Sri Lanka was in many ways a good place for such work, the developing civil war on the island and its distance from our bases in Europe

Figure 1.10 *Tulip,* a 10 m ocean-going sloop used in the first long-term studies of living sperm whales off Sri Lanka in 1982–1984, here seen leaving Greece in late 1981.

and North America made us look for more convenient research locations. Jonathan started work in the Azores (Gordon and Steiner 1992), and later the West Indies (Gordon et al. 1998), while, together with colleagues Tom Arnbom, Vassili Papastavrou, and Linda Weilgart, I began a long-term study of the sperm whales off the Galápagos Islands in 1985. This work later broadened geographically to include other waters of the South Pacific, especially those off mainland Ecuador and northern Chile.

We were followed by others who started studies of individually identified sperm whales in a variety of parts of the globe. The longest and most detailed have been those of males at high latitudes, especially off Andenes in Norway (e.g., Ciano and Huele 2000) and Kaikoura, New Zealand (e.g., Jaquet et al. 2000). Although there have been other studies of living female sperm whales at lower latitudes (e.g., Kahn 1991; André 1997; Jaquet and Gendron 2002), rather disappointingly, except for our work in the eastern tropical Pacific and Jonathan Gordon's off the Azores, none have yet fully developed into long-term sustainable research projects of individually identifiable animals of the type that have proliferated on bottlenose dolphins (Connor et al. 2000b) and humpback whales (Clapham 2000). However, there are prospects for such projects in a few areas, including Greece, the Canary Islands, and the Gulf of California, where several research seasons have been completed.

That long-term research on living sperm whales has not grown to the extent of that on killer whales, humpback whales, right whales, and bottlenose dolphins is, to a large extent, a result of the expense inherent in studying mobile oceanic animals over long time periods. The success of the Andenes and Kaikoura operations is due partially to the fact that in these locations animals consistently come close to shore and partially to synergistic relationships with the whale-watching industries that have developed there: scientists may either use the whale-watch vessels as platforms to obtain data, as at Andenes, or may receive funding from government agencies that need data in their attempts to regulate the industry sensibly, as at Kaikoura. Sperm whale watching has spread to other locations, including the Azores, Dominica in the West Indies, and the Gulf of St. Lawrence (International Fund for Animal Welfare 1996). Quite a few humans have now been able to watch sperm whales leading their normal lives in a portion of their natural habitat; many others can do so vicariously through films and television programs.* As a result, our collective conception of these animals has changed radically, from the raw material for fine candles and the "monstrously ferocious" *Moby-Dick* to fascinating animals, in some ways (such as their dives) very strange to us and in others (their mammalian suckling behavior) quite familiar.

Whale-watching, filming, and greater public appreciation have promoted sperm whale science in a variety of different ways, providing opportunities, platforms, funds, and an overall supportive atmosphere. It is not only the long-term research on identified individuals that has benefited. There have been important studies using new and sophisticated acoustic methods (Goold and Jones 1995; Møhl et al. 2000), genetic techniques (Lyrholm et al. 1996; Richard et al. 1996a), and high-tech tags (Watkins et al. 1999). In this book I use the results from such studies as well as those from the old whaling-based work to fill out the picture of sperm whale society that is emerging from long-term research on living animals.

---

* Such programs include National Geographic's *Sea Monsters: Search for the Giant Squid* (1998), Dieter Paulman's *On the Tracks of Moby Dick* (1992), and Rick Rosenthal's *Sperm Whales: Back from the Abyss* (1996).

Not all modern developments have been positive, however. As I will discuss in section 9.2, sperm whales are threatened by increasing levels of chemical pollutants, fishing gear, shipping, garbage, and noise in their habitat. And, in 2000, the Japanese resumed sperm whaling, ostensibly "for scientific purposes."

1.7   What Is Social Structure, and How Is It Studied in Sperm Whales?

"Society," "social structure," and "social organization" are terms that are used freely, loosely, and fairly interchangeably in common speech or writing to refer to emergent properties arising from the interactions between members of a population. Scientifically, we must be more precise. There have been a number of attempts to define these terms, some giving them separate meanings and some making them synonymous. I find the most useful to be that of Hinde (1976).

Hinde suggested that social structures are based on *interactions* between individuals, and that the *relationship* between any two animals could be considered the content, quality, and temporal patterning of these interactions. The *social structure* of a population, then, is the content, quality, and temporal patterning of the dyadic (or other) relationships it contains. So, to study animal societies, we must first examine interactions, use these to describe relationships, and then build a model of social structure. However, for sperm whales and many other animals that are hard to study, interactions cannot be systematically observed. Instead, we use *associations,* observations of animals in circumstances in which interactions are likely to occur (Whitehead 1997a).

Thus, the fundamental elements of our study of sperm whale social structure are methods for finding and following the animals (passive acoustic and visual location and tracking); a way of identifying individuals (photoidentification); techniques for classifying animals into age-sex classes (sexual size dimorphism with visual or photographic estimates of length and genetic sexing) and estimating their genetic relatedness (molecular techniques); measures of association (spatial or temporal proximity based on observations of behavior and vocalizations); and techniques for modeling social structure (lagged association rates, association permutation tests). These techniques are described in the appendix. Important additional information is drawn from observations of rare but important events (see the prologue) and inferences from life history, anatomy, and other fields of sperm whale biology.

This is the approach used in my attempt to describe sperm whale social structure in chapter 6.

1.8   What Is Culture, and How Is It Studied in Sperm Whales?

Culture can be defined in many ways (see Rendell and Whitehead 2001b). From an evolutionary perspective, a useful formulation of culture is "information acquired from members of the same species through some form of social learning which causes similarities in behavior among members of a population or sub-population" (Rendell and Whitehead 2001a). Cultural inheritance, then, is a complement to genetic inheritance—a conclusion that leads to theoretical investigations of the properties of a dual-inheritance system (Cavalli-Sforza and Feldman 1981; Boyd and Richerson 1985). Cultural inheritance has some similarities to genetic inheritance—for instance, it can lead to evolution through natural selection—but also some differences. The most obvious are that an animal can receive culture from a broader range of conspecifics than the two parents who provided its genes, that acquired cultural characteristics can be passed on, and that culture can evolve much more quickly than genes. These attributes have important consequences for the evolution of cultural transmission and the evolution of particular cultures, as well as for gene-culture coevolution (e.g., Cavalli-Sforza and Feldman 1981; Boyd and Richerson 1985).

Although most theoretical work on cultural evolution has focused on humans, there is a growing recognition that culture is not an exclusive property of humans (e.g., de Waal 1999). There have been two principal approaches to the study of nonhuman culture, one focusing on transmission mechanisms, usually imitation,* largely through experimental manipulation, and the other based on ethnography: the examination of patterns of behavioral similarity and variation between individuals that cannot be explained by genetics or ecological differences plus individual learning. One of the most prominent examples of the latter approach is the catalog of chimpanzee cultures produced by Whiten et al. (1999). With an animal that is as hard to

---

* Some scientists, and especially experimental psychologists, restrict culture to information or behavior transmitted through imitation or teaching. Thus, as teaching is widely considered to be very rare in nonhuman animals (Caro and Hauser 1992), there has been a focus on imitation as the root of culture (e.g., Galef 1992; Tomasello 1994). In contrast, other scientists, including some psychologists (e.g., Whiten and Ham 1992), note that imitation itself does not have an agreed-upon definition, is not necessarily more sophisticated than other forms of social learning such as goal emulation, and has not been shown to be used in the transmission of many elements of human culture (see Rendell and Whitehead 2001a).

manipulate as the sperm whale, I am largely restricted to the ethnographic approach, looking at behavioral differences between places, times, and, especially, social entities.

To establish that a behavioral pattern is culturally determined, I must rule out both genetic determination and individual learning under different ecological conditions. Two elements of sperm whale society make the study of culture easier than it is for some terrestrial species, but one makes it harder. The first advantage is that sperm whale social units may contain several unrelated matrilines (Mesnick et al. 1999; Mesnick 2001; see section 6.6), so genetic determination is unlikely to explain similarities in the behavior of members of units or differences between them. Second, many units are sympatric, often intermingling to form groups, so differences between units are probably not the result of different ecological circumstances. But this intermingling is also a disadvantage: it makes it hard to assign a behavioral pattern observed at sea to a unique social unit.

My initial approach when looking for sperm whale culture was to examine the behavior of different sperm whale social units, looking for elements that are consistent over time and that vary between social units. This approach restricts the elements of sperm whale culture that can be studied. For instance, cultures that are specific to different places and cultural differences within social units are missed. However, the approach has generated some unexpected results, and in particular, has revealed a new level of population and social organization: the acoustic clan (see section 7.2).

1.9   What Do We Know and What Don't We Know about Sperm Whale Societies?

There is much we do not know about sperm whales and, in particular, their social behavior. Readers may be disappointed with the lack of information on interactions between individuals. Unfortunately, human eyes and artificial sensors have so far given us little more than a glimpse, and not a very systematic glimpse, of the second-to-second details of the animals' social lives. We have not yet worked out appropriate ways of collecting behavioral data from sperm whales in some circumstances. On 8 April 2000, off the coast of Chile, our boat was surrounded by at least twenty sperm whales socializing actively in many small clusters. The clusters formed and split; the whales rolled around one another; we were surrounded by vigorous behavior of many types; and we heard strange sounds through the air and water. But we had no way to record this diversity of fascinating behavior in any scientifically useful manner. The brightest area is that of vocalizations, for which we can get a clear record of the sounds made by a particular group, although we can rarely say who made which sound.

This lack of small-scale observations is partially offset, however, by multi-year data sets indicating the social relationships of individual animals over large spatial and temporal scales, with studies spanning over 15 years and covering the 11,000 km width of the Pacific. These data principally come from our studies off the Galápagos Islands and in other parts of the South Pacific. However, recent research by scientists working in other areas with different goals, protocols, and techniques is producing many useful results. From these sources, as well as data collected in conjunction with the whaling industry, I will construct a model of the social structure of the sperm whale (see chapter 6). Like all models, it will be wrong in detail, but, I hope, correct enough in its fundamentals that it can be used as a focus around which to consider social evolution in the peculiar habitat of the deep and open ocean (see chapter 8).

An important caveat to virtually all our data on sperm whale social structure is that they come from populations that have been affected by whaling. We can only speculate as to the effects of whaling on sperm whale societies. In the final chapter (chapter 9), I consider these uncertainties, as well as the future of this most remarkable animal in the face of past exploitation and growing indirect threats from humans.

1.10 Summary

1. The sperm whale is the most phylogenetically distinct of all the seventy-five-odd species of odontocetes.

2. The sperm whale is the largest of the odontocetes, and possesses the largest brain on Earth.

3. In the sperm whale, the most sexually dimorphic cetacean species, mature males are one-and-a-half times the length and three times the mass of females.

4. The spermaceti organ, which takes up 25–33% of the sperm whale's body, is an extremely powerful, highly directional sonar system.

5. Female sperm whales live a highly social life, largely separate from the much more solitary males. Reproductive rates are very low.

6. Two large-scale, worldwide hunts targeting sperm whales, which peaked in 1750–1850 and in 1945–1980, have been the subjects of a large volume of popular and scientific literature.

7. I define associations between sperm whales based on photographic identifications of pairs of known individuals in circumstances in which interaction between them is likely to occur. Records of associations are used to measure dyadic relationships, which in turn are the foundations for a model of social structure.

8. In searching for sperm whale culture, I look for patterns of behavior across social entities that are consistent over time and vary among sympatric groups.

9. Our model of sperm whale social structure is short on small-scale detail, but draws from temporally and spatially extensive data sets.

2 The Oceanic Habitat of the Sperm Whale

They are also seldom or never seen on "soundings," that is where the bottom of the sea can be touched by the deepest sea-line.
—Beale 1839, 189

2.1 Spatio-Temporal Structure of the Ocean

The sperm whale is an animal of the depths, but its habitat is much more than a tall water column. The horizontal and temporal aspects of its open-ocean environment are also remarkable. Vast, fluid, dense, and connected, the ocean presents challenges to its inhabitants quite different from those faced by organisms of terrestrial habitats. There is no animal that is ecologically successful in both realms, except modern humans. For scientists, too, the pelagic ocean* and the land have become dichotomous. Oceanographers use the differential equations of fluid dynamics to describe a habitat, while terrestrial landscape ecologists plot two-dimensional resource and organism distributions with Geographic Information Systems, and their results are rarely comparable. However, John Steele has contrasted some of the most fundamental attributes of terrestrial and pelagic systems (Steele 1985, 1991). In the following paragraphs I summarize his arguments.

Over large time scales (a century or more), physical measures, such as temperature, show a pattern of "red noise"; that is, the variance in the measure, after regular cycles (diurnal, lunar, and seasonal) are removed, is roughly proportional to the square of the time scale being considered (fig. 2.1A). Atmospheric-oceanic coupling causes similar "red" patterns in terrestrial and in marine habitats over centuries and millennia. Over scales of less than about a hundred years, however, terrestrial variance levels off—a pattern described as "white noise"—while that in the ocean remains red, continuing to fall as time scales become smaller. Thus, on land, physical conditions can be predicted with about equal accuracy over short and medium time scales, up to about a century. In contrast, while the ocean environment is extremely stable over short time scales (because of thermal buffering and for other reasons), it becomes less and less predictable as time lags increase from a month to a year to decades (Steele 1985; Boyd 2001). Phenomena such as El Niño (techni-

* "Pelagic" refers to the open ocean, away from land and the bottom. The "mesopelagic" zone consists of the waters 200–1,000 m below the surface and the "bathypelagic" zone, those roughly 1–3 km deep. Thus the sperm whale is principally a mesopelagic predator.

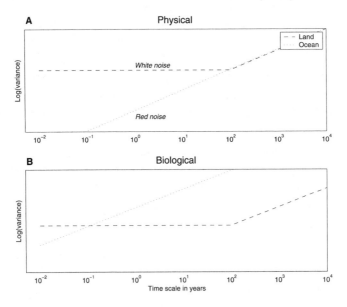

Figure 2.1 (A) Scales of variation in physical measures, such as temperature, in terrestrial and marine systems (based upon Steele 1985). (B) Scales of variation in biological measures, such as biomass of any species in any area, for terrestrial and marine systems (based on a liberal interpretation of Steele's arguments—the time scale at which the amounts of variation in the two systems cross is particularly uncertain).

cally, the "El Niño-Southern Oscillation" phenomenon), the Pacific Decadal Oscillation, and the North Atlantic Oscillation make dramatic long-term changes in physical and biological oceanographic features over huge areas of ocean (McKinnell et al. 2001).

In horizontal space, there are similar contrasts. Considerable variation in terrestrial habitats is caused by plants, rocks, and puddles over small spatial scales and by climatic differences and geological processes over larger scales. In contrast, environmental variation in the ocean, with a few important exceptions such as oceanic fronts, is normally very small at horizontal ranges of meters, or even a few kilometers, but on a scale of hundreds of kilometers, water masses differ, often quite dramatically.

What do these differences mean for organisms? While short-lived and long-lived terrestrial organisms face similar levels of unpredictable environmental variation, that is not the case for marine organisms. Small, short-lived organisms, which constitute the great bulk of biomass at low trophic levels in the ocean, generally live in remarkably uniform physical environments, which favor physiological and ecological specialization. However, because of variation at larger scales, the conditions that a particular organism experiences in any place during its short lifetime may change dramatically over succeeding generations, leaving its descendants with unfavorable conditions. Small oceanic organisms often deal with such large-scale temporal variation by long-distance dispersal of early life stages, thus taking advantage of large-scale spatial variation and the relatively predictable elements of the physical ocean, such as currents.

This specialization-plus-dispersal strategy and the short life histories of so many of the organisms near the base of marine food webs means that their populations are particularly variable, changing by many orders of magnitude over scales of months, years, or decades. By contrast, terrestrial populations are either reasonably stable, vary in predictable ways with seasonal cycles (as do many insects and plants), or are driven into fairly regular cycles by features of ecosystem dynamics (as in the spruce budworm and tree defoliation cycle; Steele 1985). Consequently, marine ecosystems are much more likely than their terrestrial counterparts to experience switches between alternate ecological states, sometimes called "regime shifts" (Benson and Trites 2002). Examples include switches between haddock (*Melanogrammus aeglefinus*) and herring (*Clupea harengus*) as the dominant fish species (Steele 1985), or, more generally, between piscivores and planktivores (Benson and Trites 2002). So, ironically, a physical environment that is stable over small spatial and temporal scales has spawned a highly unstable biotic environment over larger scales (fig. 2.1B).

A large, long-lived, slowly reproducing animal living in the open ocean, such as the sperm whale, is thus faced with surviving, reproducing, and competing in a habitat that is particularly unpredictable, compared with the terrestrial habitat its ancestors left, over time scales of much less than a lifetime. Unlike smaller marine animals, it cannot use high reproductive rates and passive dispersal to overcome downturns in its environment. How, then, does the sperm whale survive in such a capricious environment, and how did it prosper before the arrival of humans? The answer, I think, involves a combination of two attributes:

> *Large size.* A larger animal generally has a higher ratio of energy storage capacity (which is approximately proportional to body mass) to metabolic rate (which is approximately proportional to body mass$^{0.75}$; Kleiber 1975), and so can survive starvation for longer, than its smaller neighbor. A female sperm whale should be able to live on the lipid stores in her blubber for about 3 months (Whitehead 1996b), whereas a smaller cetacean may be able to do so for only a week or two.

Active movement. In chapter 3, I show that sperm whales make substantial horizontal movements and suggest that these movements are adaptive in allowing the whales to survive pronounced downturns in the environment in any particular place.

2.2   Oceanographic Habitat of the Sperm Whale

Latitude and Temperature

"The spermaceti whale has been seen," wrote Beale in 1839, "and even captured, in almost every part of the ocean between the latitude of 60° south and 60° north" (188). Beale understated the geographic range of the sperm whale by a full 10° (1,000 km) in each hemisphere (Gosho et al. 1984). It is one of the most widely distributed animals on Earth, rivaled among mammals only by killer whales (*Orcinus orca*) and modern humans. Sperm whales are found near the ice edge at both poles and at all latitudes in between (fig. 2.2).* They have made their way into deep semi-enclosed seas, including the Gulf of Mexico, the Sea of Cortez, the Sea of Japan, and the Mediterranean, but not, as far as I know, the Black Sea or the Red Sea, both of which have shallow entrances (Rice 1989).

The two sexes have strikingly different distributions (Gulland 1974), with females using just a subset of the waters where males are regularly found. Temperature marks the clearest boundary for females: they are normally restricted to areas with sea surface temperatures greater than about 15°C, whereas males, and especially the largest mature males, can be found in waters bordering pack ice with temperatures close to 0°C (Rice 1989). Latitude is closely related to mean sea surface temperature, so that the thermal limits on female distribution correspond approximately to the 40° parallels (50° in the North Pacific) (see fig. 2.2).

Depth

The distributions of the sexes are also distinguished by water depth, the most characteristic attribute of sperm whale habitat (Caldwell et al. 1966). Females rarely enter the shallow waters above continental shelves. The groups

* For detailed information on the distribution of sperm whales around the world, see the descriptions by Berzin (1972, 152–78) and Rice (1989), as well as the charts of Maury (1851) and Townsend (1935; see fig. 2.8).

Distribution of sperm whales

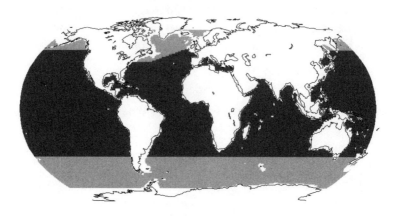

Figure 2.2 Approximate distribution of sperm whales. Dark shading represents the general habitat of females and young animals; light shading represents the higher-latitude regions used only by mature males. (Data from Rice 1989.)

that we followed off the Galápagos Islands spent only about 1.6% of their time in waters less than a kilometer deep, although there are extensive shallower areas very close to the waters that they were moving through (figs. 2.3, 2.4, and 2.5). Similarly, off Chile, the groups of females that we followed were found in water less than 1 km deep only about 0.3% of the time (figs. 2.6 and 2.7). Likewise, in most other low-latitude areas, sperm whales rarely move onto the continental shelf, although, as off the Galápagos, they may be found along steep drop-offs (e.g., Clarke 1956; Gordon 1987a, 15–17; André 1997, 99). An exception is the Sea of Cortez, where 34% of Jaquet and Gendron's (2002) encounters with groups of females and immatures were in waters less than 1 km deep. However, in water deeper than a kilometer, bathymetry does not seem to be an important direct determinant of the distribution of female sperm whales (Gordon et al. 1998). Off the Galápagos Islands, the time spent in water of any depth greater than 1 km is roughly proportional to the amount of habitat present at that depth (see fig. 2.5), and we have tracked sperm whales through some of the deepest waters in the eastern Pacific (>8 km) off northern Chile (see fig. 2.7). It seems that female sperm whales do not often feed on the bottom, and thus that water depths below about 1 km are of only secondary consequence to them.

Male sperm whales, too, are largely creatures of the deep, but, like sea surface temperature and latitude, depth is less of a constraint for them than it is for females. Off British Columbia and South Africa, male sperm whales are

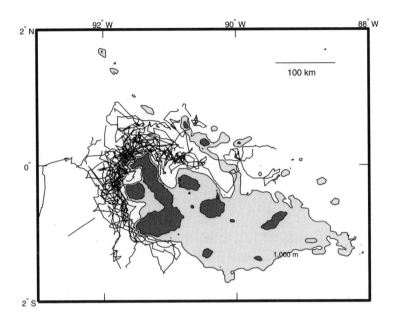

Figure 2.3 Tracks taken while following groups of sperm whales off the Galápagos Islands, 1985–1995. Positions from SATNAV or GPS are joined if less than or equal to 6 hours apart. The landmass of the islands is represented by dark shading; water less than 1 km deep is represented by light shading.

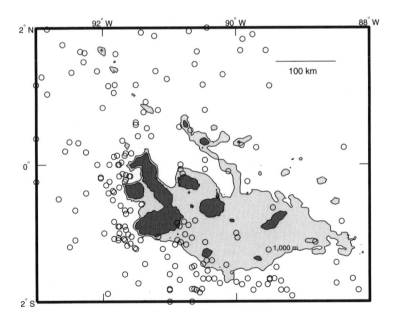

Figure 2.4 Recorded positions of whalers working off the Galápagos Islands between 1830 and 1850 on days when sperm whales were sighted. The landmass of the islands is represented by dark shading; water less than 1 km deep is represented by light shading. The accuracy of the whalers' navigation was variable, with errors of up to 100 km occurring (see Hope and Whitehead 1991 for details).

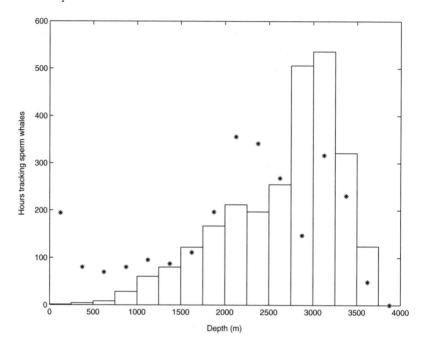

Figure 2.5 Distribution of depths used by groups of sperm whales tracked during studies between 1985 and 1995 off the Galápagos Islands (bars), together with the distribution of depths in the general Galápagos area (2° S–2° N, 88°–93° W; asterisks; from bathymetry of William Chadwick, Oregon State University).

quite consistently found closer to shore, and thus in shallower waters, than females (Best 1999; Gregr et al. 2000). In some parts of the world, such as the western North Atlantic (Mitchell 1975; Whitehead et al. 1992), males regularly use waters as shallow as 200 m. Most notably, off Long Island, New York, male sperm whales have been repeatedly found in waters 41–68 m deep (Scott and Sadove 1997).

The Grounds

Although sperm whales may use just about any ice-free deep water, they are most common in certain geographic areas, which the whalers called "grounds" (fig. 2.8). These grounds tend to be a few hundred to a few thousand kilometers across. Sperm whale grounds are usually, but not universally, areas of high primary and secondary productivity (Gulland 1974; Jaquet and Whitehead 1996; Jaquet et al. 1996) (fig. 2.9). For instance, historic sperm whale abundance, primary productivity, and secondary productivity are all

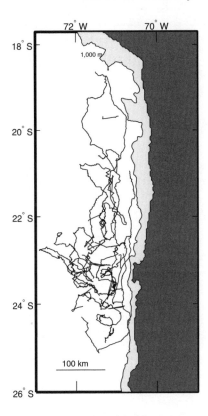

Figure 2.6 Tracks taken while following groups
of sperm whales off northern Chile in 2000.
The landmass of South America is represented
by dark shading, and water less than 1 km deep
is represented by light shading.

high along the equator in the Pacific and in the Humboldt Current off the
west coast of South America, including the waters shown in figure 2.6. However, sperm whales were also caught, and are currently found, in some relatively unproductive waters, such as the "Charleston Grounds" in the Sar-
gasso Sea, between the southeastern United States and Bermuda, and near
the Society Islands, including Tahiti, in the central Pacific (Jaquet 1996b).
The habits of sperm whale prey—largely mesopelagic cephalopods (see section 2.3)—may account for such discrepancies; for instance, sperm whales
might concentrate to feed on squid spawning in unproductive waters (Jaquet
et al. 1996). The distributions of male sperm whales at high latitudes seem
more uniform and less well delineated into grounds that those of the females,
but there are some generally preferred areas (Gosho et al. 1984).

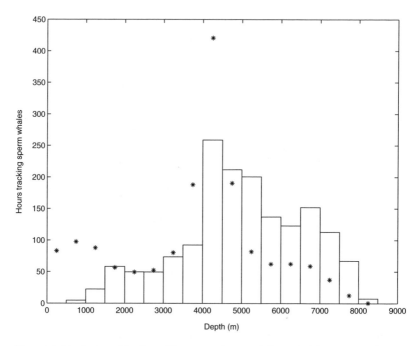

Figure 2.7 Distribution of depths used by groups of sperm whales tracked during 2000 off northern Chile (bars), together with the distribution of depths in the general area (15°–26°45′ S, 70°–73°22′ W; asterisks).

The Oceanography of Sperm Whale Distribution

A number of scientists have searched for oceanographic correlates of sperm whale distributions over scales smaller than the ground, and have come up with general confusion (Jaquet 1996b). For instance, Berzin (1972, 184–89) suggested that sperm whales concentrate at downwellings where currents converge and water sinks, bringing nutrients to the deeper parts of the ocean and so fueling life in the deep waters where sperm whales feed. Some studies of sperm whale distributions seem to support this idea (Jaquet 1996b). In contrast, other studies, and other data sets, indicate a correlation of sperm whale abundance with upwelling zones, where rising waters bring nutrients to the surface, promoting the growth of phytoplankton (e.g., Caldwell et al. 1966; Smith and Whitehead 1993).

Jaquet (1996b) has tried to resolve such apparent discrepancies by invoking the concept of scale. Her work (see also Jaquet and Whitehead 1996, 1999; Jaquet et al. 1996; Jaquet and Gendron 2002) suggests that, in the warmer waters of the Pacific, there is a positive correlation between sperm

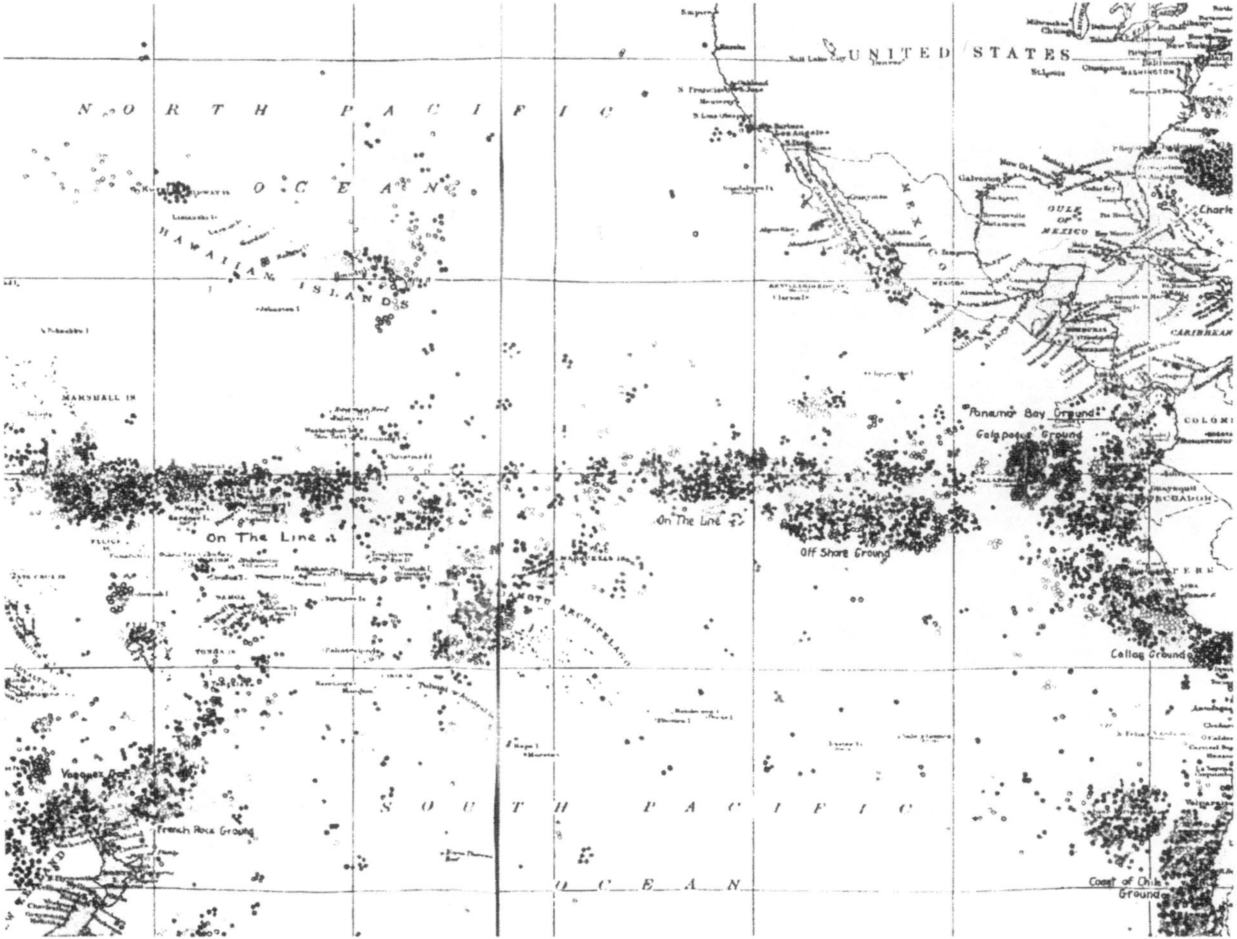

Figure 2.8 An extract from Townsend's (1935) charts of the positions at which nineteenth-century American whalers killed sperm whales.

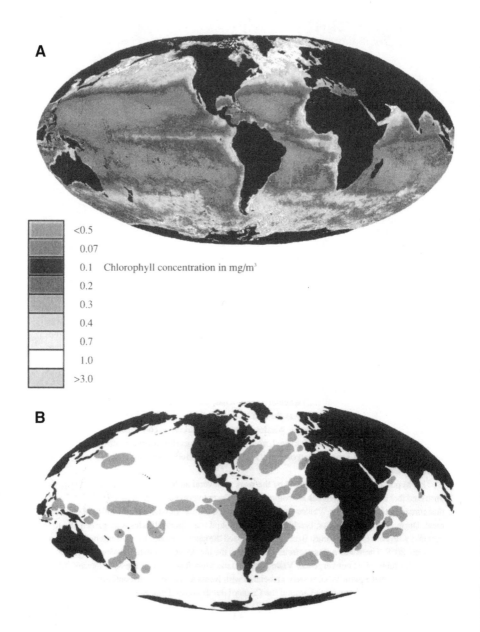

A

<0.5
0.07
0.1   Chlorophyll concentration in mg/m³
0.2
0.3
0.4
0.7
1.0
>3.0

B

Figure 2.9 Composite images of (A) ocean phytoplankton pigment concentration (Coastal Zone Color Scanner data from November 1978 to June 1986) and (B) major areas of sperm whale distribution. (From Jaquet 1996b.)

whale abundance and primary productivity, but only over large scales of a few hundred kilometers. Sperm whales are associated with areas of high plankton biomass (e.g., see fig. 2.9), but there are spatial lags of hundreds of kilometers and temporal lags of months as this production works its way up the food chain and down the water column (Jaquet 1996b). Over smaller scales of kilometers or tens of kilometers and days, sperm whale distributions do seem to be related to strong oceanographic features, such as the continental shelf break (e.g., Gregr and Trites 2001; Waring et al. 2001), oceanic fronts where water masses meet (e.g., André 1997, 155–59), cyclonic eddies (Biggs et al. 2000), and warm-core rings spinning off the Gulf Stream (Waring et al. 1993; Griffin 1999; fig. 2.10), probably because these features concentrate sperm whale prey.

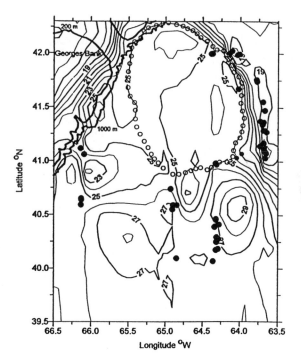

Figure 2.10 Sperm whale distribution in relation to a warm-core ring spinning off the Gulf Stream in the North Atlantic Ocean. Open circles represent the approximate boundary of warm water; solid circles represent sperm whale sightings. Also shown are 200 m and 1,000 m depth contours, as well as con-tours of sea surface temperature. (From Griffin 1999.)

Variation in Sperm Whale Abundance

From my Galápagos logs:

> 30 March 1985: It's a huge group, forming huge subgroups, up to 40 [sperms] or so, with . . . other distant subgroups.
> 22 April 1995: Another no-whale day . . . I don't believe there are sperm whales.

Sperm whale distributions in some areas (particularly at low latitudes) show a great deal of variation over time scales of months and years. Most sperm whale scientists have been disappointed at one time or another to find no, or very few, sperm whales on known grounds. Kahn, for instance, encountered no sperm whales during a survey around Fortune Bank, Seychelles, which was "named by the whalers for its generous supply of sperm whales" (Kahn 1991, 78). When we traversed the famous and extensive "Offshore" and "On the Line" grounds of the equatorial eastern Pacific in 1992, we heard only one sperm whale (Jaquet and Whitehead 1996). The density of sperm whales off the Galápagos Islands varied considerably between years during both the operations of Yankee whalers between 1830 and 1850 (CV = 0.48), and our studies between 1985 and 1995 (CV = 0.73) (fig. 2.11).

In contrast to this picture of temporal variation in the density of female

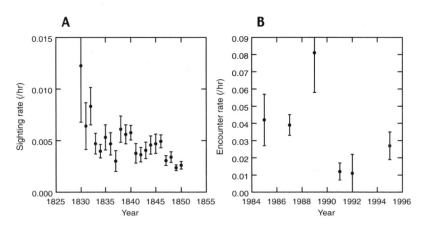

Figure 2.11 Abundances of sperm whales close to the Galápagos Islands (1°30′S–1°30′ N, 89°–92°30′ W). (A) Rates of sighting by whalers between 1830 and 1850 (see Hope and Whitehead 1991 for methods). (B) Encounters during acoustic surveys between 1985 and 1995 (from Whitehead et al. 1997). Bars show approximate standard errors.

sperm whales at any one location (a few hundred kilometers across), the abundance of male sperm whales in some high-latitude areas seems to change much less. For instance, densities of male sperm whales off Kaikoura, New Zealand (42° S), were quite stable both between and within years, with no statistically significant differences in abundance, and little variation (CV = 0.16) in the numbers of individuals identified, among fifteen seasons of fieldwork in both summer and winter (Jaquet et al. 2000). The reliable occurrence of males has allowed high-latitude locations such as Kaikoura and Andenes, Norway, to develop more sophisticated whale-watching industries than have emerged for sperm whales in tropical waters. The generally better weather and the presence of females (which have more interesting behavior; e.g., see section 5.6) at low latitudes seems to be outweighed by the predictability of males in colder waters.

2.3 The Diet of the Sperm Whale: The Walnut, the Pea, and the Half-Pound Steak

[The sperm whale's diet] is comparable to a 90kg man aiming to sample food the size of a walnut and swallowing anything between the size of a pea and a half pound [225 g] steak.
—Clarke et al. 1993

What Does the Sperm Whale Eat?

The stomachs of many thousands of dead sperm whales have been opened and examined by scientists (Kawakami 1980). Most of these whales were killed by the whaling industry (see appendix, section A.1), but some were stranded (e.g., Clarke and Young 1998). Scientists examine two types of material found in stomach contents: flesh remains of incompletely digested prey items (especially "buccal masses" for cephalopods) and indigestible remnants of prey (especially cephalopod beaks). Recently, a new source of information on sperm whale diet has been added: collections of feces from living animals. Initial analyses of feces concentrated on identifying squid beaks (Smith and Whitehead 2000; see appendix, section A.2), but it may also be possible to use DNA to identify prey (H. Poinar, personal communication).

All these methods are subject to biases (table 2.1). For instance, smaller, more gelatinous prey are digested more quickly than large, muscular animals (Clarke 1980; Clarke et al. 1988; Clarke et al. 1993), leading to a bias to-

Table 2.1.   Significance of Potential Biases in Different Methods of Studying Sperm Whale Diet

	Method[a]		
Bias	Fecal Samples: Beaks	Stomach Contents: Beaks	Stomach Contents: Flesh Remains
Large beaks sink	XXX	0	0
Very small beaks are missed	XX	X	0
Differential excretion of beaks	X	XX	0
Differential vomiting of beaks	X	XX	0
Loss of squid head during ingestion	X	X	X
Differential digestion rates of flesh	0	0	XXX
"Prey of prey"	X	X	0

[a]Significance of potential biases as assessed by Smith and Whitehead (2000): 0 = unimportant, X = slightly important, XX = important, XXX = very important.

ward large prey in analyses of flesh remains if there are substantial post-mortem times (see box 2.1). In contrast, analysis of squid beaks in fecal samples generally favors smaller prey types, as larger squid beaks sink more quickly and so are harder to collect. Furthermore, if different types of beaks are vomited or excreted more frequently, they will be underrepresented in studies of indigestible material in stomachs (Smith and Whitehead 2000). The importance of these biases will vary with the actual diet of the animals, and their magnitudes are unknown. However, if results using methods with biases of opposite expected directions are similar, then perhaps we may have some confidence in them.

It is clear from studies of stomach contents that in most areas of the world the principal prey of sperm whales are cephalopods (Kawakami 1980). However, in the northern North Atlantic and North Pacific, as well as near New Zealand, fish form a substantial part of the diet, and off Iceland, fish seem to be more important than cephalopods. The fish species eaten vary considerably between areas and individuals and over time. Sperm whales consume members of at least 55 genera and 68 species of fish (Kawakami 1980), but particularly significant groups are sharks and rays, as well as many types of teleosts. The fish eaten by sperm whales are mainly medium to large (about 0.3–3 m in length) bottom-dwelling animals (Rice 1989). Fish are a much larger part of the diet of males than of females, a finding that agrees with the differences in their distributions, males being more likely than females to be in shallow waters (see section 2.2) and to dive to the bottom (see section 3.2).

Sperm whales occasionally ingest a range of other animals, including mysids, crabs, salps, krill, lobsters, tunicates, jellyfishes, sponges, starfishes, sea cucumbers, seabirds, and seals, as well as inanimate objects such as wood, coconuts, mollusk shells, kelp holdfasts, stones, sand, tin plate, plastic, and fishing gear, as well as one quite well-documented human corpse (Berzin 1972, 206–9; Kawakami 1980; Clarke et al. 1988; Rice 1989; Clarke et al. 1993; Best 1999). Some of these items are likely to have been eaten intentionally (probably some of the large mysids, crabs, tunicates, and sea cucumbers), exploited as "secondary prey items" during normal foraging (Best 1999). Others (particularly the smallest, such as salps) may be taken incidentally while ingesting more typical prey, while objects such as plastic bags and fishing gear are probably mistaken for digestible prey. Males seem to be more likely to ingest these unusual items than females (Clarke et al. 1988), following the pattern of their generally wider niche breadth (described in section 2.2).

But it is the cephalopod—primarily the squid (but also the occasional octopus)—that is the mainstay of most sperm whale diets. Many types of cephalopods are found in the stomachs or feces of sperm whales (table 2.2; figs. 2.12 and 2.13). For simplicity, and because some areas of cephalopod taxonomy are uncertain, table 2.2 describes diet diversity at the family level, but it is worth noting that in most of these families a number of species are eaten. These squids vary in size from the small chiroteuthids (sometimes < 100 g; the "pea") and histioteuthids (~400 g; the "walnut") to the architeuthids, or giant squids (up to 400 kg; a 100 kg architeuthid is equivalent to the "half-pound steak"; fig. 2.12). They also vary in activity, from the active ommastrephid *Dosidicus gigas* of the Humboldt Current, which has been described as "a living horror from the deep" (Duncan 1941), to the placid, gelatinous histioteuthids. However, we know rather little about many of the sperm whale's prey, and a substantial proportion of our knowledge of several species comes from studies of sperm whales themselves (Clarke 1987). For instance, M. R. Clarke (1980) makes plausible inferences on the schooling behavior of different squid species based on the numbers of fresh animals co-occurring in sperm whale stomachs (see below). Largely absent from the sperm whale diet are surface-living cephalopod species (Berzin 1972, 197), but given this restriction, almost all evidence indicates that sperm whales have a wide dietary range. One study, however, presents a radically different picture: that of R. Clarke and his colleagues (1988) on the sperm whales off Peru and Chile. Their view of the sperm whale as an animal with only one important food species is considered in box 2.1.

Table 2.2. Cephalopod Families in the Diet of Sperm Whales

Family	Description[b]	Photophores	Approx. mass (g)[c]	Estimated Percentage of Cephalopod Diet (by Mass)[a]									
				Azores	Iceland	Chile, Peru	Chile, Peru	New Zealand	Hawaii	S. Africa: ♀	S. Africa: ♂ < 12 m	S. Africa: ♂ > 12 m	Galápagos
Architeuthidae	Giant squids, neutrally buoyant and antitropical	Y/N	24,000	11.0	3.4	0	0	0	26.8	3.4	4.3	1.8	0
Ommastrephidae	Swift and muscular, may live on bottom	Y/N	8,000	4.7	0.0	31.6	100.0	5.6	30.6	15.5	7.1	2.0	0
Lepidoteuthidae	Scaled squids, active, mid- to deep waters	N	2,000	4.1	0	0	0	0	2.2	1.1	0.9	0.3	0
Pholidoteuthidae	Now thought to be a genus within the family Lepidoteuthidae, may be associated with bottom	N	1,700	3.4	0.0	0	0	0	1.2	1.4	0.5	0.1	0.1
Vampyroteuthidae	Gelatinous, of deep waters with webbed tentacles	Y	1,000	0.0	0	3.3	0	0	0.0	0.2	0.0	0.0	0
Psychroteuthidae	Muscular, of cold waters, mid- to deep waters	Y	1,000	0	0.0	0.1	0	0	0	0	0	0	0.1
Octopoteuthidae	Medium to large, gelatinous, several species	Y	1,000	42.2	25.3	1.9	0	0	14.1	27.3	34.1	11.6	11.3

Family	Description												
Histioteuthidae	Slow-swimming, gelatinous, ammoniacal, 18 species found in mid-waters	Y	800	29.6	46.5	56.3	0.0	1.0	11.0	20.5	20.5	2.6	48.0
Ancistrocheiridae	Mesopelagic warm waters	Y	700	0.6	0.1	2.6	0.0	0	0.3	12.9	14.6	3.9	40.0
Alloposidae	Gelatinous, often associated with sea floor	N	600	0.7	7.5	0	0	0	0	0.0	0.0	0.0	0.2
Onychoteuthidae	Wide size and depth range, muscular	Y/N	500	0.6	5.9	0.1	0	89.7	0.9	4.4	6.1	59.7	0.1
Cycloteuthidae	Mostly mesopelagic, tropical, somewhat gelatinous	Y	500	1.2	0.4	0	0	0	4.4	3.3	1.6	0.3	0
Octopodidae	Bottom-living octopuses, not gelatinous	?	300	?	0	0	0	0	0	0	0	0	0
Mastigoteuthidae	Weakly muscled, somewhat gelatinous, mid- to deep-water species often near bottom	Y	200	0.0	0	0	0	0	0.3	0.2	0.1	0	0

(continued)

Table 2.2. (continued)

Family	Description[b]	Photophores	Approx. mass (g)[c]	Estimated Percentage of Cephalopod Diet (by Mass)[a]									
				Azores	Iceland	Chile, Peru	Chile, Peru	New Zealand	Hawaii	S. Africa: ♀	S. Africa: ♂ <12m	S. Africa: ♂ >12m	Galápagos
Gonatidae	Abundant, well muscled and gelatinous, deep waters, often associated with bottom	Y	200	0.2	0.1	1.4	0	0	0.9	0.0	0.0	0	0
Cranchiidae	Surface-dwelling, ammoniacal, sluggish	Y	200	1.3	0.6	0.5	0	0	7.7	6.5	6.3	16.3	0
Chiroteuthidae	Gelatinous, slow-moving, deep-sea/benthic	Y	100	0.1	0.0	1.0	0	0	0.0	1.9	1.8	0.1	0.2
Unidentified				0	10.1	1.1	0	3.7	0	1.3	0.6	0.3	0
Reference				Clarke et al. 1993	Clarke and MacLeod 1976	Clarke et al. 1976	Clarke et al. 1988	Gaskin and Cawthorne 1967	Clarke and Young 1998	Clarke 1980	Clarke 1980	Clarke 1980	Calculated from Smith and Whitehead 2000
Method[d]				BS	BS	BS	FS	FS	BS	BS	BS	BS	BD

[a] 0.0 indicates members of family found but not thought to account for more than 0.1% of diet; 0 indicates members of family not found.

[b] Descriptions of cephalopod families from the "Tree of Life" (http://www.soest.hawaii.edu/tree/cephalopoda/cephalopoda.html) and other sources.

[c] Sizes of species within a family vary considerably. The given "approx. mass" is a rough median of the species most commonly used by sperm whales, from the sizes of animals found in sperm whale stomachs.

[d] B, studies of cephalopod beaks; S, studies of stomach contents; F, studies of flesh remains; S, studies of stomach contents; D, studies of defecations.

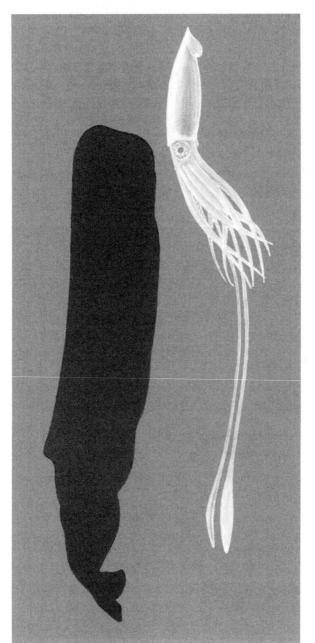

Figure 2.12 The sperm whale and its largest prey, the giant squid *Architeuthis*, drawn to scale. (Copyright Emese Kazár.)

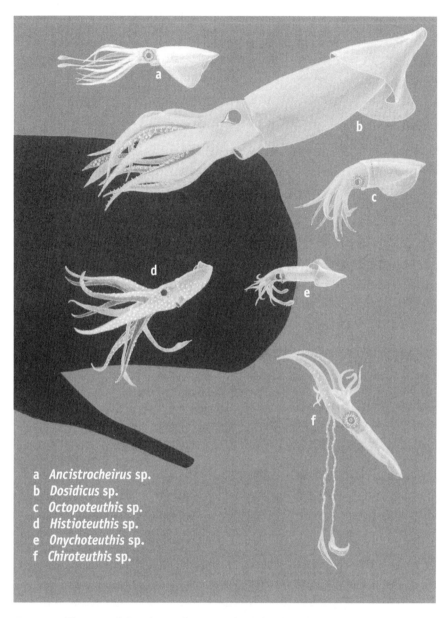

a *Ancistrocheirus* sp.
b *Dosidicus* sp.
c *Octopoteuthis* sp.
d *Histioteuthis* sp.
e *Onychoteuthis* sp.
f *Chiroteuthis* sp.

Figure 2.13 The sperm whale and some of its principal cephalopod prey, drawn to scale. (Copyright Emese Kazár.)

Box 2.1 **An Alternative View of the Sperm Whale Diet: The Picky Eater**

Robert Clarke and his colleagues* (Clarke et al. 1988; Clarke and Paliza 2000) have a view of the sperm whale diet that differs radically from that of most scientists. The conventional wisdom, summarized in section 2.3, is that the animal is a generalist, eating a wide variety of foods from the deep waters of the ocean. In contrast, Robert Clarke and his colleagues argue that the diet of the sperm whale in the southeastern Pacific is virtually monospecific, and by extension, that this may also be the case in other oceans.

This view arises from the results of a study in which Robert Clarke and his colleagues (1988) described the flesh remains from 2,403 stomachs of sperm whales caught by whalers off Peru and Chile between 1959 and 1962. Of these remains, 99.4% were from the large ommastrephid squid *D. gigas,* 0.4% from *Ancistrocheirus leseuri,* and 0.1% from histioteuthids. When the sizes of the different species are considered, the dominance of *D. gigas* is even more dramatic, with the other species contributing less than 0.1% of the sperm whale's total diet (see table 2.2). Hence, Robert Clarke and Obla Paliza (2000) concluded that "sperm whales off Chile and Peru feed practically exclusively on *D. gigas.*"

These findings contrast with the results of all other research on sperm whale diet (see table 2.2), and specifically with those of an earlier analysis of squid beaks from some of the same Chilean and Peruvian sperm whale stomachs by Malcolm Clarke et al. (1976). Malcolm Clarke and his colleagues concluded that, while *D. gigas* constituted a substantial part of these whales' diet (32% of estimated mass), histioteuthids were more important (56%), and that the whales were eating members of at least eighteen squid species. Robert Clarke et al. (Clarke et al. 1988) dismiss this study on two principal grounds. First, because sperm whales frequently lose the heads, and thus the beaks, of *D. gigas* while eating them, the abundance of this species is underestimated in studies of beaks alone. Second, the beaks of smaller squid found by Malcolm Clarke and colleagues (1976) in the Chilean and Peruvian whales, as well as those that we found in sperm whale feces off the Galápagos Islands (Smith and Whitehead 2000), actually came from the prey of *D. gigas,* not of the sperm whales (the "prey-of-prey phenomenon") (see table 2.1).

Box 2.1 continued

This controversy is not restricted to the haunts of *D. gigas* in the eastern Pacific. Other large ommastrephid squids—for instance, *Ommastrephes bartrami* in the North Pacific (Kawakami 1980)—are eaten by sperm whales, and if the assumptions of Robert Clarke et al. (1988) are right, these may be much more important, and the smaller squids, such as the histioteuthids, much less important, in sperm whale diets than usually supposed.

Despite the obvious confidence of Robert Clarke and his colleagues in their monospecific diet and prey-of-prey perspectives, their views have not been widely accepted. For instance, members of the Sperm Whale Subcommittee of the International Whaling Commission's Scientific Committee noted that Robert Clarke et al.'s (1988) results "were atypical . . . smaller species will be digested more quickly and this percentage may not be an accurate representation of the diet composition" (International Whaling Commission 1987). Malcolm Clarke et al. (1993) added that, in their study of beaks from Chile and Peru (Clarke et al. 1976), the smaller species (histioteuthids, chiroteuthids, and ancistrocheirids) were sometimes represented by flesh remains and so must have been important in the diet of sperm whales, not just as "prey-of prey." However, in my view, the low representation of members of these species in flesh remains is best explained by differential digestion rates (Smith and Whitehead 2001): smaller squid will be digested more rapidly than larger ones, and gelatinous squids will be digested more quickly than muscular forms (Clarke 1980; Clarke et al. 1988; Clarke et al. 1993; Best 1999). *D. gigas* is much larger and more muscular than the other squids found in the stomach contents (or feces) of sperm whales in the southeastern Pacific, and so will be overrepresented among the flesh remains in the stomachs of whaled animals. The effect will be particularly severe if postmortem times are long (Berzin 1972, 204), as tends to be the case in the shore-based whaling industry that provided the Chilean and Peruvian carcasses. In Chile, postmortem times were 12–36 hr (Aguayo 1963, quoted in Clarke et al. 1988), while in Peru they ranged to over 48 hr (Clarke et al. 1988).

For these reasons, I believe that the extreme position taken by Robert Clarke and his colleagues (1988) should be rejected, and that sperm whales of the southeastern Pacific have a diverse diet like those elsewhere. Some of their points are valid, however, and need to be incorporated when we consider sperm whale diet. For instance, the presence of "prey-of-prey" may bias estimates using beaks from stomachs or fe-

Box 2.1 continued

ces (as previously noted by Malcolm Clarke in 1962), and the abundance of *D. gigas,* and possibly other species, will be underestimated in such studies if their heads are not always eaten.

* One of the confusing aspects of the debate over the sperm whale's diet is the similarity of the names of the two principal protagonists: M. R. Clarke and R. M. Clarke. In an attempt at clarification, I use their first names here: Malcolm Clarke and Robert Clarke. Further confusion arises from the presence of Obla Paliza as co-author of each of the lead papers on the two sides of the debate (Clarke et al. 1976; Clarke et al. 1988).

The majority of squids, and other prey items, taken by the sperm whale are small compared with its body size (perhaps about 0.01% of its body mass) and not particularly nutritious; thus, sperms must eat many of them. M. R. Clarke (1980) estimated that females eat about 750 squid per day and males 350 (but usually larger ones). Both our observations (see section 5.4) and the stomach contents data strongly suggest that the sperm whale forages fairly continuously, being a "nibbler" rather than a "meal eater" (Best 1999). Thus, R. Clarke et al.'s (1988) consideration of sperm whale feeding using the concept of the "full meal" is probably inappropriate.

The diet of the sperm whale varies considerably with geographic location (fig. 2.14; see table 2.2). The diet in any one area is also quite diverse, with animals using about ten to twenty-five species of cephalopods reasonably frequently (Clarke 1980). Even by the standards of the unpredictable oceanic biotic environment (see section 2.1), squids have notoriously variable abundances (Arnold 1979), so it is not surprising that the diet of sperm whales in any place varies considerably with time (Kawakami 1980; Smith and Whitehead 2000) and between individuals (e.g., Clarke et al. 1993). In general, compared with females and smaller males, adult male sperm whales prey upon larger squid species, and larger members of the same species, although there is considerable overlap (Clarke 1980; Clarke et al. 1993; Best 1999; see table 2.2).

Characteristics of Sperm Whale Prey: Behavior and Aggregation

The cephalopods eaten by sperm whales vary not only in size, but also in nutritional quality and ease of capture. The smaller squids are generally more gelatinous, neutrally buoyant, slow-swimming, and bioluminescent than the larger, more powerful animals such as *D. gigas* (Clarke et al. 1993). Thus, gram for gram, the larger animals probably provide a better meal, but a more

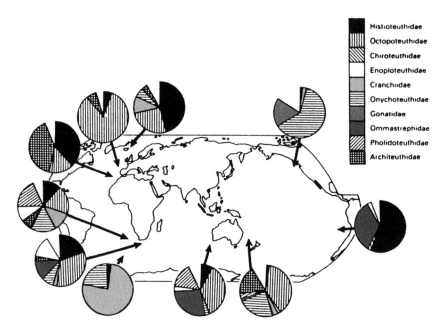

Figure 2.14 Composition by mass of squid families in the diets of sperm whales caught in various regions of the world, estimated from beaks found in the whales' stomachs. (From Clarke 1987, fig. 11.3.)

challenging capture, than the small cephalopods that are the staple food for most sperm whales most of the time (Clarke et al. 1993; see table 2.2). While chiroteuthids have limited escape responses and are easily visible, catching an adult *D. gigas,* large, agile, and invisible in the dark deep ocean, is probably challenging (Clarke 1980).

From the behavioral perspective, the aggregative nature of squids is important, as the size and stability of groups of predators is often thought to depend on the size of prey aggregations (Connor 2000). Cephalopods are notoriously difficult to study in the wild, so we have limited information on their aggregations. Perhaps the best strategy is to examine the data from sperm whale stomachs. Table 2.3 presents data from M. R. Clarke (1980) on the numbers of fresh squid of particular species found in stomachs together. These animals are likely to have been caught on the same foraging dive or on sequential dives (Clarke 1980), and so would probably have been within a kilometer or two of one another (see section 3.4). Only about one to five members of most species are caught per dive, suggesting that these animals are not usually found by sperm whales in very large aggregations. The more bottom-dwelling species, listed at the end of table 2.3, seem more solitary than those found in midwater.

Table 2.3.   Numbers of Buccal Masses of Squid Species in Complete Stomachs Sampled at Durban, South Africa, 1962–1969

Species/genus[a]	No. of Stomachs	Total Buccal Masses	Mean Buccal Masses in Stomach	Maximum Buccal Masses in Stomach
Todarodes	43	200	4.65	16
Octopoteuthis	38	247	6.50	30
Histioteuthis bonnellii	24	96	4.00	14
Histioteuthis miranda	25	68	2.72	20
Ancistrocheirus	29	78	2.69	12
Moroteuthis robsoni	19	32	1.68	6
Taningia	11	11	1.00	1
Taonius megalops	15	20	1.33	4

SOURCE: Clarke 1980, table 52.

[a]Only species found in at least 10 stomachs are included.

Moving to larger scales, we continue to lack direct information on the sizes of squid concentrations. However, once again, it is possible to make inferences from studies of sperm whales—in this case, of their movements. Off the Galápagos Islands, groups of sperm whales with high feeding success zigzagged back and forth over areas about 20 km across, suggesting patches of prey of this scale, whereas off northern Chile the patches seemed larger, perhaps 60 km across (see section 3.4; see also Jaquet and Whitehead 1999).

Cephalopods often aggregate to spawn and may be an especially easy and profitable prey at this time (Clarke 1980). That different species have different times and modes of spawning may be useful to the sperm whales, and may be partially responsible for their wide dietary range.

2.4   The Sperm Whale and Its Competitors

There are other animals that feed principally on the larger organisms of the deep ocean waters, and so are potential competitors of the sperm whale. These animals include some large cephalopod species, fishes, and other marine mammals. For the marine mammals, Pauly et al. (1998) have provided a useful, but approximate, summary of diet, dividing it into eight categories. Those species with a substantial overlap with sperm whales are listed in table 2.4. It is noteworthy that, despite the rather crude characterization of diet used, no species has a very close match to the sperm whale: the closest are the pygmy sperm whale (*Kogia breviceps*) and southern elephant seal (*Mirounga leonina*), with 70% overlaps. In contrast to their rather low dietary overlap

Table 2.4. The Diets of Potential Marine Mammal Competitors of Sperm Whales

| Species | | Benthic Invertebrates | Squids[a] | | Fish | | Other | Percentage Overlap with Sperm Whales |
			Small	Large	Small Pelagics	Meso-pelagics		
Physeter macrocephalus	**Sperm whale**	**0.05**	**0.1**	**0.6**	**0.05**	**0.05**	**0.15**	**100**
Berardius bairdii	Baird's beaked whale	0.1	0.3	0.25	0.1	0.1	0.15	65
Mesoplodon bidens	Sowerby's beaked whale	0	0.25	0.3	0.05	0.2	0.2	65
Mesoplodon layardii	Strap-toothed whale	0	0.3	0.4	0	0	0.3	65
Mesoplodon hectori	Hector's beaked whale	0	0.4	0.4	0	0	0.2	65
Ziphius cavirostris	Cuvier's beaked whale	0.1	0.3	0.3	0	0.15	0.15	65
Hyperoodon ampullatus	Northern bottlenose whale	0.15	0.35	0.35	0	0.05	0.1	65
Kogia breviceps	Pygmy sperm whale	0.05	0.35	0.4	0	0.1	0.1	70
Kogia simus	Dwarf sperm whale	0.1	0.4	0.4	0	0.05	0.05	65
Peponocephala electra	Melon-headed whale	0	0.35	0.35	0.1	0.1	0.1	65
Globicephala macrorhynchus	Short-finned pilot whale	0	0.3	0.3	0.1	0.1	0.2	65
Mirounga leonina	Southern elephant seal	0.05	0.4	0.35	0.05	0	0.15	70

SOURCE: Data from Pauly et al. 1998.

NOTE: Shown are the estimated distributions of the diets of those species whose overlap with sperm whales was greater than 60% (see Krebs 1989, 381 for a definition of the percentage overlap index of niche overlap). Proportions of "large zooplankton" and "higher vertebrates" were given as zero by Pauly et al. (1998) for all these species.

[a]Squids from families with mantle lengths greater and less than 0.5 m were classified as "large" and "small" respectively.

with the sperm whale, each of the other species listed in table 2.4 has an over-lap of at least 90% with one of the other species on the list. The distinctive-ness of the sperm whale's diet, and in particular its fondness for large squid, seems to hold, even with more refined taxonomic distinctions among food items. For instance, the deep-living squid *Mesonychoteuthis hamiltoni* seems to be the major dietary item of Antarctic sperm whales (77% of cephalopod mass, according to Clarke 1987), but has yet to be associated with any other predator, including the southern elephant seal, another deep-diving preda-tor of these waters (Clarke 1987; Green and Burton 1993).

Based on the information in table 2.4 and other considerations, the most likely mammalian competitors of the sperm whale would seem to be the pygmy and dwarf sperm whales (*Kogia* spp.), beaked whales (*Berardius, Hy-peroodon, Mesoplodon,* and *Ziphius* spp.), elephant seals (*Mirounga* spp.), and pilot whales (*Globicephala* spp.). For some areas of the ocean, there are estimates of the numbers of sperm whales and these potential competitors. One such area is the Antarctic, and M. R. Clarke (1987) attempted to com-pare squid consumption by sperm whales and by some of their potential competitors in the Antarctic waters. He estimated that sperm whales in the Antarctic eat about 11.8 million metric tons (Mt) of cephalopods per year, considerably more than any of the other species considered—the nearest ri-val was the southern elephant seal, with 4.5 Mt. These assessments should be considered very cautiously, however, as they are based on most uncertain assumptions, particularly population sizes and diets (Clarke 1987).

In table 2.5, I have attempted a similar exercise for other areas of the world, but I have listed only biomasses and have not attempted to convert them into tonnage consumed. Although most population estimates are not very precise, and are probably biased because animals under water are not censused, they give some idea of the relative biomasses of the different species. In only one location—off the northeastern United States—does a potential competitor, in this case the long-finned pilot whale (*Globicephala melas*), have a greater es-timated biomass than the sperm whale. Of the total biomass of the guild of squid-eating mammal species listed in table 2.5, the sperm whale's biomass is an estimated 66% in the eastern tropical Pacific, 42% along the continental shelf edge off the northeastern United States, 94% in the Gulf of Mexico, 62% off California, and less than 67% in the Antarctic.

Thus, the sperm whale has comparatively little dietary overlap with other marine mammal species and generally dwarfs its closest competitors in terms of biomass. What about other taxa? Several large squid species, such as *D. gigas,* have diets that overlap with those of sperm whales—especially in their consumption of smaller squids—but, as these animals are themselves a

Table 2.5. Approximate Numbers and Biomass (in Metric Tons) of Sperm Whales and Some of Their Potential Competitors in Areas for Which There Are Estimates of Cetacean Numbers

Group	Mass per Individual (t)	Eastern Tropical Pacific		Shelf-Edge Region of Northeastern US (summer)		Northwestern Gulf of Mexico		California		Antarctic	
		N	Biomass	N	Biomass	N	Biomass	N	Biomass	N	Biomass
Sperm whale	20 (40 Antarctic)	22,700	454,000	215	4,300	313	6,260	756	15,120	15,000	600,000
Pygmy/dwarf sperm whales	0.35	11,200	3,920	0	0	107	37	870	305	0	0
Beaked whales (Mesoplodon, unidentified)	1	25,300	25,300	0	0	124	124	1,572	1,572	?	
Beaked whales (Ziphius)	3	20,000	60,000	0	0	14	42	1,621	4,863	0	0
Beaked whales (Berardius)	9	0	0	0	0	0	0	38	342	?	
Beaked whales (Hyperoodon)	4.7	0	0	9	42	0	0	0	0	?	
Pilot whales	0.9	160,200	144,180	6,823	5,800	215	194	0	0	?	
Elephant seals	0.5	0	0	0	0	0	0	≈6,000[a]	≈3,000	600,000	300,000
Reference	Hain et al. 1985; Clarke 1987; Mead 1989a; Wade and Gerrodette 1993; Whitehead and Mann 2000	Wade and Gerrodette 1993		Hain et al. 1985		Davis and Fargion 1996, 70		Barlow 1995		Clarke 1987; Butterworth et al. 1995	

[a] This figure is very approximate and was calculated from an approximate population of 60,000 elephant seals in the North Pacific in 1991 (Sydeman and Allen 1999), of which 10% are assumed to be in the study area used for the cetacean surveys (Barlow 1995; Stewart and DeLong 1995).

major sperm whale food, they are perhaps unlikely to be important competitors, although this is not clear. There are also a number of fish species that eat the same mesopelagic and bathypelagic squids as the sperm whale, probably most notably the tunas and the swordfish (*Xiphias gladius*) (e.g., Hernández-García 1995). These fishes, while occasionally found in sperm whale stomachs, are not themselves a major food source for sperms. There are no good estimates of the stocks of most such fish species, but in the Pacific, the total potential annual yield of the highly migratory species (including tuna, swordfish, and pelagic sharks) is estimated to be about 3 Mt (National Marine Fisheries Service 1999, 18–20), and a ballpark biomass of tuna is on the order of 20–30 Mt (J. Hampton, personal communication), roughly the biomass of a million sperm whales. Thus, these species collectively may exceed or rival sperm whales in the amount of prey taken from the ocean, but the diet overlap between any of these species and the sperm is probably quite small, and I suspect (but do not know) that they are not important competitors.

Despite the assertions of R. Clarke et al. (1988; see box 2.1), the sperm whale seems to have a very catholic diet. But is this diet remarkable when compared with the diets of its principal competitors? In figure 2.15 the number of squid genera among the lower beaks of cephalopods found per stomach is plotted against the total number of lower beaks examined from the

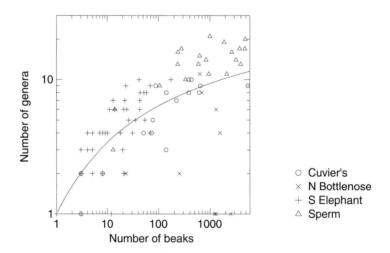

Figure 2.15 Number of cephalopod genera identified from stomachs (each data point represents one analyzed animal) plotted against the number of beaks examined for four species of deep-diving mammals, with a logistic curve (solid line) fitted to all data. (From Whitehead et al. 2003.)

stomach. We have done this for the sperm whale and three of its potential competitors, the northern bottlenose whale (*Hyperoodon ampullatus*), Cuvier's beaked whale (*Ziphius cavirostris*), and the southern elephant seal (Whitehead et al. 2003). Naturally, the number of genera found tends to increase with the number of beaks examined, but it seems that there are differences among the species in the rate of this increase. To quantify these differences, we fitted a logistic model to the data* that estimated the rate of increase of squid genera with the number of beaks examined (the parameter k). This parameter was higher for sperm whales ($k = 1.25$) than for the other species (northern bottlenose whale, $k = 0.21$; Cuvier's beaked whale, $k = 0.68$; southern elephant seal, $k = 0.90$), and these differences were statistically significant (Whitehead et al. 2003). Thus, sperm whales appear to have wider niche breadths than these other deep-diving species, especially the northern bottlenose whale. In section 8.4, I relate this contrast to differences in these animals' movement patterns.

2.5 The Ecological Role of the Sperm Whale

So, in terms of biomass and niche breadth (see table 2.5 and fig. 2.15), sperm whales seem to overshadow their potential competitors, but what does this amount to in global terms? How much do sperm whales take from the ocean? This can be estimated quite easily if we know the numbers of whales and their feeding rates. M. R. Clarke (1977) assumed a population of 501,000 males and 757,000 females worldwide, with females eating 0.15 t/d and males 0.3 t/d, and thus an annual total of 96 Mt/yr. He considered this estimate conservative. His assumed food consumption rates may be rather low (Lockyer [1981] suggests 0.4 t/d for a physically mature female and 1.5 t/d for a physically mature male), while his population sizes are optimistic. Assuming Lockyer's (1981) consumption rates and my population estimate of 361,400 animals globally (see section 4.2), of which 15% are physically mature males, I calculate a global consumption estimate of 74.5 Mt/yr. Thus, it seems reasonable that the world sperm whale population is eating on the order of 100 Mt/yr. This figure is comparable to the total annual catch of all human marine fisheries in recent years (70–85 Mt/yr between 1984 and 1998; FAO 2000).

One of the particularly interesting things about this comparison is that

---

* Model:

$$\text{Log(Number of genera)} = \frac{k \times \text{Log(Number of beaks)}}{[1 + k \times \text{Log(Number of beaks)} / \text{Log(Total number of genera)}]}$$

the overlap between the species used by sperm whales and those fished by humans is very small. Of the sperm whale's primary cephalopod prey, only *D. gigas* is the target of a substantial fishery. Thus, the sperm whales use a vast source of protein that voracious and technologically sophisticated modern humans have so far failed to tap. The reasons for our failure are partly technological—we cannot catch deep-water squid with anything like the efficiency of sperm whales (Clarke 1977)—and partly a matter of taste: unlike sperm whales (presumably!), we do not usually like them.

Thus the indications are that sperm whales are often a major element of their ecological guild (Clarke 1977). So, it is quite likely that sperm whales, through their foraging, had, and to a lesser extent still have, a substantial effect on oceanic biological processes. However, given our poor knowledge of the ecology of the mesopelagic ocean, this remains conjecture.

It has also been suggested that sperm whales may aid nutrient cycling in the ocean by feeding at depth and defecating at the surface (Clarke 1977), thus bringing limiting nutrients, especially nitrates, back into the photic zone. Katona and Whitehead (1988) looked at this suggestion in more detail. In a steady state, with no net horizontal transfers of nutrients, we showed that sperm whale feces cannot provide more than about 8% of the nitrogen used in primary production, and that usually the figure will be much less than this, estimating it to be about 0.04% off the Galápagos Islands. However, in special circumstances—for instance, if squid migrate into, and concentrate in, nutrient-poor waters for mating or other reasons—then a high feeding rate by sperm whales could induce a phytoplankton bloom (Katona and Whitehead 1988). Additionally, sperm whale feces could prolong a transient phytoplankton bloom, or induce a secondary, smaller bloom, following a major phase of production (Katona and Whitehead 1988).

2.6   Feeding Success of the Sperm Whale

Studying Feeding Success

Scientists working with the whaling industry had two principal methods of assessing sperm whale feeding success: tallying the fresh items in sperm whale stomachs (e.g., Clarke 1980) and measuring fatness (e.g., Clarke et al. 1988). The first method indicates recent success within the few hours before capture, while the second indicates long-term average success over a few weeks or months.

In recent years we have used the rate at which sperm whales defecate as they dive (defecations per fluke-up observed) to indicate the feeding success of living animals (Whitehead et al. 1989; Smith and Whitehead 1993; Whitehead 1996b). This method has several justifications (Whitehead 1996b). First, deep-diving mammals generally shut down physiological systems that are not immediately essential during their dives (Elsner 1999), and so might be expected to defecate almost always at the surface. Second, off the Galápagos Islands, defecation rates were negatively related to sea surface temperature, which itself is negatively correlated with productivity (Whitehead et al. 1989). Finally, most, though not all, defecations are coincident with dives. Defecation rates probably indicate feeding success integrated over a few hours (as maximal rates are about 0.2 defecations/dive, and diving rates are about 1.33 dives/hr; see section 3.2) and with a lag of a few hours.

All of these measures indicate substantial variation in feeding success. For instance, stomach contents off South Africa varied from 0 to 130 kg (Best 1999), blubber thickness off Peru from about 5 to 12 cm (Clarke et al. 1988), and daily defecation rates (mean of all whales followed during the day) off the Galápagos Islands from 0 to more than 0.5 defecations/dive (Whitehead 1996b).

Geographic Variation

Results from all three methods of assessing feeding success have been used to suggest that sperm whales do better in some areas than in others. The proportion of empty stomachs was 3% in the Bering Sea and 41% off Japan (Kawakami 1980). Off the west coast of South America between 1959 and 1962, Clarke et al. (1988) found an increase in fattening and fatness in sperm whales caught at more southerly whaling stations, which they attributed to a greater abundance of food at higher latitudes within the Humboldt Current. And defecation rates were significantly higher off the Seychelles in 1990 (0.16 defecations/dive) than in the eastern tropical Pacific between 1985 and 1991 (0.06 defecations/dive), a difference that Kahn et al. (1993) relate to density-dependent effects following exploitation. Moving to smaller scales, over a 3-month study in 1989, sperm whales off the Galápagos Islands had a significantly higher defecation rate in one 1° × 1° square, west of the islands, than in neighboring squares (Smith and Whitehead 1993). However, significant differences between 1° × 1° squares were not found in other study years (Smith and Whitehead 1993).

I do not think that any of these comparisons should be taken to mean that sperm whales necessarily do consistently better in one part of the world or ocean area than in another. As shown below, feeding success in any area

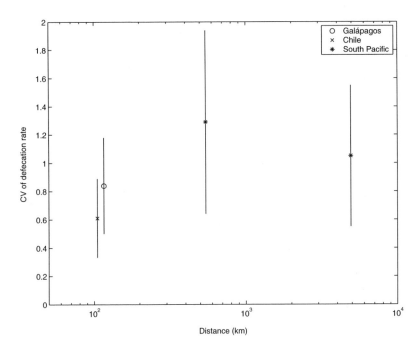

Figure 2.16 Estimated coefficients of variation (CV) in defecation rates (per fluke-up) with distance for female and immature sperm whales near the Galápagos Islands (1985–1995), off northern Chile (2000), and during a 1992–1993 survey of the South Pacific (methods as in Whitehead 1996b). Bars show estimated standard errors.

varies enormously over a variety of time scales, especially long ones, and no studies have yet gone on long enough to remove this temporal variance. Instead, I think a better question to ask is, how does sperm whale feeding success vary with spatial scale over small temporal scales? This question addresses spatial variation in feeding success from a sperm whale's perspective: how will its feeding success change as it moves around the ocean?

In figure 2.16, the estimated coefficients of variation in defecation rates in different areas are plotted against distance for our studies in the South Pacific (using the methods described in Whitehead 1996b). By moving about 100 km, the sperm whales changed their defecation rates by an average of about 60% of the long-term mean, but over ranges of several hundred, or several thousand, kilometers the change was more like 120%. However, these defecation rates were measured from just one boat in any study, which could not be in different places at the same time. Therefore, spatial variation is confounded with temporal variation at scales of about 3 d for 100 km, 10 d for 500 km, and 80 d for 5,000 km. The variation in any place over time scales of about 3 d has a CV of about 0.7 (fig. 2.17), so that animals would not ex-

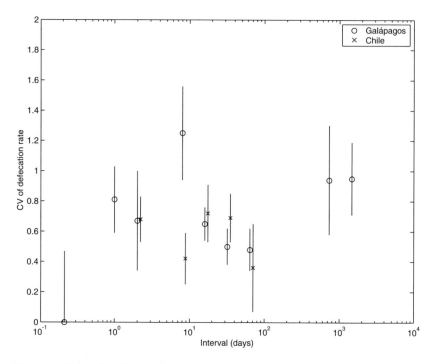

Figure 2.17 Estimated coefficients of variation (CV) in defecation rates (per fluke-up) with time interval for female and immature whales near the Galápagos Islands (1985–1995) and off northern Chile (2000) (methods as in Whitehead 1996b). Bars show estimated standard errors; only estimates with SE < 0.5 are shown.

pect to find appreciably different feeding success by moving 100 km compared with staying in one place. However, over larger scales of hundreds and thousands of kilometers, the spatial variation seems considerably greater than the variation due to the confounding time scale (figs. 2.16 and 2.17). Thus, by moving these kinds of distances, animals may be able to alter their feeding success considerably.

Temporal Variation

The ocean is a cyclical environment, and its cycles (tidal, diurnal, lunar, and annual) strongly affect the foraging of many—probably most—oceanic animals. Examples range from the strongly diurnal foraging of many zooplankton (Lalli and Parsons 1993, 100–103) to the reaction of fur seals to lunar cycles (Horning and Trillmich 1999) to the annual foraging migrations of baleen whales (Clapham 2000). But what about sperm whales? Starting at the smallest scales, I attempted to correlate sperm whale defecation rates off the Galápagos Islands with tidal cycles, but found no relationship (Whitehead 1996b).

There have been a number of examinations of diurnal foraging patterns in sperm whales. Most studies using food remnants in the stomach failed to find any consistent relationship between stomach fullness and time of day (Okutani and Nemoto 1964; Clarke 1980; Best 1999). Similarly, there was no strong diurnal pattern in the defecation rates of Galápagos sperm whales: low measured rates in the early morning and late evening could be related to difficulties in sighting defecations at low solar elevations (Whitehead 1996b). In contrast, in the Antarctic, Matsushita (1955, cited in Best 1999) found sperm whale stomachs to be fullest in the early morning (03:00– 08:00) and late evening (22:00–24:00). Perhaps this difference is related to Matsushita's ability to cover much more of the day than the other studies, or to geographic differences in the strength of diurnal cycles (Best 1999). The female sperm whales that we follow show a clear diurnal peak in non-foraging behavior in the late afternoon (see section 5.3), but this would probably lead to a reduced defecation rate at some time during the night, when we cannot observe it. Thus, while most studies have failed to find any diurnal pattern in the sperm whale's feeding success, and it is clear that sperm whales forage at all times of day, there are hints of some diurnal variation in at least some areas.

Variation in foraging rates with the lunar cycle is clear for Galápagos fur seals (*Arctocephalus galapagoensis*), which feed within 100 m of the surface (Horning and Trillmich 1999), but there is no sign of it for the sperm whales that use a deeper part of the water column in the same geographic area (Whitehead 1996b). Similarly, there was no signal of a lunar cycle in the contents of sperm whale stomachs examined in South Africa (Clarke 1980).

A number of scientists have looked for seasonal variation in sperm whale feeding success (e.g., Clarke 1956; Clarke et al. 1988; Whitehead 1996b; Best 1999 and references therein). In no case has consistent seasonal variation been found between years. Apparent seasonal patterns (e.g., Clarke et al. 1988; Best 1999) come from studies in which the data from just a few years are lumped across years and then compared by season. Seasonal patterns produced by such analyses can originate from aseasonal temporal variation in feeding success, such as that shown in figure 2.17.

Thus, there is no convincing evidence that environmental cycles drive sperm whale foraging, except possibly diurnally. Instead, we can look for aperiodic variation in foraging success. That this variation is substantial is shown in figure 2.18, in which the defecation rates for our Galápagos studies are plotted. In 1987, an El Niño year of warm water temperatures, it was

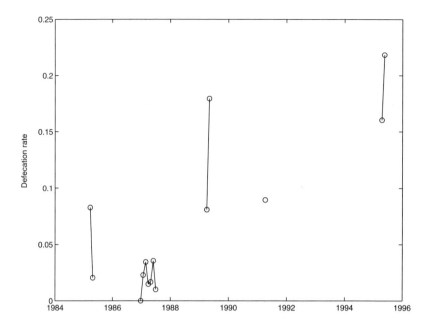

Figure 2.18 Temporal variation in feeding success of sperm whales off the Galápagos Islands as indicated by mean defecation rates (per fluke-up). Each open circle represents a 32-day period.

consistently low over 6 months, whereas in the cooler years of 1985, 1989, 1991, and 1995, the whales seemed to do much better. Off the Galápagos Islands, defecation rates were clearly inversely related to sea surface temperatures (fig. 2.19; Smith and Whitehead 1993). This pattern may result from changes in productivity or currents, as both are related to sea surface temperature in these waters (Smith and Whitehead 1993). Similarly, Berzin (1972, 202) described considerable variation between years and seasons in the fullness of stomachs from animals caught by Soviet whalers, variations he ascribes to changing ecological conditions.

In figure 2.17, the aperiodic temporal variation in the feeding success of sperm whales off the Galápagos Islands and northern Chile is summarized by plotting the estimated coefficients of variation in defecation rates against time interval between measurements. Changes in feeding success over periods of a few hours seem to be small, but reach about 60% of the long-term mean for periods of a day to several months. Over 2 years or more, the variation in defecation rates off the Galápagos Islands rises significantly (Whitehead 1996b) to approximate the long-term mean. A similarly high variation (CV = 1.42) was found between defecation rates measured off mainland Ecuador in 1992 and in 1993. Thus, the general oceanographic pattern of in-

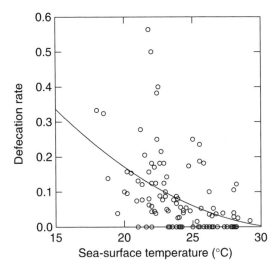

Figure 2.19 Feeding success of sperm whales off the Galápagos Islands as indicated by defecation rates (per fluke-up) plotted against 06:00 sea surface temperature on research days. (A quadratic regression line is shown; Spearman correlation coefficient $r_S = -0.493$, $P < .001$.)

creasing variation with increasing temporal scale (Steele 1991; see fig. 2.1) also seems to hold for the prey of sperm whales: the sperms are faced with substantial and prolonged food shortages in any area on a seemingly unpredictable schedule. A caution: These results come from the eastern tropical Pacific, an area with exceptionally unpredictable interannual environmental variation, mainly because of the El Niño phenomenon. It may be that sperm whales in other parts of the world have more stable or predictable food availability over these kinds of time scales.

Males and Females

Using stomach contents or defecation rates, it is possible to compare the feeding success of different classes of animals present at any place and time. A consistent pattern has been observed: in those places where mature males and females are found together, larger males have lower feeding success than females or smaller males, and the larger the males, the lower their success (table 2.6; Clarke et al. 1988; Whitehead 1993; Christal 1998; Best 1999). The effect is quite substantial, with the largest males feeding at perhaps 50–80% of the rate of females and immatures (table 2.6). This contrast could be due either to the large males having other interests while around females or

Table 2.6. Feeding Success of Different Classes of Sperm Whales Measured in the Same Area during the Same Period

Location	Measure	Females	Small Males	Medium Males	Large Males	Reference
South Africa	Flesh remains	66.4%	69.6% (<12 m)	58.0% (12–14 m)	56.4% (>14 m)	Best 1999
Peru, Chile	Feeding ratio[a]	0.58	0.62 (immature, pubertal)	0.47 (mature)	0.32 (physically mature)	Clarke et al. 1988
Galápagos (1989, 91)	Defecation rate (/fluke-up)	0.12	—	—	0.06	Whitehead 1993, 1996b
Galápagos (1995, 97)	Defecation rate (/fluke-up)	0.31	—	0.07	—	Christal 1998, 140

[a]The feeding ratio is defined as the estimated mass of squids in the first and second stomach divided by the mass of the whale.

to the smaller whales outcompeting them for available food. It is important to remember that these comparisons were made when the males were with females at relatively low latitudes, and that the males' feeding success may be higher during the probably large part of the year when they are in colder waters, away from females.

2.7   Intraspecific Competition

As sperm whales have rather little diet overlap with other species and usually eclipse their potential competitors in terms of biomass (see section 2.4), it seems that interspecific competition is not usually a key regulator of sperm whale populations. In contrast, the slow, "*K*-selected" population parameters of the sperm whale (see section 4.3) strongly indicate that intraspecific competition is important in population-level processes (see Begon and Mortimer 1986, 165–66). What evidence is there for intraspecific competition?

Feeding is generally the primary domain for intraspecific competition. Intraspecific competition in feeding can be studied in several ways, some of which are not available to sperm whale scientists. We cannot currently watch sperm whales directly competing for food or, to take the broadest approach, study their population dynamics in detail as they recover from whaling. But intermediate approaches are at least worth considering.

If intraspecific competition is important, then feeding success should be negatively correlated with population density. The huge natural variation in feeding success (see fig. 2.17) makes such a comparison difficult, but there might be some potentially useful information in our long-term data sets.

However, the correlation between measures of sperm whale abundance (the rate of encounters with groups of females and immatures) and feeding success (the defecation rate) in the Galápagos data for different years is small, positive, and not statistically significant ($r = .16$, $P = .8$, $n = 5$; data from Whitehead et al. 1997). A positive correlation between whale abundance and feeding success might be a consequence of feeding-dependent movement patterns (see section 3.4), if animals quickly leave areas with little food and linger in those where it is plentiful. As mentioned above (see section 2.6), our attempt to examine the density dependence of feeding success using defecation rates measured from different areas (Kahn et al. 1993) probably reached too far.

Another approach is to look for changes in population parameters as the population size changes. There have been attempts to do this using twentieth-century whaling data by comparing a parameter in the earlier stages of exploitation with its value later on, after the population had been reduced. But this methodology is problematic with sperm whales for several reasons. First, because of the largely unknown effects of the open-boat whalers of the nineteenth century, it is unclear what the population level was, relative to the pre-whaling numbers, in 1950 when modern whaling for sperm whales hit its stride (Whitehead 1995b; see section 4.4). Second, biological data were often not collected, or not collected well, in the earliest stages of modern whaling, and for a slowly reproducing animal like the sperm whale, some density-dependent effects would be expected to appear in the earliest stages of population reduction (Fowler 1984). And, finally, the possibility that a reduction in the density of mature males reduces the pregnancy rate of females (Clarke et al. 1980; Whitehead 1987; see section 9.1) should counteract the expected increase in pregnancy rates as populations are reduced.

Despite these difficulties, two studies have uncovered what seem to be density-dependent effects in sperm whale population parameters. Best et al. (1984) found that pregnancy rates off South Africa increased by about 15% over 10 years of exploitation, and Kasuya (1991) showed that male sperm whales, but not females, increased their growth during the 1970s. Over this period, the proportion of mature males longer than 16.8 m rose from zero to over 20%, which Kasuya attributed to a reduction in feeding competition as whaling reduced the population during the 1940s and 1950s, when these males were juveniles. Thus, these two studies provide indirect, but quite strong, evidence that intraspecific feeding competition is important for sperm whales.

It seems unlikely that any food resources, or oceanic areas where resources are plentiful, could be economically defended against other sperm whales (see section 2.9). Given the mobility of deep-ocean animals, it is similarly difficult to imagine that one animal could seriously interfere with the feeding of another, unless it swam directly in front. (In fact, as will be argued in section 6.7, sperm whales may benefit from information provided by other sperm whales feeding nearby.) Thus, intraspecific competition for sperm whales probably occurs at the level of resource depletion caused by other sperm whales feeding in the same large general area during the previous months; in such "scramble competition," increasing densities of animals lead to lower feeding success for all.

2.8   Natural Enemies

The sperm whale is not at the top of the marine food chain, and while we usually think of it as a predator, in some cases sperm whales may also be prey. The most serious predator of the sperm whale, *Homo sapiens,* has been a threat only since 1712, and so is not a "natural enemy." The relationship between humans and sperm whales is discussed briefly in section 1.6.

Of the pre-1712 predators on sperm whales, one stands out: the killer whale (*Orcinus orca*) (fig. 2.20). Jefferson et al. (1991) list six observations of killer whales attacking sperm whales, and I know of four more recent accounts (Brennan and Rodriguez 1994; two in Pitman et al. 2001; and an apparent attack off the Galápagos Islands observed by G. Merlen [personal communication]). In only one of these cases, off California in 1998, was a fatality observed, and it was likely that more than one sperm whale died in that incident (Pitman et al. 2001). In addition, Best et al. (1984) describe a sperm whale calf that stranded alive showing clear evidence of a recent killer whale attack, and Visser (1999) reports a dead 9.8 m female sperm whale that washed up off New Zealand with extensive wounds from killer whales. In neither instance is it certain that the sperm whales were killed by killer whales, although, especially in the second case, this seems likely.

In the accounts of killer whale attacks, with one exception, the sperm whales seemed to defend themselves successfully (see section 5.9 and box 5.3), even though most of the groups contained calves, which may have been the focus of the killer whales' attentions. However, the California group in which animals were killed did not contain a calf. Killer whales and sperm whales are actually more often observed together but not behaving as predator and prey. Jefferson et al. (1991) list thirty-three such "non-predatory interactions," and in our fieldwork in the eastern tropical Pacific we have seen sperm whales and killer whales in the same area on four occasions, but on

Figure 2.20 Killer whales attacking sperm whales off the Galápagos Islands during 1985.

only one of these was there any sign of attack by the killers (see box 5.3). The defensive responses of the sperm whales are described in section 5.9.

Additional evidence for killer whale attacks on sperm whales comes from observations of tooth scars, probably made by killer whale teeth, on their bodies (Shevchenko 1975, cited in Jefferson et al. 1991; Dufault and Whitehead 1998; fig. 2.21). Shevchenko (1975, cited in Jefferson et al. 1991) estimated that 65% of Southern Hemisphere sperm whales had marks of killer whale teeth, and 4% of the 305 South Pacific sperm whales flukes examined by Dufault and Whitehead (1998) possessed such marks. We cannot be certain that all these marks were made by killer whales, and in no case were they fatal.

Some attacks are deadly, however, as shown graphically by the observed attack off California. Additionally, Yukhov et al. (1975) found remains of sperm whales in the stomachs of killer whales from the temperate and tropical Pacific, although these might have been scavenged.

Besides the killer whale, what else might be dangerous to a sperm? There are reports of pilot whales (*Globicephala* spp.) and false killer whales (*Pseudorca crassidens*) harassing or attacking sperm whales (Palacios and Mate 1996; Weller et al. 1996). In none of these cases were there fatalities, al-

Figure 2.21 Marks from the teeth of killer whales on sperm whale flukes.

though pieces of flesh were observed in the water after the false killer whale incident, suggesting that this species may cause some of the marks and scars on sperm whales' bodies (Palacios and Mate 1996). It seems probable that these species rarely, if ever, kill sperm whales. So what is going on in these incidents, in which the sperm whales seem to show reactions similar to those when they are faced by the palpably much more dangerous killer whales or humans (see section 5.9)? We do not know. Suggestions include the possibilities that the false killers and pilots may be attempting to drive sperm whales away from common food resources, trying to find sick animals, "snacking" on flukes and fins, trying to induce the sperm whales to vomit food—a form of kleptoparasitism—or simply playing (Palacios and Mate 1996; Weller et al. 1996).

Large sharks are sometimes observed around sperm whales, particularly when newborn calves are present, and scars from shark bites have been found on calves (Best et al. 1984). Sperm whale remains have been found in the stomachs of sharks, but these were probably scavenging on carcasses killed by the whaling industry rather than preying on live animals (Rice 1989).

Thus, with the vivid exception of the attack off California (Pitman and Chivers 1999; Pitman et al. 2001), there is no concrete evidence that killer whales, other cetaceans, or sharks kill sperm whales. It seems, then, that while killer whales, and possibly the other species, do pose a threat, sperm whales can usually counter it. Thus predation is unlikely to be a significant influence on the mortality of adult sperm whales. However, the potential for predation, especially on calves, may have been important in sperm whale evolution. Even a small risk of calf mortality will be strongly selected against in a species, such as the sperm whale, with an extremely low reproductive rate (see section 4.3).

Although all of these potential enemies are smaller than the sperm whale, they are still of comparable size. Frequently, though, very small organisms pose the greatest threat to an animal. Sperm whales are infected by a range of microbes, helminth parasites, ectoparasites, and epizoites (see reviews by Rice 1989; Lambertsen 1997), but none of them seem to be a substantial cause of mortality (Rice 1989). Lambertsen (1997) reviewed the natural diseases of the sperm whale, concluding that only two had been identified that had the potential to cause death: myocardial infarction associated with coronary atherosclerosis, and gastric ulceration associated with anisakid nematodes, which are very common in sperm whale stomachs. However, no deaths from either of these causes have been documented.

2.9   Attack, Defense, and Escape in the Ocean

A habitat not only provides an animal with a set of prey and predators, but also serves as the stage on which an individual feeds, is fed upon, and interacts with members of its own species. Like the set of a play, but much more profoundly, the structure of the habitat underpins these actions.

Important structural features of the open ocean include its lack of refuges, the many avenues of approach in a three-dimensional medium (Norris and Schilt 1988), and the potential for sophisticated maneuvering in a medium that offers resistance in all directions over three dimensions, but no directional force field such as gravity (fig. 2.22). A few of the sperm whale's prey— those of the benthos—may be able to hide from it in mud, sand, or crevices, but most have no such resource. Some mesopelagic animals might be able to descend beneath the sperm whales' diving range, perhaps as far as 2 km (see section 3.2), but this strategy would probably remove them from their own preferred feeding habitat, so it would make sense only when sperm whales were present, and it is unlikely that they could outpace a sperm whale in a prolonged dive. Thus, common defensive measures may often be unavailable to the sperm whale's prey, so that they must rely on close-range maneuver-

Figure 2.22 A group of sperm whales under water. (Photograph by Flip Nicklin, courtesy of Minden Pictures.)

ing, which is likely to be more effective in three dimensions (Norris and Schilt 1988), or confusion to escape (see section 8.10).

Similar issues face sperm whales when they are attacked by killer whales. They may be able to outdive the killers, but must return to the surface to breathe. This strategy is unlikely to work for calves, which do not seem to be able to dive for as deep or as long as their mothers (see section 3.2). Even for older sperms, deep dives will not be an effective form of defense if, as seems likely, killer whales can track sperms under water using echolocation. Thus, to defend themselves against predators, sperm whales must rely largely on their own fighting abilities as well as those of their group members (see section 5.9). This means that, especially for groups with calves, communal defense and vigilance against threats—allowing appropriate defensive formations to be adopted before the predator closes in—become especially significant.

Animals may also wish to fend off close approaches by members of their own species, such as unwanted suitors or competitors for the attention of a mate. As with predators, there are few, if any, refuges from these unwanted conspecifics in the deep ocean. Three dimensions give the instigator of an interaction advantages and disadvantages. Take the example of a breeding male: compared with his counterpart in a two-dimensional medium, he may find approaching a female relatively easy (as there are so many avenues of approach), but guarding her against other males becomes hard (as they can also approach her easily). Forcing his attentions on her is also very difficult (she can easily maneuver away), and prolonged herding against her will is impossible. By acting in consort, several males may improve their chances of guarding, forcing, or herding a female, but these actions will still be expensive, and alliances have their own costs—most notably the sharing of mating opportunities. Thus, mating systems that do not involve guarding, herding, or forced copulations, but rather incorporate elements such as sperm competition, male dominance hierarchies, or female choice, would seem more likely to evolve in such an open habitat (see section 6.11).

Mating is not the only source of intraspecific conflict. Food resources or territory are defended by many mammals, but this becomes very difficult in the open, three-dimensional sea. I know of no case of territorial defense in the open ocean, and few of resource defense by pelagic cetaceans (but see Baird 2000; Nolan et al. 2000), although such cases might be hard to identify.

Another intraspecific interaction that is believed to influence the evolution of social structure in some species is infanticide (Van Schaik and Kappeler 1997). Females may kill other females' infants if resources are limited (Wolff 1997), and males may kill infants fathered by other males to induce estrus in the mother and improve their own mating opportunities (Packer and Pusey 1983). In the open ocean, feeding competition between any two animals will generally be of minor consequence, so infanticide by females is unlikely (Wolff 1997). Large male sperm whales could probably kill infants quite easily, and mothers would have a difficult time defending their off-spring, but, given the likely long time delay between infant mortality and subsequent conception, and the openness of the habitat, the likelihood that an infanticidal male would father the mother's next offspring is small. Thus, I think that infanticide is unlikely to be a common phenomenon in sperm whales (see sections 6.11 and 8.10).

So, overall, the open, three-dimensional habitat of the sperm whale would seem to disfavor exclusive control by an individual, or even a group of in-dividuals, over any element of the environment, whether it be geographic territory, food resources, or mates. A potential exception is in relationships between breeding males. A large male sperm whale may not be able to con-sistently keep other males away from a female, but he can easily attack another male and has the ability to inflict serious injuries (Kato 1984; Car-rier et al. 2002). Thus, in areas containing only a few males, dominance hierarchies could be set up based on actual or perceived fighting ability (see section 6.11).

2.10 Summary

1. The deep and open ocean is a habitat with great biological variation over medium to large temporal and spatial scales. Pelagic animals, such as sperm whales, must be able to deal with this variation.

2. Sperm whale habitat is usually deep (>1 km), in the vicinity (within a few hundred kilometers) of productive waters, and often near steep drop-offs or strong oceanographic features.

3. Compared with males, female sperm whales have a more restricted habitat (in terms of both sea temperature and depth) and more variable den-sities both spatially and temporally. Male sperm whales generally eat larger prey species, and larger members of the same species, than females.

4. Sperm whales feed on a wide variety of mesopelagic and bathypelagic prey, principally squids, but also other invertebrates and fish. The modal prey item is a rather slow-moving, easily caught, 400 g histioteuthid squid, but the whales also eat smaller animals, as well as much larger species that provide a more challenging capture. Most sperm whale prey do not seem to form large schools.

5. Sperm whales show rather little prey overlap with other mesopelagic and bathypelagic predators, and usually have a higher biomass than their potential competitors. Compared with some of their principal potential mammalian competitors for mesopelagic squid, sperm whales have a wider niche breadth.

6. The world's sperm whale population, although depleted by whaling, may take about 75 million metric tons of food from the ocean per year. This amount is similar to that taken by all human marine fisheries.

7. The feeding success of sperm whales shows great variation in any place over time scales of months and more, and at any time between places separated by a few hundred or thousand kilometers. There is little evidence for any consistent diurnal, seasonal, or geographic variation in sperm whale feeding success. However, large males have lower feeding success than females on the breeding grounds.

8. Changes in life history measures over the course of whaling suggest that intraspecific competition is an important factor in sperm whale population dynamics.

9. Apart from humans, killer whales seem to be the major natural enemies of the sperm whale. Although they rarely kill sperms, the threat of killer whale predation, especially on the young, may have been an important factor in sperm whale evolution.

10. In the three-dimensional medium of the open ocean, there are no refuges, and approaching other animals is generally easier than in two dimensions. However, guarding another animal, or forcing copulations, may be difficult. Thus it is probably hard for any animal, or group of animals, to maintain exclusive control of geographic territory, food resources, or mates.

3 The Sperm Whale:
On the Move through an Ocean

3.1 Why Movement Is Crucial

In a habitat where resources are highly variable in space and time, movement patterns are crucial determinants of how an animal relates to its environment (Turchin 1998, 1–7; Stevick et al. 2002). Animals that can range widely and wisely will generally have higher mean, and less variable, resource acquisition rates than others whose movements are more constrained or less well attuned to environmental variation. Prudent nomads will also have a greater effect on the densities and distributions of renewable resources, tending to reduce their temporal and spatial variation. Movement is also important in social life, being a significant factor in the quantity and quality of an animal's relationships, and so a potentially strong influence on social and cultural evolution (Rendell and Whitehead 2001b). Finally, geographic population structure is a result of large-scale movement patterns, so these movements patterns are vital input to management and conservation decisions (e.g., Soulé and Gilpin 1991). Their importance was demonstrated by the long and inconclusive debates about sperm whale stock structure within the International Whaling Commission's Scientific Committee as it tried to provide useful management advice during the 1970s and early 1980s with only few, and indirect, data on large-scale movements (Donovan 1991).

Among mammals, the sperm whale is unusual in that its vertical movements are of a similar order of magnitude to its horizontal ones over time scales of up to a few hours. As we shall see, over a 50 min dive cycle, a foraging sperm covers a span of about 3 km horizontally and 0.5 km vertically. The movement patterns of this species vary geographically, environmentally, and culturally (see sections 3.4 and 7.3). Given the unpredictable spatio-temporal structure of oceanic biomass (see section 2.1), they are likely to be important determinants of fitness. For all these reasons, this chapter examines sperm whale movement patterns.

3.2   Vertical Movements: The Sperm Whale's Dive

That head upon which the upper sun now gleams, has moved amid the world's foundations.
—Melville 1851, 417

The most characteristic movements of sperm whales are their deep dives, which they make to forage. Sperm whale dives are discussed from the perspective of a forager in section 5.4; here I will describe their general characteristics. Deep dives are usually preceded by a fluke-up (fig. 3.1), and they usually last about 30–45 min, although sometimes the whale stays under water for well over an hour. Dives are usually separated by periods of about 7–10 min spent breathing at the surface (see section 5.4; table 5.6). These long, deep foraging dives are a major feature of the behavior of sperm whales, taking up about 62% of an animal's life.*

Studies using a variety of methods give a fairly consistent picture of the whales' foraging dives. They usually descend to about 300–800 m, and occasionally to 1–2 km (table 3.1; fig. 3.2). However, there are some exceptions. The dives of males are sometimes constrained to less than 200 m by very shallow water (e.g., Whitehead et al. 1992; Scott and Sadove 1997). At the other extreme, the presence of a fresh bottom-dwelling shark in the stomach of a large male sperm whale captured over a flat bottom greater than 3,193 m deep is strong, but not conclusive, evidence that sperm whales may sometimes descend more than 3 km below the surface (Clarke 1976).

Very young sperm whales do not seem to make long foraging dives (Best et al. 1984; Gordon 1987a, 193–94; Papastavrou et al. 1989). While they can descend several hundred meters (see table 3.1), the calves usually return quickly to the surface.

Sperm whale dives are frequently, though not always, nearly vertical during the initial descent (Watkins and Schevill 1975; Gordon 1987a, 184; Papastavrou et al. 1989). Dive rates have been measured using depth sounders, sonars, and tags. Both ascent and descent rates are usually in the range of 30–100 m/min, or 1.8–6.0 km/hr (table 3.2), but descent rates tend to decrease as the whale approaches its maximum depth (Gordon 1987a, 179; Papastavrou et al. 1989), probably reflecting a change from vertical to horizontal

---

* The whales are foraging about 75% of the time (section 5.3), dive for approximately 37.5 minutes, and are at the surface for about 7.5 minutes between dives (see section 5.4), giving an overall proportion of time on dives of $0.75 \times 37.5/(7.5 + 37.5) = 0.625$.

Figure 3.1 Fluke-up of a sperm whale at the start of a deep foraging dive off Kaikoura, New Zealand. (Photograph by Linda Weilgart, courtesy of Linda Weilgart.)

movement. Much higher speeds, up to 760 m/min, were calculated by Lockyer (1977), but these may be in error, as she used a sonar, and sonars can give very imprecise estimates of ascent and descent rates when not used with nearly vertical tilt. In contrast, Mano (1986), who seems to have used methods similar to those of Lockyer, reports comparatively low ascent and descent rates (table 3.2). The reasons for this discrepancy are not clear, but Mano may have included the more horizontal central parts of the dive in his calculations.

With mean ascent and descent rates of about 65 m/min and a mean dive depth of approximately 500 m, a sperm whale spends roughly 15 min ascending and descending, with about 15–30 min of a 30–45 min dive left for foraging at depth. Thus, the sperm whale's dive often has a U-shaped profile. There are many exceptions to this pattern, however (Gordon 1987a, 179–82; Lockyer 1977; Mano 1986; Watkins et al. 2002). For example, a 12 m male radio-tracked by Watkins et al. (2002) in the Caribbean "stopped" at intermediate depths for periods of minutes. Some earlier writers (e.g., Beale 1839, 35) believed that sperm whales hang motionless while at depth, but recent evidence shows that they principally move horizontally, with some

Table 3.1. Dive Depths of Sperm Whales

Sex/age Class[a]	No. of Dives	Mean (or Range)	Maximum	Method	Reference
M	—		> 1,000 m	Line taken out	Caldwell et al. 1966
F/I		554 m	—	Line taken out	Caldwell et al. 1966
M	—		1,500 m	Line taken out	Beale 1839, 164
M	1	—	> 3,193 m	Bottom-living shark in stomach	Clarke 1976
?		—	1,145 m	Entanglement in cables	Heezen 1957
F/I < 8 m	11	495 m	650 m	Sonar during chasing	Lockyer 1977 (from table 2 using dives > 200 m)
F/I 8–9.5 m	93	457 m	550 m	Sonar during chasing	Lockyer 1977 (from table 2 using dives > 200 m)
F/I 9.5–12 m	467	438 m	950 m	Sonar during chasing	Lockyer 1977 (from table 2 using dives > 200 m)
M > 12 m	57	475 m	1,150 m	Sonar during chasing	Lockyer 1977 (from table 2 using dives > 200 m)
F	12	316 m	581 m	Sonar during chasing	Mano 1986 (from appendix 1 using dives > 200 m)
M < 12 m	3	272 m	340 m	Sonar during chasing	Mano 1986 (from appendix 1 using dives > 200 m)
M > 12 m	2	402 m	473 m	Sonar during chasing	Mano 1986 (from appendix 1 using dives > 200 m)
?		—	1,827 m	Sonar	Rice 1978, cited in Rice 1989
F/I	48	355 m	450 m	Depth sounder[b]	Papastavrou et al. 1989
M	2	265 m	270 m	Depth sounder[b]	Papastavrou et al. 1989
F/I		300–800 m	—	Depth sounder[b]	Gordon 1987a, 192–93
C	2	305 m	320 m	Depth sounder[b]	Gordon 1987a, 192–93
M	10	314 m	415 m	Depth sounder[b]	Whitehead et al. 1992
M: 12 m	93	990 m	1,330 m	Radio tag	Watkins et al. 2002
M: 11 m, 15 m	18	611 m	1,185 m	Tracked with transponder	Watkins et al. 1993
M	—		< 68 m	Bottom depth	Scott and Sadove 1997

[a]M = males; F = females; I = immatures; C = calves.

[b]Depth sounder records may underestimate dive depths, as deep animals are less likely to be picked up.

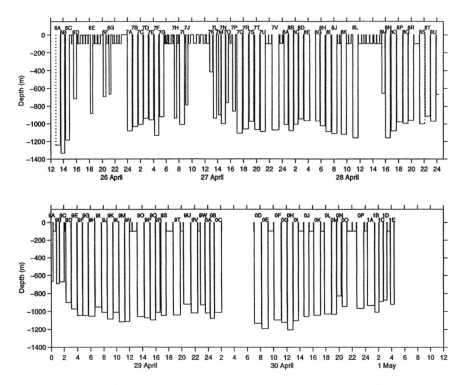

Figure 3.2 Sequence, relative duration, and depth of dives recorded during the tracking of a male sperm whale off Dominica, West Indies, using a radio tag. Dives are displayed by hour and date of occurrence. "Shallow dives" are shown as 100 m dives to distinguish these from surface activities and deep dives. Dotted lines indicate incomplete data, and blanks show periods with missing signals. (From Watkins et al. 2002.)

Table 3.2. Descent and Ascent Rates of Sperm Whales

Descent Rate (m/min)	Ascent Rate (m/min)	Method	Reference
ca. 87–132	ca. 89–141	Sonar during chasing	Lockyer 1977 (from tables 5 and 6)
28.1	26.3	Sonar during chasing	Mano 1986 (table 6)
ca. 40–100	—	Depth sounder	Papastavrou et al. 1989
ca. 45–195	—	Depth sounder	Gordon 1987a, 179
ca. 40–80	—	Depth sounder	Whitehead et al. 1992
ca. 62	—	Acoustic array	Watkins and Schevill 1975
91.2	84	Radio tag	Watkins et al. 2002 (excluding dives with "stops")
57.5	55.6	Tracked with transponder	Watkins et al. 1993
60	—	Acoustic array	Thode et al. 2002
60	75	Suction-cup tag	Madsen et al. 2002

vertical excursions (Fristrup and Harbison 2002; Gordon 1987a, 187; Lockyer 1977; Mano 1986; see fig. 5.14). Sometimes their movements under water seem fairly direct and straight (e.g., Mullins et al. 1988), while at other times they can make considerable turns in the horizontal plane (Gordon 1987a, 187; Lockyer 1977; Mano 1986). Measurement of horizontal speeds through the water at depth using sonars or acoustic tracks is error-prone because of unknown currents in the waters through which the whale is swimming and for other reasons. However, my impression from the available information, and from my own experiences during months spent tracking sperm whales, is that when at depth they usually move horizontally at speeds similar to their speeds at the surface and when ascending and descending—that is, at roughly 4 km/hr. We should soon have better data from speed-recording tags attached to sperm whales.

Thus far, I have been describing the sperm whale's deep dive, whose principal purpose is foraging.* However, sperm whales, especially when disturbed, may make "shallow dives," usually with maximum depths of 50–300 m. These dives are usually not preceded by fluke-ups, the descents do not seem to be vertical, and they are generally of shorter duration than foraging dives. Many of the dives observed from whaling vessels by Mano (1986) and Lockyer (1977) were of this type, and we sometimes seem to trigger shallow dives when our research vessel approaches within about 50 m of the whales. However, shallow dives can occur when sperms are disturbed by other events, such as the sudden appearance of other cetaceans or pinnipeds, and are also made by seemingly undisturbed animals at times (e.g., Watkins et al. 2002). The purpose of these shallow dives seems to be to remove the sperm whale from the surface for a time and to make a horizontal displacement, often away from the source of a disturbance. Shallow dives by undisturbed whales are quite rare, however, and most sperm whale diving is preceded by a fluke-up. The whale's intention seems to be seeking food rather than avoiding disturbance.

3.3   Costs and Benefits of Vertical Movement

Why dive? The sperm whale's dives consume time and energy. But there is more to it than this. The species has made a considerable evolutionary in-

---

* However, some, or perhaps many, of the deep dives reported by Lockyer (1977) and Mano (1986) during their work from whaling vessels may have been attempts to escape capture.

vestment in anatomical structures and physiological systems that give it the capacity to spend much of its life hundreds of meters below the surface (see review by Elsner 1999 of adaptations of deep-diving mammals). This investment has resulted in many trade-offs. Because they dive, and have evolved to dive, sperm whales have less time and energy for other activities, and are less good at some of them (such as fleeing from predators), than animals that have not evolved extreme diving ability.

For these ecological and evolutionary investments to pay off, the food resources available to sperm whales at depth must be substantial—and, presumably, they are. Part of the benefit of specializing on these deep-water food sources may be the relative lack of competitors, or the advantages that sperm whales have over their potential competitors. Compared with non-air-breathing predators of the mesopelagic and bathypelagic zones, such as fishes and squids, sperm whales have metabolic supremacy, as the waters in which they feed are particularly deficient in oxygen. And their large size may give them better diving abilities than many of the other deep-diving marine mammals, although they are certainly rivaled by some of the beaked whales (Ziphiidae) and the elephant seals (*Mirounga* spp.), which make dives of similar depth and duration (Stewart and DeLong 1995; Hooker and Baird 1999). Then there is the spermaceti organ, the sperm whale's extremely powerful sonar (see section 1.4), which is likely to be especially efficient in the deeper waters where background noise is lower, and refraction less of a problem, than at the surface. The strange and clumsy-looking sperm whale of the surface becomes the supreme predator of the depths.

3.4 Horizontal Movements of Females and Immatures

Movements over Time Scales from Minutes to Days

While speeds at depth are hard to measure, speeds at the surface are easily estimated, often by comparing the movements of whales with those of the observer's vessel. The available estimates suggest average speeds between about 3.5 and 6.0 km/hr (table 3.3). Speeds measured during whaling operations, both open-boat and modern, tend to be higher—perhaps because the whales were at least partially responding to the chasing. The most comprehensive data set comes from our 1985 and 1987 scan samples off the Galápagos Islands (Whitehead 1989a), in which the mean speeds of whales observed at the surface were 3.6 km/hr (SD 1.9 km/hr; $n = 1,312$ observations) in 1985 and 3.3 km/hr (SD 1.1 km/hr, $n = 2,510$ observations) in

Table 3.3.   Estimated Instantaneous Speeds of Sperm Whales through the Water at the Surface

Usual Speed (km/hr)	Maximum Speed (km/hr)	Method	Reference
4.7–8.0	22.3–39.0	Summary of observations of whalers. Maximum speeds are when towing whaleboat	Caldwell et al. 1966
5.6–7.4		Observations from modern whalers	Gaskin 1964
5.8		Aerial observations	Gambell 1968
3.7–5.6		Observations from sailboat—females/immatures	Gordon 1987a, 98
2.3–6.0		Observations from sailboat—females/immatures	Whitehead 1989a

Table 3.4.   Estimated Speeds of Sperm Whales over the Bottom over a Few Dive Cycles (Periods of About 0.5–2 hr)

Mean (km/hr)	Range (km/hr)	Observations	Reference
4.6	0.9–8.3 (most)	Males, Nova Scotia	Whitehead et al. 1992
2	0.4–4.1	Males, Kaikoura, New Zealand	Gordon et al. 1992, 31
1.9		Males, Kaikoura, New Zealand (mean distance traveled divided by mean length of dive cycle)	Jaquet et al. 2000
3.8	0–13.9	Females/immatures tracked off Galápagos Islands, displacement between satellite navigator fixes	Whitehead 1989a
2.6, 3.5		Two males radio-tracked off Dominica	Watkins et al. 1999
4.3	0–12.2	Females/immatures tracked off Chile in 2000, displacement between GPS fixes 1 hr apart	H. Whitehead, unpublished data
	1.6–3.0	Two whales, Madeira	Ohlsohn 1991, cited in Gordon et al. 1992, 31

1987. Estimates of displacements over periods of 0.5–2 hr—that is, over a few dive cycles—generally agree with the surface estimates of about 2–5 km/hr, supporting the impression that the sperms travel at similar speeds whether at the surface or at depth, although there seem to be some differences among study areas (table 3.4).

Assessments of maximum speeds are very variable. Some whalers believed that sperms could travel up to 39 km/hr for very brief periods (Caldwell et al. 1966). However, if such speeds are real, they are likely to be feasible for a minute or less. Maximum sustained speeds over more than a minute or two seem to be on the order of 18–22 km/hr (Caldwell et al. 1966).

Over time periods of an hour or more, movement can be described more rigorously. In several analyses, we have used our data from tracking and identification of sperm whales to describe how they move through the ocean (Dufault and Whitehead 1995c; Whitehead 1996b, 2001a; Jaquet and Whitehead 1999). In this section, I will go further by adding data from our tracking of sperm whales off Chile during 2000 and considering the results using an analytical framework based on that set out by Turchin (1998). The two movement measures used are displacement—the straight-line distance between the initial position of a whale or group of whales and its position after a certain time interval—and the change in heading between the beginning and the end of the time interval.

First, I will look at movements over a few hours to a few days using the positions of the research vessel as we tracked groups of female and immature sperm whales off the Galápagos Islands, off northern Chile, and in other parts of the South Pacific. The position of the boat can be taken as a fair measure of the position of the group, with a few provisos: the boat could sometimes be 1–2 km from the center of the group, a group could be spread over a kilometer or two (see section 6.4), estimates of position were not very accurate until we started using GPS in 1992, and our position recording system became much more accurate and reliable in 2000.* For these reasons, I did not use the tracking data to look at movements over time scales of less than 3 hr before 2000 or less than 1 hr in 2000. At the other extreme, the identity of the group being tracked sometimes changed inadvertently, especially overnight (Whitehead 2001a), so I restricted the analysis to tracking over less than 2 d.

I plotted the root-mean-square (Rms)[†] horizontal displacement of the tracked group against time lag, so that for any time lag τ the plot shows the average straight-line distance between the position of a group at any time and its position τ time units later. Horizontal displacement is a straight-line distance and does not take into account any wiggles in the track. The overall displacements calculated using this method are plotted in figure 3.3. Over an hour, the female and immature sperm whales moved an average of about 4 km, which is in agreement with estimates of the speeds of sperm whales

---

* Between 1985 and 1991 we used the SATNAV navigation system, which gave about 15 fixes over a 24-hour period on an irregular schedule, with each fix accurate to about 500 m. From 1992 onward we used a GPS system with continuous fixes accurate to 200 m, but until 1995 these were only transcribed each hour. In 2000, we downloaded fixes every 5 min throughout the day from a new GPS unit to a computer; at this time, the precision of the GPS system was improved to about 20 m.

† I use root-mean-square displacements rather than mean displacements, as squared displacements are a theoretically more significant measure than simple displacements (Turchin 1998, 135) and emphasize the longer movements. However, the two measures will be similar unless the distribution of displacements is strongly skewed, and it is not in our data.

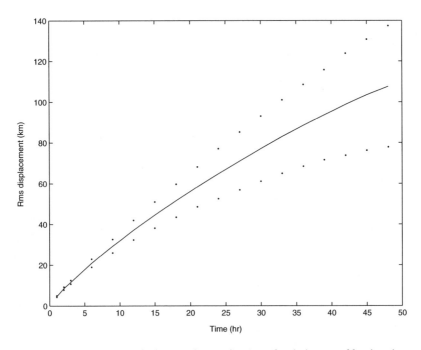

Figure 3.3 Root-mean-square displacement between locations of tracked groups of female and immature sperm whales plotted against time lag, for studies in the South Pacific (values for time lags of 1–2 hr only from studies off northern Chile in 2000, when locations were recorded more accurately). Dots above and below the line indicate estimated 95% confidence intervals for mean displacement (obtained from the mean displacements calculated separately for different studies and monthly trips off northern Chile).

over these time periods from other sources (see table 3.4). Over 6 hr, they have traveled about 20 km (in a straight line). Over 24 hr, the distance between the locations of our sperm whales was about 70 km, and over 2 d, roughly 110 km.

The curve in figure 3.3 bends to the right because the sperm whales have an increasing probability of changing their heading over longer time lags; if they always moved perfectly straight, the displacement-time plot in figure 3.3 would also be perfectly straight. I have used the data from our tracks to quantify these changes in heading, and they are shown in figure 3.4. Over 1 hr, a group changes its heading by about 5°, but after 6 hr the change is more like 50°, and over intervals of 36 hr or more, the headings of most groups are unrelated (with a median change in heading of about 90°).* De-

---

* These median changes in heading are rather less than the mean changes for our Galápagos Islands studies in 1985 and 1987, as plotted in Whitehead (1989a). This difference may be explained by the mean change being much more affected by occasional large changes in direction than the median.

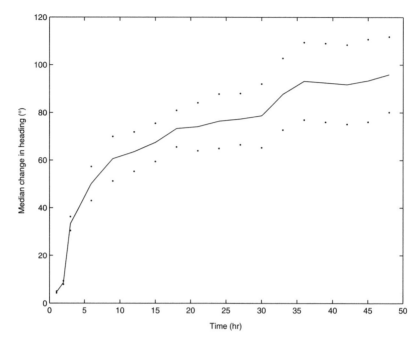

Figure 3.4 Change of heading of tracked groups of female and immature sperm whales plotted against time lag, for studies in the South Pacific (values for time lags of 1–2 hr are only from studies off northern Chile in 2000, when locations were recorded more accurately). Dots above and below the line indicate estimated 95% confidence intervals for median change in heading (obtained from comparing the median changes calculated separately for different studies and monthly trips off northern Chile).

spite the large amount of tracking data (215 d) that went into these calculations, the estimates of mean displacement and median change in heading are disappointingly imprecise (as shown by the approximate confidence intervals in figs. 3.3 and 3.4) over periods of longer than about 12 hr. The reason for this is that the movements of sperm whales over 12–24 hr vary very considerably (Whitehead 1996b; Jaquet and Whitehead 1999).

One source of such variation is the feeding success of the whales (Whitehead 1996b; Jaquet and Whitehead 1999). Over any time scale up to 2 d, groups that have had little or no feeding success, as indicated by defection rates (see section 2.6), move, on average, about one-and-a-half times to twice as far as those with the greatest feeding success, and groups with intermediate feeding success tend to have intermediate displacements (fig. 3.5). These differences were not due to variation in instantaneous speed through the water under different foraging conditions: the whales moved at similar speeds—about 4 km/hr—whether they found food or not (Whitehead

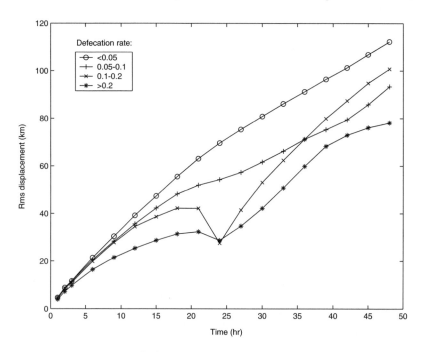

Figure 3.5 Root-mean-square displacement between locations of tracked sperm whales plotted against time lag, for four rates of feeding success (0.0–0.05, 0.05–0.10, 0.10–0.20, > 0.20 defecations per dive), during studies off the Galápagos Islands and northern Chile (values for time lags of 1–2 hr are only from studies off northern Chile in 2000, when locations were recorded more accurately).

1996b; Jaquet and Whitehead 1999). Instead, the sperms, behaving in an intuitively sensible fashion, tended to move in straight lines when there was little to eat, but frequently doubled back on themselves over scales of about 10–30 km when conditions were good, as is indicated in figure 3.6 by the plots of median change in heading for different levels of feeding success. That heading changes much more when feeding is successful is also shown by randomly selected 24 hr tracks from both the Galápagos Islands and northern Chile (figs. 3.7 and 3.8). Over periods of 18–24 hr, the groups that we followed that were finding a lot of food tended to be retracing their steps rather than continuing in the same direction (as shown by median changes in heading of > 90° in fig. 3.6).

But, as might be expected, there are other factors that influence movement patterns. Displacement rates are shown for different study areas in figure 3.9. The whales that we tracked moved most widely in the western Pacific, least off the Galápagos Islands, and showed an intermediate pattern

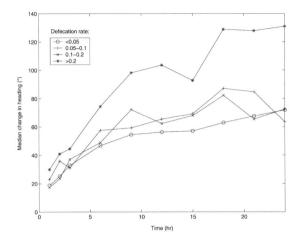

Figure 3.6 Median change in heading of tracked sperm whales
plotted against time lag, for four rates of feeding success (0.0–0.05,
0.05–0.10, 0.10–0.20, > 0.20 defecations per dive), during studies
off the Galápagos Islands and northern Chile (values for time lags
of 1–2 hr are only from studies off northern Chile in 2000, when
locations were recorded more accurately).

in the waters off the west coast of South America. And, when comparing our
largest data sets, from Chile and the Galápagos Islands, a difference in response to changes in feeding conditions seems apparent. Even given their
generally lower movement rates, the whales off the Galápagos Islands
showed a particularly dramatic response to good foraging, staying within
relatively small areas, just 20 km or so across, for periods of at least a day
(fig. 3.10; see also fig. 3.7). Off Chile, the doubling back when conditions
were good usually lasted only a matter of hours (see fig. 3.8). These differences are probably related to the different food types and oceanographic conditions in the two areas. Perhaps there are denser, smaller, or longer-lasting
food patches off the Galápagos Islands. The waters around the Galápagos Islands have much greater bathymetric structure over scales of about 20 km
than those off northern Chile (compare the 1,000 m contours in figs. 2.4
and 2.6). Off northern Chile, where the coast and depth contours run principally in a north-south direction and the potentially productive waters of
the Humboldt Current form a strip about 150 km wide oriented parallel to
the coast (see fig. 2.6), groups of whales tended to move north-south when
feeding success was low and east-west when it was high (fig. 3.11). This observation suggests that good feeding conditions for the sperms formed latitudinal bands across the Humboldt, extending perhaps 60 km in a north-
south direction.

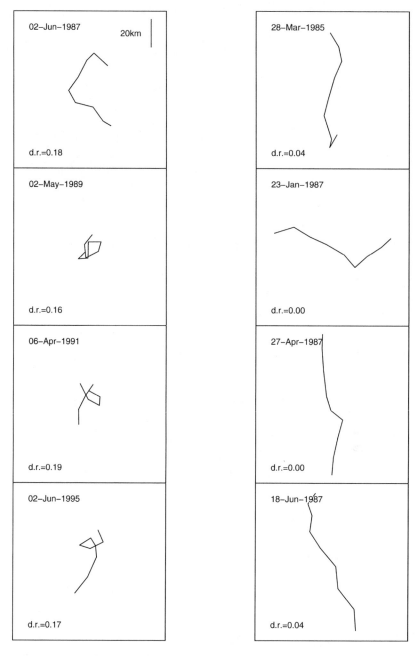

Figure 3.7 Twenty-four-hour tracks of sperm whales off the Galápagos Islands, chosen randomly from days with good (left) and poor (right) feeding conditions ("d.r." is defecation rate for that day, in feces observed per fluke-up; north is up the page; resolution = 3 hr; boxes are 1° × 1° or 111 km square).

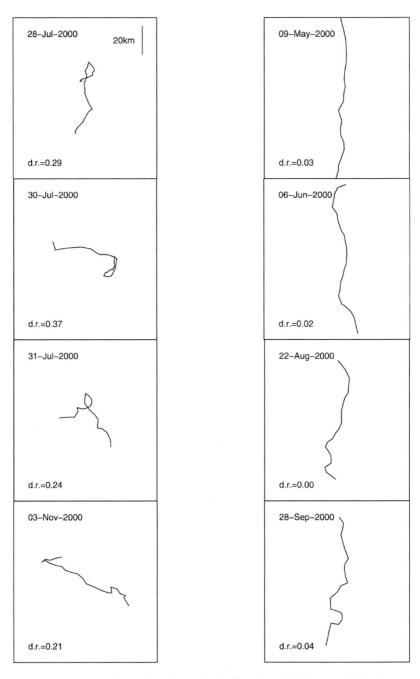

Figure 3.8 Twenty-four-hour tracks of sperm whales off northern Chile chosen randomly from days with good (left) and poor (right) feeding conditions ("d.r." is defecation rate for that day, in feces observed per fluke-up; north is up the page; resolution = 1 hr; boxes are 1° × 1° or 111 km square).

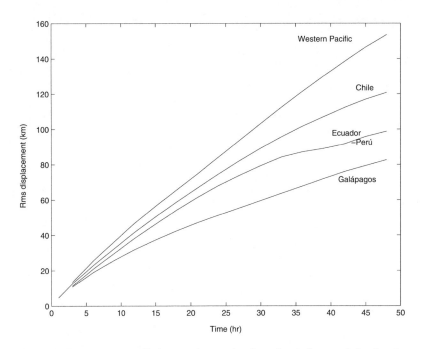

Figure 3.9 Root-mean-square displacement between locations of tracked sperm whales plotted against time lag, during studies off the Galápagos Islands, Ecuador-Peru, Chile, and in the western South Pacific.

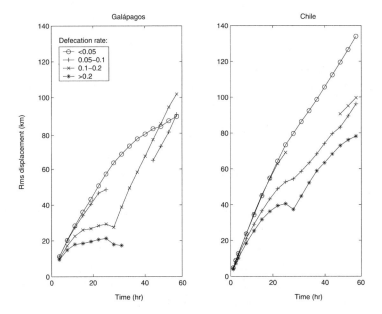

Figure 3.10 Root-mean-square displacement between locations of tracked sperm whales plotted against time lag, during studies off the Galápagos Islands and northern Chile for four rates of feeding success (0.0–0.05, 0.05–0.10, 0.10–0.20, > 0.20 defecations per dive). Missing values indicate that fewer than six displacements were available for that combination of area and defecation rate.

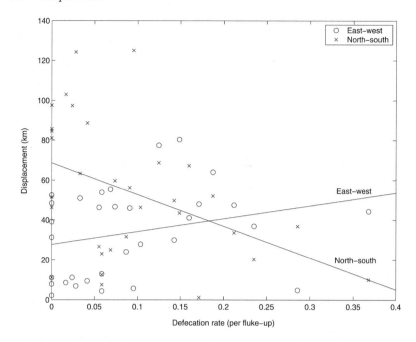

Figure 3.11 Twenty-four hour displacements of groups of female and immature sperm whales off northern Chile in 2000 in east-west and north-south directions with defecation rates. Regression lines for the east-west and north-south displacements have significantly different slopes ($P = .004$).

One final, but probably vital, source of variation in movement patterns is related to cultural identity. As will be shown in chapter 7, female and immature sperm whales can be divided into clans based on their vocal repertoires. During our studies off the Galápagos Islands, most of the whales we studied came from two clans, which had quite different movement patterns (for further details, see section 7.3).

Movements over Time Scales from Weeks to Years

As the time scale increases beyond a couple of days, the methodology for looking at horizontal movements changes. We do not have many tracks of the same group of whales over more than 3–4 d, and so far, tracks of sperm whales using radio or satellite tags are very few (but see Watkins et al. 1999 for tracks of males over a few days). Instead, we have used photoidentifications to look at movements of individually identified whales. Initially, we "connected the dots" to see where individuals did and did not seem to go (Dufault and Whitehead 1995c; fig. 3.12). It seems that females and immatures identified off the Galápagos Islands quite frequently moved to the wa-

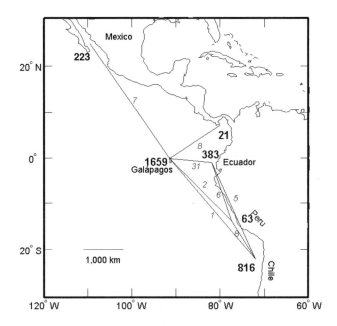

Figure 3.12 Movements of photoidentified female and immature sperm whales (photographic quality $Q > 3$) in the eastern tropical Pacific between areas off the Galápagos Islands, Chile, mainland Ecuador, Panama, Peru, and the Gulf of California (Mexico). Boldfaced numerals indicate the number of animals identified in each location; italic numerals indicate the numbers moving between locations. We also have photoidentifications from the western South Pacific (114 animals), but there are no matches between this area and any other. (Data for Gulf of California courtesy Nathalie Jaquet; data for Chile are not fully checked yet.)

ters off mainland Ecuador (1,000 km away), occasionally to those off Panama and Peru (1,500–2,000 km distant), and rarely to those off Chile and in the Gulf of California (> 3,000 km away). We have no evidence for movements of more than 6,000 km into the western South Pacific. However, such plots clearly depend heavily on where and when photoidentifications are collected. So I developed a likelihood methodology for analyzing movements using opportunistically collected individual identifications, which accounts for effort, and applied it to our South Pacific photoidentification data set (Whitehead 2001a).

In figure 3.13, I have added the results of this work to those from the tracking data to give estimates of the root-mean-square displacement of individual female or immature sperm whales in the South Pacific over time scales from an hour to decades. The displacements from the two methods do not agree perfectly because almost all the photoidentification displacements

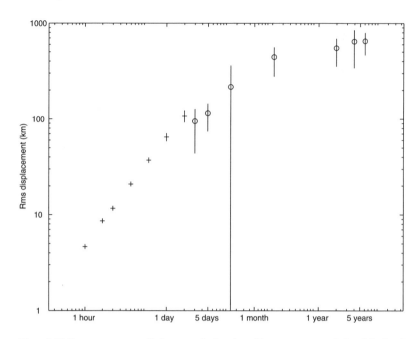

Figure 3.13 Root-mean-square displacement for female and immature sperm whales of the South Pacific over time scales from 1 hour to 6 years, from tracking data (plus signs; from fig. 3.3) and photoidentifications (open circles; data from Whitehead 2001a). The vertical bars show one standard error.

for time scales longer than 3–4 d come from the Galápagos Islands studies, where tracking displacements were lower (see fig. 3.9)—if only the Galápagos tracking displacements are used, the two curves match almost exactly. The photoidentification displacements are clearly less accurate than those based on tracking, but do show that over about 5 d the average Galápagos female is displaced about 100 km, over a month perhaps 300 km, and over several years approximately 650 km. The leveling off of the displacement-time curve over time scales of years at 650 km suggests that female and immature sperm whales have home ranges spanning on the order of 1,450 km* in any direction. The longest of the movements shown in figure 3.12, from the Galápagos Islands to the Gulf of California, is 3,400 km, but it seems that maximal displacements of about 1,000–2,000 km are more normal, as indicated by the more general movement analysis in figure 3.13.

These results agree with those from the marking of sperm whales using metal tags that were recovered from carcasses during processing (see appen-

---

* I found by simulation that the mean distance between two randomly chosen points within a circle of diameter 1 unit was 0.45 units. Thus the estimated diameter of a circular home range that is used uniformly and within which randomly chosen positions are 650 km apart is 650/0.45 = 1,450 km.

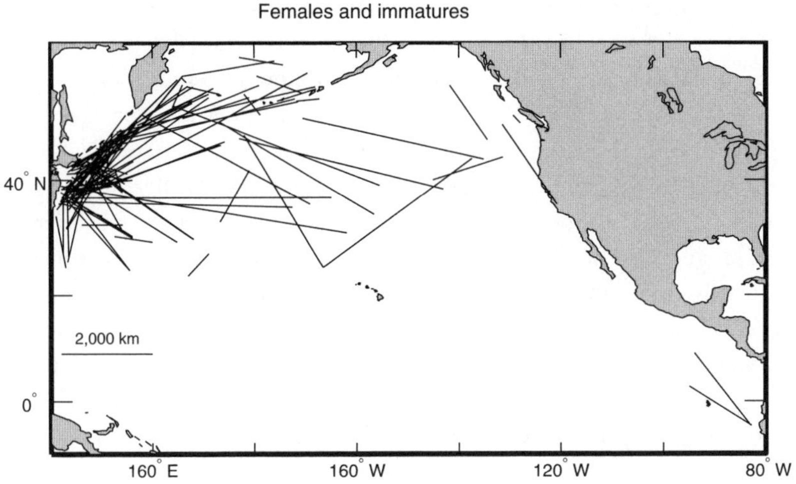

Females and immatures

Figure 3.14 Movements of female and immature sperm whales in the North Pacific between marking (with metal tags) and capture for those animals who lived for at least 10 days following marking. These data are heavily influenced by the highly nonrandom spatial and temporal distributions of marking and whaling effort. Some of the short movements off Japan are omitted. (Data from the figures and tables of Kasuya and Miyashita 1988.)

dix, section A.1). Movements of tagged females and immatures in the North Pacific are plotted in figure 3.14. There is great variation in movements, with some animals being caught very close to where they were marked and others moving several thousand kilometers.* In the Southern Ocean, twenty-two females who lived for at least 1 month after marking had a mean displacement of 690 km between the positions of marking and capture (Best 1979), which seems consonant with the North Pacific data shown in figure 3.14. These statistics clearly depend on the distributions of marking and whaling effort, which have not been taken into account. It should be possible to analyze movements from artificial marking data using a technique that corrects for effort, such as that which I carried out using photoidentifications, especially in the North Pacific, for which copious data are available (Kasuya and Miyashita 1988).

The scale of the movements of the average whale (as depicted in fig. 3.13) is vital input to models of population and social structure, but for some purposes we need information on extreme movements. The few animals that move particularly long distances have disproportionate effects on population

* We cannot calculate meaningful mean displacements for movements of tagged female and immature whales in the North Pacific from the data presented by Kasuya and Miyashita (1988), as they omitted short movements from their plots.

structure, affect the flows, and the geographic structure, of genetic and cultural traits, and may be important in repopulating an area after it has been depleted. Despite their importance, these extreme movements are hard to document, and so scientists often work backward, estimating the rates of migration between areas from patterns of genetic diversity (see section 4.1). For instance, Lyrholm and Gyllensten's (1998) study showed differences in maternally inherited mitochondrial DNA between sperm whales in different ocean basins, but no significant departure from homogeneity within any ocean. This finding suggests that females at least sometimes move across substantial parts of an ocean basin, but rarely cross between basins. The extreme movements of female sperm whales that were photoidentified—3,400 km (N. Jaquet et al., unpublished data; see fig. 3.13)—and marked—5,100 km (Kasuya and Miyashita 1988; see fig. 3.14)—are consistent with this hypothesis.

Seasonal Movements

So far, I have discussed movements of female and immature sperm whales following a nomadic model, in which the groups move around the ocean in response to variation in resources. However, since at least the time of Colnett (1798, 147), who proposed that females came to the Galápagos Islands to calve, there have been suggestions that their movements might be more regular, perhaps seasonal. Berzin (1972, 179–83) considered the data available and concluded that the evidence for regular migrations was somewhat contradictory. Evidence from South African whaling operations suggests that female groups may move toward the equator in winter, although the patterns are not particularly clear (Best 1969; Gambell 1972).

In 2000, we studied female and immature sperm whales off the coast of Chile between 18° and 25° S on what might be assumed to be a major migration route, the Humboldt Current (see fig. 2.6). While the groups we tracked were usually moving northward during the southern autumn, as predicted by a seasonal migration hypothesis, they were also moving northward in the southern spring, the reverse of the expected pattern (fig. 3.15).

In our other major study area, off the Galápagos Islands on the equator, a seasonal migration scenario is rather more complex, with the possibility that "Northern Hemisphere animals" replace "Southern Hemisphere animals" in the northern winter (Rice 1977). A prediction of such a scenario is that animals should tend to appear at the same season in different years. Because sperm whales travel in fairly permanent social units (see section 6.5), I looked at the appearance of units in different seasons. On 20 pairs of

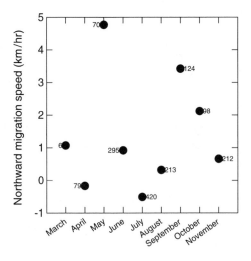

Figure 3.15 Mean northward speed of groups of female and immature sperm whales tracked off northern Chile in 2000. Positive speeds indicate general northward movement; negative speeds indicate southward movement. A seasonal migration hypothesis predicts that speeds should be positive between about March and June and negative between about September and November. Hours spent tracking in each month are given beside each plot.

occasions, the same unit was observed off the Galápagos Islands during the same calendar month, but in different years (e.g., "March 1985" and "March 1989"). However, this is almost exactly the expected number, 20.4, if particular units were no more likely to appear in any one calendar month than another, given the number of units sighted in each calendar month and year. Thus there is no evidence that certain animals replace other animals in these equatorial waters on a regular seasonal schedule.

Using the logbooks of the Yankee whalers working off the Galápagos Islands between 1830 and 1850, we have looked for seasonal changes in abundance of sperm whales around the islands (Hope and Whitehead 1991). There was no significant seasonal variation in the rate at which sperm whales were sighted by these whalers in a 10° square around the islands, but during March–April, sighting rates were about 50% higher close to the islands than farther away, suggesting that the whales tended to move in toward land at this time of year.

In the temperate western Pacific, surveys found that sperm whales were "not strikingly concentrated during the breeding season" (Barlow and Taylor 1998).

A Model of the Movement of Females and Immatures

Overall, then, while there may be some seasonal variation in sperm whale abundance in different areas, probably in relation to seasonally abundant food supplies, the females do not seem to make pronounced seasonal migrations. Instead, their movements are best described by models of nomadic animals responding to changes in food abundance by moving. The results that I have summarized indicate several of the elements in this model:

1. An overall, and fairly constant, speed through the water of about 4 km/hr

2. A fairly straight track (with median changes in heading of $<20°$/hr) when feeding success is poor, but a more recurving track (with median changes in heading of $>20°$/hr) when success is good, leading to 24 hr displacements of about 70 km when conditions are poor and 25 km when they are good

3. Constraints imposed by oceanography, particularly shallow waters and unproductive ocean areas

4. Cultural aspects of movement, with animals adopting the distinctive movement pattern of their clan

5. An overall home range for most animals spanning about 1,450 km in any direction

6. Occasional longer-distance movements by a few animals, sometimes spanning at least 5,000 km, but very rarely extending into a different ocean basin

It is likely that, over all time scales, the movements of female sperm whales are far from random: the whales move when their senses or knowledge or culture suggest that feeding will be better elsewhere. Much of their knowledge is likely to be communal.

3.5   Movements of Males

We have few data on the movements of mature and maturing males. This uncertainty is one of the biggest gaps in our understanding of sperm whale behavior. The movements of males have an important bearing on the mating strategies of both sexes, on how genes spread through the population, and on how populations respond to whaling—especially modern whaling, which has concentrated on the largest males. Male movement is more complex than that of females; it includes different ranging patterns within breeding and feeding grounds as well as movements between the two. Although our knowledge of male movement is limited, we do know that males depart from the female model both by sometimes covering larger stretches of ocean and by remaining within, or repeatedly returning to, much smaller habitats at other times.

I do not know of any measurements of the instantaneous speeds of males through the water, but when we have observed them with groups of females and immatures, the whales all move along together, and so must have similar speeds—about 4 km/hr. Away from females, the speeds of males over a dive cycle or two are similar to this, although there are some geographic differences. The males off Kaikoura, New Zealand, move comparatively slowly, while those on the Scotian Shelf move rather faster (see table 3.4). These differences probably relate to variation in foraging techniques with prey type and oceanographic conditions. Jaquet et al. (2000) found that Kaikoura males moved farther during a dive cycle in summer (mean of 1.9 km) than in winter (mean of 1.3 km), a difference that they also related to differences in foraging conditions.

Expanding to time scales of a few days and using repeat photoidentfication, we do not have sufficient data for males to carry out a movement analysis corrected for effort like the one shown for females in figure 3.13. But identifications of the same mature male from photographs taken 1–5 d apart during our studies in the eastern tropical Pacific show a range of displacements similar to that of the females, with a Rms displacement over this time span of 80 km, very similar to that of the females and immatures (fig 3.16; cf. fig. 3.13). Three males tagged by Watkins et al. (1999) off Dominica in the West Indies had displacements of 58 km in 1.9 d, 17 km in 4.6 d, and 73 km in 21 d, which are also consistent with those from our photo-identifications of females and immatures (see fig. 3.13).

At time scales of weeks to years, the picture of male movements becomes much more variable. At one extreme, some animals spent weeks or months within the canyon at Kaikoura, New Zealand, which spans about 30 km (Childerhouse et al. 1995; Jaquet et al. 2000). Some remained there, or returned there repeatedly, in up to eleven summer and winter seasons (Jaquet et al. 2000). Males in other high-latitude feeding areas, such as Bleik Canyon off Andenes, Norway (which also spans about 30 km), show similar long-term site fidelity (Ciano and Huele 2000). An individual male sperm whale, who has been named Tryphon, has returned to a small area in the St. Lawrence Estuary, Canada, almost every summer since 1991.*

However, off both Andenes and Kaikoura, other male sperms pass briefly

---

* http://www.Whales-online.net/eng/FSC.html?sct=1andpag=1.

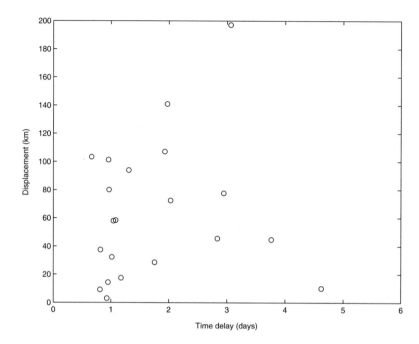

Figure 3.16 Displacement of photographically identified mature male sperm whales over time lags of 12 hours–5 days during our studies off the Galápagos Islands and mainland Ecuador. Only the first identification of each male on each day is used.

through the canyons, never to be identified again (Ciano and Huele 2000; Jaquet et al. 2000). Recoveries by whalers of artificial tags shot into male sperms show some very substantial displacements, including a mean of 1,600 km for seventeen males in the Southern Ocean with more than a month between marking and recapture—over twice the mean for females (Best 1979)—and a mean of 1,300 km (Rms 1,700 km) for twenty mature males (≦14 m) killed in the North Pacific (data from Kasuya and Miyashita 1988). The movements of these North Pacific males are shown in figure 3.17. The North Pacific males, as well as those from the Southern Ocean (see Brown 1981), sometimes move substantial distances east-west as well as north-south. Extremes include a male marked off Nova Scotia who was killed 7 years later on the other side of the Atlantic off Spain, 4,300 km to the east (Mitchell 1975), another male marked south of Mexico and killed 3 months later 4,850 km away to the northwest off British Columbia (Kasuya and Miyashita 1988), and—the record—a male marked off the North African coast at 22° N who crossed the equator and was killed off South Africa at latitude 33° S 4.5 yr later and 7,400 km away (Ivashin 1967).

Mature males

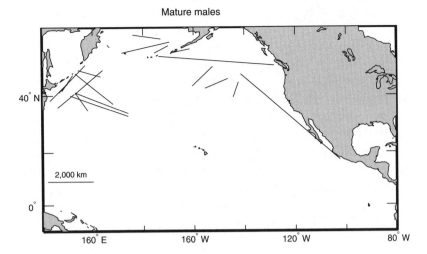

Figure 3.17 Movements of mature male (≥ 14 m) sperm whales in the North Pacific between marking (with metal tags) and capture for those animals who lived for at least 10 days following marking. These movements are heavily influenced by the highly nonrandom spatial and temporal distributions of marking effort and whaling. (Data from Kasuya and Miyashita 1988.)

A lack of any significant difference in the distributions of nuclear genes between sperm whales in different oceans (Lyrholm et al. 1999) suggests that males, at least occasionally, breed successfully in a different ocean from the one in which they were born.

There is reasonable evidence, based on whaling data (Best 1979), for north-south seasonal migrations of large males along the South African coast. However, off Kaikoura, New Zealand, where individually identified animals are studied in both winter and summer, little pattern of seasonal migration is apparent (Jaquet et al. 2000). Males are found at similar densities in both seasons, and there is no significant tendency for there to be "winter" whales and "summer" whales, although, on a much smaller scale, there are differences in the distributions of animals within the canyon in summer and in winter (Jaquet et al. 2000).

At low latitudes, there is some evidence of seasonality in the density of mature males on the breeding grounds. Their density correlates roughly with the presumed peak receptive periods of the females, but, at least off the Galápagos Islands (Hope and Whitehead 1991) and northern Chile (H. Whitehead, unpublished data), and probably elsewhere, large mature males are present on the breeding grounds throughout the year.

These and other results suggest a model for the movements of mature or maturing male sperm whales containing the following elements:

1. A general gradual, and probably irregular, movement to higher latitudes with age

2. Periodic migrations between lower-latitude feeding grounds and higher-latitude feeding grounds beginning at about age 27 (Best 1979)

3. Variable patterns of movement while at high latitudes, with some animals staying resident in small areas for long periods or making repeated returns to particularly favored locations, while others seem more nomadic (although these could be animals en route between their preferred areas)

4. While at low latitudes, nearly continuous movement within a breeding area a few hundred kilometers across as the mature male visits and revisits the resident groups of females (see section 6.11), punctuated perhaps by larger-displacement shifts to new breeding grounds

5. At the smallest scale, movement through the water at about 4 km/hr during both breeding and feeding

Of these, element 2 is the most uncertain. We do not know the duration or frequency of male migrations between feeding and breeding areas. We do not know the geographic relationship between a male's breeding and feeding grounds, nor how they are positioned in relation to the home range of his maternal social unit. We do not know whether he repeatedly uses the same breeding grounds in subsequent breeding migrations, nor whether he returns to the same feeding areas between them. Do males from feeding grounds in different hemispheres, or different oceans, mix and compete when they are with females near the equator? Are there males who feed in the Antarctic during some interbreeding periods and in the Arctic in others? My suspicion, based on the little evidence that is available, is that behavior is variable in all these respects, with males having individually distinctive movement patterns over these large spatial and temporal scales.

3.6   Movements of Other Oceanic and Terrestrial Animals

At least for the females, we have the beginnings of a model as to how sperm whales move around the ocean, and although there is little information on the large-scale temporal patterning of male migrations, we do know the spatial scales that they use. In this section, I look at these movements from a comparative perspective. Are sperm whale movements exceptional by the standards of either animals that use the same habitat or those that live in very different environments?

At the smallest temporal scale, the sperm whale's speed through the water of about 4 km/hr (see table 3.3) is comparable to average speeds of between

2 and 7 km/hr for a variety of pinnipeds, the sea otter (*Enhydra lutris*), and baleen whales (Costa and Williams 1999), and only a little above the 1.7–3.0 km/hr measured for the Humboldt penguin (*Spheniscus humboldti*) (Culik et al. 2000). Deep-diving marine mammals, whether pinnipeds or cetaceans, show speeds similar to those of sperms: 3.3–6.0 km/hr for the northern elephant seal (*Mirounga angustirostris*) (Le Boeuf et al. 1992) and about 3.9 km/hr for the northern bottlenose whale (*Hyperoodon ampullatus*) (Arch 2000, 53). However, many of the smaller odontocetes tend to move faster, averaging perhaps 9 km/hr (Costa and Williams 1999).

The sperms' horizontal displacements over intervals of hours or days are also not unusual for pelagic mammals, large pelagic fishes, and even penguins: displacements of 70 km in a day and 300 km over a month seem to be in the range of those measured for other pelagic species (table 3.5). An exception is the short displacements of only a few kilometers over 24 hr by northern bottlenose whales when they are within the highly structured environment of the Gully submarine canyon off Nova Scotia (Hooker et al. 2002). This apparent outlier is discussed in section 8.4. It also seems that the extent of the sperm whale's home range is also not particularly exceptional for large pelagic mammals (table 3.5).

In contrast, inshore odontocetes may stay within areas just a few tens of kilometers across for months or years (Connor et al. 2000b), and the movements of terrestrial mammals are even more limited. The monthly displacements of about 25 km per month (see table 3.5) of the savannah elephants (*Loxodonta africana*) studied by Thouless (1995) are for a particularly mobile population of a particularly mobile terrestrial mammal species.

So, overall, while sperm whales clearly use large areas of ocean, they are not unusual among pelagic animals in this respect. But pelagic animals as a group are much more mobile than nonavian terrestrial or inshore vertebrates.

The other principal conclusion from our analyses—that movements depend on feeding success—is not unexpected and is mirrored by results from other animals. For instance, northern right whales (*Eubalaena glacialis*) doubled back and forth within patches of plankton (mainly copepods), but moved in fairly straight lines between them (Mayo and Marx 1990).* The plankton patches spanned on the order of 100 m, rather than the tens of kilometers for the sperm whales' areas of good feeding. Traveling right whales had a median change in direction of 3.5° per 10 m (~15 s), but within food patches it was 10.2° per 10 m (Mayo and Marx 1990). These measurements

---

* This comparison was suggested by Stevick et al. (2002).

Table 3.5. Displacements of Large Marine Animals (and One Terrestrial Animal) over Time Scales of One Day and One Month and Approximate Spans of Home Ranges

Species	Displacement (km) ~1 day	~1 month	Range span (km)	Reference
Sperm whale (mature male)	~20–90	~20–400	~2,500	This chapter
Sperm whale (female/immature)	~30–70	~300	~1,450	This chapter
Gray whale (*Eschrichtius robustus*)	~85 (c,e)		~7,000	Mate and Harvey 1984
Blue whale	67, 101 (a)			Laquerist et al. 2000
Fin whale	36 (a)	~280 (a)		Watkins et al. 1996
Northern elephant seal	73–103 (a)		~4,500	Stewart and DeLong 1995
Southern elephant seal (*Mirounga leonina*)	31 (a)		~900–3,000	McConnell and Fedak 1996
Northern bottlenose whale	4 (d)			Hooker et al. 2002
Bottlenose dolphin	<24 (a)			Mate et al. 1995
Atlantic white-sided dolphin (*Lagenorhynchus acutus*)	71, 84 (a)			Mate et al. 1994b
Humboldt penguin (*Spheniscus humboldti*)	10–60 (a)	900, 60 (a)		Culik et al. 2000
Tuna, billfish	26–65 (b)		~3,000–10,000 (b)	Joseph et al. 1988, 10–11
African elephant (*Loxodonta africana*)		10–50 (a)		Thouless 1995

NOTE: Letters in parentheses indicate methods used: a, displacements between satellite tag locations; b, physical tags; c, radio tags; d, photoidentifications; e, on migration.

can be compared with the median changes in heading of about 18° per hour for Chilean sperm whales in poor feeding conditions and 30° per hour in good feeding conditions (see fig. 3.6). The temporal and spatial scales are very different for the two species, but the patterns are the same, and are similar to those of a range of other animals in a variety of habitats (see references in Mayo and Marx 1990).

3.7   Costs and Benefits of Horizontal Movement

Although water offers considerably more resistance to movement than air, this disadvantage is offset by the nearly neutral buoyancy of a marine mammal and the end of the battle against gravity. Consequently, the physiological costs of movement are similar for truly marine mammals, including whales, and terrestrial mammals (Williams 1999). Using the allometric relationship of Williams (1999), the minimum "cost of transport" for a 10 t sperm whale would be 0.6 joules/kg/m (i.e., it costs 0.6 joules to move each kg of sperm whale 1 meter). This is roughly the same cost that 10 t "runners" and "fliers" would pay, if there were such animals (extrapolating from the regressions of Tucker 1975). However, Williams (1999) notes an important difference in the sources of these costs: in runners and fliers, about 12–14% of the cost of transport is metabolic, whereas in marine mammals, with their higher basal metabolic rates, the proportion is about 40%. In other words, a bat or goat multiplies its energetic costs by about 8 when it moves, whereas for a whale the factor is only 2.5. Thus the marginal costs of movement are much less for marine mammals, which partially explains why they spend much more of their lives on the move, and so travel much greater distances, than terrestrial mammals (Connor 2000).

Why 4 km/hr? This may be the speed that minimizes the cost of transport. For bottlenose dolphins (*Tursiops truncatus*) and gray whales (*Eschrichtius robustus*), the cost of transport is minimized at about 7 km/hr, and for killer whales (*Orcinus orca*) at about 10 km/hr (Williams 1999). The less hydrodynamic shape of the sperm, together with the trade-offs it may have made through its evolution into an accomplished deep diver, may account for the difference.

But there is another factor involved: the rate of finding and acquiring food will increase with increasing speeds. Perhaps a speed of 4 km/hr optimizes the difference between the expected energy gain per unit time, which will increase with speed—almost linearly with the sparse and sluggish prey that make up most of the sperm whales' diet—and the energy expended in movement per unit time, which will also increase with speed, but more quickly, approximately as speed squared. Therefore, maximal net energy intake will be achieved by increasing speed as prey density increases until the predator starts to reach satiation levels (Ware 1975; box 3.1). This formulation does not seem to match the results for sperm whales, who travel at a similar speed whether in good or poor feeding conditions (see section 3.4). However, the formulation ignores two crucial factors: first, that sperm whales, because of their blubber layers, can store energy over long periods, and so these energetic considerations should be integrations over at least months; and second, that the faster the speed, the more likely that the animal will change habitats from poor to good feeding conditions, or vice versa. In a model that includes these factors, animals have optimal net energy gain when their speeds through the water in good and bad environments are equal (box 3.1), which is what is found with sperm whales, as well as right whales (Mayo and Marx 1990). However, it clearly makes sense to maintain a straighter track when conditions are poor.

Box 3.1 **How Fast Should a Sperm Whale Swim?**

Sperm whales use energy when traveling, but also gain it by encountering food. If their movement costs are proportional to some power, β, of their speed, v, z represents basal metabolism, and ρ is the energy gain per unit distance traveled, then the net rate of energy gain per unit time when traveling is $\rho v - z - \alpha v^\beta$, where α is proportional to the drag coefficient. In this formulation, net energy intake is maximized when

$$v = \left(\frac{\rho}{\beta\alpha} \right)^{\frac{1}{\beta-1}}$$

As β is usually close to 2, this suggests that if net energy input is to be optimized, speed should increase as prey density increases.

However, sperm whales can store energy, and by moving faster, they are more likely to move from good to poor conditions, or vice versa, so a more sophisticated formulation is called for. This model estimates the optimal speed of a sperm whale trying to maximize net energy intake averaged over long time periods while swimming through an ocean with unpredictable foraging conditions. Imagine two environments, A and B. For an animal swimming at speed v, the energy input per unit time is $\rho_A v$ in environment A and $\rho_B v$ in environment B. As the animal swims along, conditions change from environment A to environment B at a rate of λv per unit time, and from environment B to environment A at a rate of μv per unit time (this is consistent with the idea of patches of prey).

If the speed of the animal in environment A is v_A, and in environment B, v_B, then the proportion of time spent in environment A is

$$\frac{\mu v_B}{\mu v_B + \lambda v_A}$$

and similarly for environment B. Then the net expected energy gain per unit time is

Box 3.1 continued

$$E = \frac{\mu v_B \, (\rho_A v_A - z - \alpha v_A{}^\beta)}{\mu v_B + \lambda v_A} + \frac{\lambda v_A \, (\rho_B v_B - z - \alpha v_B{}^\beta)}{\mu v_B + \lambda v_A}$$

$$= \frac{v_A v_B \times [\rho_A \, \mu + \rho_B \, \lambda - \alpha(\mu v_A + \lambda v_B{}^{\beta - 1})]}{\mu v_B + \lambda v_A} - z$$

It can be shown, either by differentiation or numerically, that E is maximized when

$$v_B = v_A = \left[\frac{\rho_A \mu + \rho_B \lambda}{\beta \alpha \, (\mu + \lambda)} \right]^{\frac{1}{\beta - 1}}.$$

Therefore, to maximize net energy gain, whether in a good or a poor environment, the whales should always swim at the same speed.

But movement can entail more than energetic costs. One disadvantage of constant active traveling is that it makes maintaining social relationships more difficult. Sperm whales seem to have developed methods of coping with this problem (see chapter 6). Over larger temporal scales, substantial movements have another set of benefits and costs. As noted above, these movements allow sperm whales to avoid, or mitigate, periods of poor feeding conditions, but with a larger range come increasing difficulties in keeping track of local environmental conditions and making adaptive movements. Long-distance movements may bring animals into contact with unfamiliar prey, predators, or social circumstances. The horizontal ranges of about 1,500 kilometers that are indicated for female sperm whales and other large pelagic animals may represent a balance between using enough habitat so that acceptable feeding conditions can be found in some part of it at any time and being at least moderately familiar with its contents.

3.8 Summary

1. For a pelagic animal whose resources are highly variable spatially and temporally, movement is crucial to fitness. Movement patterns have a strong influence on social behavior and population structure.

2. Sperm whales generally dive to about 300–800 m for about 30–45 min, but occasionally they descend to 1,000–2,000 m and are under water for well over an hour. Young sperm whales do not make these long foraging

dives. Ascents and descents are often nearly vertical and are made at about 4 km/hr. When at depth, the whales move largely horizontally, seemingly traveling at about 4 km/hr.

3. Instantaneously measured surface speeds also average about 4 km/hr, and over an hour, groups of females displace about 4 km. But their headings change so that after 6 hr they have moved an average of about 20 km in a straight line, and a day later they have displaced about 70 km.

4. Movement patterns change with feeding success, with tracks typically being straight over many hours when food is scarce, but recurving over scales of about 10–30 km when feeding conditions are good, although speed through the water is not much changed by feeding success.

5. Sperm whale clans have distinctive movement patterns.

6. A female off the Galápagos Islands had displaced about 300 km after a month, and the ranges of females in the eastern tropical Pacific, and elsewhere, seem to span about 1,450 km.

7. Females occasionally make long-distance movements of several thousand kilometers, but rarely move between ocean basins.

8. There is only a little evidence for seasonal migrations by sperm whales, and these are probably not very pronounced or consistent.

9. Less is known of the movements of males, but their home ranges are probably larger than those of the females, although in some coastal high-latitude locations some males show substantial small-scale site fidelity.

10. The speed and scale of sperm whale movements are comparable with those of other large pelagic animals, but large ocean creatures range much more widely than almost any terrestrial mammals.

4 Sperm Whale Populations

4.1 Geographic Structure

There is but one species of Sperm Whale "in the ocean roving," and as far as any variety is visible, it exists in every zone and hemisphere.
—Scammon 1874, 77

Many opinions have been put forward concerning the stock identification of the North Pacific sperm whale.
—Wada 1980

With individual sperm whales regularly moving 1,000–2,000 kilometers, we would not expect much geographic population structure over such scales. But beyond this? The answer to this question is not only of evolutionary and ecological interest, but also an important input into any attempt to manage sperm whale populations. Scientists have approached the question of sperm whale population (or "stock") structure from a number of directions (see Dufault et al. 1999 for a recent review, upon which what follows is based). These approaches include catch and sighting distributions (see section 2.2), geographic variation in catch-per-unit-effort trends, movements of naturally and artificially marked animals (see section 3.4), comparison of life history measurements or parasitic infestations between areas, morphometric analyses, biochemical and genetic studies, and comparisons of vocal repertoire (see section 5.2). The earlier research, conducted in conjunction with the whaling industry, often suffered from a very unequal distribution of effort and a lack of standardized methods. A problem with some of these studies, perhaps especially the genetic ones, was a lack of consideration of the nonindependence of whales sampled from the same social group. More recent work, mostly using data collected from living animals, has addressed these issues, and so is more likely to reflect the true geographic structure of sperm whale populations.

In the past 15 years or so genetic techniques have become the dominant means of looking at geographic structure in animal populations (see, for instance, Sunnucks 2000). This is partly because genes are fundamental to the study of evolution and the conservation of biodiversity, but also because genetic techniques have become very flexible and powerful. Samples can come from commercial catches, bycatches, historical artifacts, strandings, biopsy darts, feces, or sloughed skin, and can then be combined in multi-marker studies (e.g., Lyrholm et al. 1999).

Genetic studies have been carried out on sperm whales over large and small geographic scales. They have used both mitochondrial, maternally inherited DNA (mtDNA) and nuclear markers, which an individual receives from both parents (see appendix, section A.5). Using nuclear microsatellites from 315 samples from around the world, Lyrholm et al. (1999) could find no significant differences between gene distributions in different oceans ($G_{ST} = 0.001$; $P = .232$), and alleles unique to particular oceans were rare, even though sperm whale microsatellites are quite diverse. Sperm whale mtDNA is much more homogenous (see section 7.4); however, Lyrholm and Gyllensten (1998) were able to find moderate, but statistically significant, differences between oceans ($G_{ST} = 0.030$; $P = .0007$). For neither the nuclear microsatellites nor the mtDNA were there any significant differences between areas within the North Pacific, the ocean with the largest sample ($P = .392$ for microsatellites; $P = .12$ for mtDNA; Lyrholm and Gyllensten 1998; Lyrholm et al. 1999). A similar study of worldwide mtDNA variation in sperm whales, using an independent set of samples, by Dillon (1996) found no significant differences among samples, either within the South Pacific or between ocean basins. Her sample sizes were smaller (30 independent social groups versus 136 for Lyrholm and Gyllensten 1998), which probably explains the discrepancy. Thus the genetic studies indicate that movements of both sexes covering substantial parts of ocean basins are quite common, and that males, but not females, fairly frequently breed in different ocean basins from the one in which they were born (a mean of at least one migrant per generation, and probably more).

Results from studies of markings, morphometrics, and parasites are rather confusing, but generally support the genetic patterns: there are few consistent differences between sperm whales in different oceans, and even fewer between those in different areas of the same ocean (Dufault et al. 1999). Given these findings and the data on movements over long time scales (see section 3.4), all the effort put into defining within-ocean stocks during the 1970s and early 1980s (see Donovan 1991) seems, in retrospect, rather misdirected. Instead, we have suggested that movement information be directly incorporated into sperm whale population models (Dufault et al. 1999; Whitehead 2000).

There is one character, however, that shows consistent within-ocean geographic variation: the repertoire of coda vocalizations (Weilgart and Whitehead 1997; see section 7.2). Groups of female sperm whales have distinctive, geographically specific (and statistically significant) repertoire features that differ between parts of ocean basins (fig. 4.1), but not between places about 1,000 km apart in the South Pacific (Weilgart and Whitehead 1997). For instance, off the Galápagos Islands, the last inter-click interval of a coda was often prolonged ("+1" codas), and short codas made up of fewer than five clicks predominated in the western Pacific, whereas in the Caribbean most codas contained more than seven clicks. These patterns can now be explained using the concept of clans. Clans are sets of social units with very similar coda repertoires (a common dialect; see section 7.2). While the clans have overlapping geographic ranges, certain clans seem to predominate in certain areas of an ocean, such as the "Regular" clan in the eastern tropical Pacific and the "4+" clan in the southeastern Pacific. So, coda repertoires show geographically based patterns at spatial scales (~5,000 km) at which genes, and genetically based characters, do not. This is presumably because a clan's coda repertoire, while seemingly fairly stable (Weilgart and Whitehead 1997; Whitehead et al. 1998; see section 7.2), evolves faster than genes mutate, so that the geographic pattern of coda repertoires has not been homogenized by the quite infrequent movements of sperm whales across substantial parts of ocean basins. Additionally, geographic habitat use also seems to be a part of the clan's cultural complex, which includes its coda dialect (see section 7.3).

So should we consider sperm whale populations only at scales within which genes seem to be homogeneous—oceans—or should we also consider the smaller scales of coda dialects? It might seem that, from the perspectives of biodiversity and the management of whaling or other anthropogenic threats, only genes matter. However, I think not. Compared with genes, the coda dialects, with their higher rates of evolution, probably give a better picture of how sperm whales move around oceans over the time scales important to managers and conservationists—for instance, following a severe reduction in population size, they can show whether or not there is likely to be substantial recolonization within a few generations. Additionally, just as selectively neutral nuclear microsatellites or mitochondrial D-loop markers are used to indicate potential patterns in more functionally important genes, coda dialects may serve as indicators for functionally significant cultural traits, such as movement patterns (see section 7.3). In species such as sperm whales in which culture may play an important role in determining

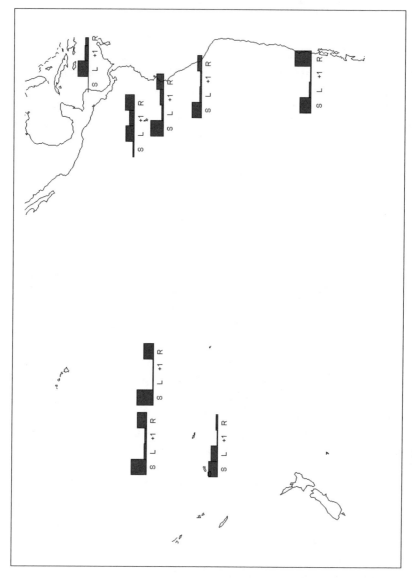

Figure 4.1 Features of regional coda repertoires within the South Pacific and the Caribbean. For each area, the bar charts show the proportional use of short (S, with < 5 clicks), long (L, with > 6 clicks), plus-one (+1, with roughly a double interval between the last 2 clicks), and regular (R, with equally spaced clicks) codas.

behavior, there is a strong argument for conserving cultural diversity for its own sake, just as we wish to preserve biodiversity.

Thus we come out of this consideration of sperm whale geographic structure with inter-ocean differences in mtDNA and regional differences in coda dialects at scales of about 5,000 km, but not much else. However, this rather weak geographic structure is overlaid upon much stronger patterns: the characteristic social units and clans of female and immature sperm whales (e.g., fig. 7.6). Indeed, according to genetic, morphological, and acoustic measures, sperm whale populations seem much more structured socially, at the level of the social unit and clan, than geographically (Whitehead et al. 1998; see chapter 7).

4.2 The Sizes of Sperm Whale Populations

With this lack of clear geographic structuring in sperm whale populations, and the whales' nearly continuous distribution in the deeper waters of the world, it is hard to delineate biologically meaningful populations at less than a global scale. So, if we are interested in numbers, we can take two approaches: consider the global population, or, if working with some smaller area, recognize that there may be substantial movement across its borders, and perhaps incomplete mixing within it. Most population studies of sperm whales have adopted the second approach. At the end of this section, however, I summarize an attempt that I have made to extrapolate from regional population estimates to obtain a feel for the global population.

Four principal techniques have been used to try to assess sperm whale populations: catch-per-unit-effort analyses, length-specific techniques, mark-recapture techniques, and surveys. They are outlined, and discussed briefly, in box 4.1.

Box 4.1 **Methods of Assessing Sperm Whale Populations**

Catch-per-Unit-Effort Techniques

During the 1970s, catch-per-unit-effort analyses were applied to the data on sperm whale catches (e.g., Ohsumi 1980). Catch-per-unit-effort techniques are, in principle, quite simple: the rate at which whalers catch whales will decrease as the population decreases. So, if

Box 4.1 continued

the catch per unit effort halves after, say, 10,000 animals have been caught, then perhaps the original population was 20,000, and the current population is 10,000 (see Caughley 1977, 16–44, for a description of these techniques).

However, various factors can complicate catch-per-unit-effort analysis when it is used on sperm whale data (Cooke 1985). These factors include recruitment and mortality (Cooke 1986b); changes with time in whaling technology (Horwood 1980) or the whaling grounds used (Smith 1980); movements of whales into or out of the whaling area (Cooke 1986b); whale schooling (Allen 1980); changing catch regulations; multispecies hunts; and falsified data (Zemsky et al. 1995). As the Scientific Committee of the International Whaling Commission gradually realized that all of these problems affected sperm whale assessments using catch-per-unit-effort analyses, disillusion grew (e.g., International Whaling Commission 1980c), and the technique was virtually abandoned in the early 1980s (Tillman and Breiwick 1983).

Age-Specific and Length-Specific Techniques

In the deliberations and calculations of the Sperm Whale Subcommittee of the International Whaling Commission, catch-per-unit-effort methodology was largely supplanted by age-specific, and especially length-specific, techniques (e.g., Cooke and de la Mare 1983). Here, instead of looking at changes in the catch rate with time, changes in the length or age distributions of captured whales were used to indicate the degree of depletion of the stock, as a highly exploited population would tend to have fewer older, or larger, animals. These methods were complex and computer-intensive.

With this new technique, some of the problems of the catch-per-unit-effort techniques (problems with changes in whaling technology and whale schooling, for instance) were no longer at issue. Others, such as falsified data (Best 1989) and movements of whales (Cooke 1986b; Whitehead 2000), remained. Furthermore, new challenges were introduced (Cooke 1986b): accurate aging methods (for age-specific methods) or an accurate age-length key (for length-specific methods) were required. After a few years of hard work, it became apparent to many that these techniques were unable to give reliable estimates (Cooke 1986b), even before the massive scale of the Soviet falsification of whaling data (Yablokov 1994) became generally known.

Box 4.1 continued

Mark-Recapture Techniques

Mark-recapture techniques have been used both with whaling data and in studies of living animals. Animals are individually identified—using either metal "Discovery"-type tags (see appendix, section A.1) or photographs of their fluke markings (see appendix, section A.3)—and then later reidentified, either by finding the tag in the carcass of a captured whale or by rephotographing the tail of a living whale. In the simplest form of this method (the "Petersen two-sample" technique), if N_1 animals are marked or photographed in the first period and N_2 are caught or photographed in the second period, of which m were marked or identified in the first period, then an intuitive population estimate is $N_1 \times N_2/m$ (see box 6.1 for a slight elaboration of this formula). There are many more complex techniques that correct for bias or allow for a number of sampling periods, recruitment to the population, mortality, and other effects (see, for instance, White and Burnham 1999).

Major problems with applying these methods to sperm whale marking data included determining whether an animal had been marked, mark shedding, nonreporting of marks in dead animals, and the geographic spreads of marking, capture, and whale movements (e.g., Cooke 1986a,b). Most of these factors, if not corrected for, will result in overestimates of population size (as the number of recaptures, m, is artificially reduced). There were rather few recoveries of sperm whale tags, and the estimates of population size obtained using this method (e.g., Ohsumi 1980) were not given much consideration by the International Whaling Commission.

With photographic identifications, we have an advantage: the "recaptured" animal is not dead and can be identified many times, giving the technique much more potential power. But this technique still presents problems, including changing marks, poorly marked animals, and a range of factors producing what is known as "heterogeneity." Heterogeneity biases population estimates downward and occurs when some animals are more likely to be rephotographed than others (Hammond 1986). Such differences among individuals could result from their movement patterns, their behavior (boat-shy or boat-curious, or more or less likely to show their flukes) or for other reasons. Among female and immature sperm whales, heterogeneity of photographic capture within groups seems to be only a small problem, although younger animals appear to be identified less frequently (Whitehead 2001b). However, some groups may behave in ways that make them more or

Box 4.1 continued

less available to photographers. Another difficulty with these techniques is that it is not often clear for what area the population is being estimated; it may be much larger than the area where photographs are being taken if animals move widely. Mark-recapture analyses using photographic individual identifications of sperm whales are currently available for the whales using the Galápagos Islands (Whitehead et al. 1997; see box 4.2), the waters off Kaikoura, New Zealand (Childerhouse et al. 1995), and the Azores (Matthews et al. 2001).

Ship or Aerial Surveys

The most intuitively simple method of assessing a whale population is to count animals in visual surveys from ships or aircraft. But there are difficulties with this method, too. These difficulties include planning coverage of a study area, estimating how much area was surveyed, considering variable sighting conditions, and whales being under water during the transit of the surveying ship or aircraft. However, there is a well-developed methodology for dealing with these issues (Buckland et al. 1993), and a number of useful estimates of the number of sperm whales in particular study areas, with confidence intervals, have appeared (see table 4.1). The principal remaining difficulty with most of these estimates is that they have not been corrected for animals under water. In technical terms, $g(0)$, the probability of sighting an animal that is present on the transect line, is usually assumed to be 1. This assumption clearly is false for sperm whales, which make long dives, and so transect population assessments in which $g(0) = 1$ is assumed represent minimum estimates.

In recent years there have been a number of attempts to assess sperm whale populations using acoustic rather than visual contacts. This method is potentially more efficient than visual surveys, as sperm whales vocalize for more of the day than they spend at the surface, sperm whale clicks are audible at greater ranges than the whales are visible, and acoustic surveys are practical under a wider range of conditions (including nighttime and fog) than visual surveys. There have been attempts at solving the major problems that need to be addressed in sperm whale acoustic censuses: estimating the range at which sperm whales can be heard and their rates of vocalizing (Whitehead and Weilgart 1990; Leaper et al. 1992; Mellinger et al. 2002). Some estimates of sperm whale numbers and densities have been derived using acoustic surveys (e.g., Leaper et al. 1992; Davis and Fargion 1996, 177–79; Barlow and

Box 4.1 continued

Taylor 1998), but, so far, visual surveys have predominated. Combined visual and acoustic surveys may be particularly productive, as by comparing visual and acoustic contacts it should be possible to estimate $g(0)$ empirically for both types of surveys (Barlow and Taylor 1998).

Catch-per-unit-effort and length-specific techniques were extensively applied to whaling data between about 1975 and 1985 by the Scientific Committee of the International Whaling Commission, but, because of the problems outlined in box 4.1, were never able to produce credible population estimates. Unfortunately, despite the acknowledgment of these problems by those involved in the assessments (e.g., Cooke 1986b), some of the estimates of sperm whale populations from catch-per-unit-effort or length-specific analyses have been reproduced in important reviews of sperm whale biology (e.g., Gosho et al. 1984; Rice 1989) and have spread from there into the more widely read literature (e.g., Berta and Sumich 1999, 428).

More recently, ship and aerial surveys and mark-recapture techniques using photoidentification have been used to assess sperm whale numbers over areas of less than an ocean basin. These methods also present difficulties, but if these difficulties are considered carefully, the resulting estimates of numbers can be used for some purposes, including, for ship and aerial surveys, a rough assessment of sperm whale densities.

In table 4.1, I present all the estimates of sperm whale population size that I know of resulting from either ship or aerial visual surveys.* Figure 4.2 shows the areas to which these estimates apply. I have corrected the estimates for animals missed because they were under water when the surveyors passed, using the estimate of $g(0)$ of Barlow and Taylor (1998), which seems to be the best available (Whitehead 2002), but is still far from perfect. In particular, it is almost certainly too large for the aerial surveys (as observers on aircraft are more likely to miss diving whales than those on ships), biasing those estimates downward. The aerial surveys represent only a very small part of the ocean covered by the estimates in table 4.1, so the effect of using a ship survey–based $g(0)$ to correct aerial surveys on the bias of the worldwide estimates discussed below is negligible.

The densities estimated in the different areas range from 0.7 to 17.0 whales per thousand square kilometers (Whitehead 2002), with an overall

* Estimates superseded by later papers with increased sampling in the same area, or better methodology applied to the same data, are omitted.

Table 4.1. Estimates of Sperm Whale Population Size, *P,* from Ship (SS) and Aerial (AS) Surveys

Area	Survey Type[a]	Population Estimate		Area (1,000 km²)	Density (Whales/ 1,000 km²)	Reference
		P[b]	CV			
Northeastern Atlantic	SS	6,013	0.32	1,226	4.9	Christensen et al. 1992
Iceland-Faeroes	SS	1,772	0.18	2,309	0.77	Gunnlaugsson and Sigurjónsson 1990
U.S. East Coast	SS	5,405	0.37	317	17.04	National Marine Fisheries Service 2000b
Northern Gulf of Mexico	SS, AS	609	0.32	258	2.36	National Marine Fisheries Service 1995
Antarctic (S of 60° S)	SS	9,540	0.17	18,548	0.51	Branch and Butterworth 2001
Eastern tropical Pacific	SS	26,053	0.24	19,148	1.36	Wade and Gerrodette 1993
Eastern temperate North Pacific	SS	24,000	0.46	7,786	3.08	Barlow and Taylor 1998
Hawaii	AS	76	0.57	81	0.94	National Marine Fisheries Service 2000a
Western North Pacific	SS	29,674	0.14	25,681	1.16	Kato and Miyashita 2000
All surveyed areas		105,671	0.13	75,354	1.40	
Total globe				316,620		

SOURCES: Whitehead 2002; Branch and Butterworth 2001.

[a]SS, ship survey; AS, aerial survey.

[b]Population size estimates (*P*) are corrected for the proportion of animals on the survey line that are counted, *g*(0).

weighted average of 1.4 whales/10^3.km^2 (CV = 0.13). The highest estimated densities are in the northwestern Atlantic between the edge of the continental shelf and the Gulf Stream, and the lowest are in the Antarctic. For smaller areas, the location of the boundary of the study area seems very important: the area off Norway, with one of the highest densities of sperm whales, abuts and overlaps the area off Iceland, with one of the lowest.

The estimates given in table 4.1 cover approximately 24% of the sperm whale's deep-water habitat (see fig. 4.2). I have used three methods to try to scale these estimates up to obtain a global estimate of sperm whale numbers (Whitehead 2002):

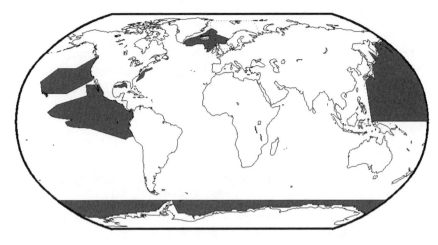

Figure 4.2 Study areas for which there is a sperm whale population estimate from ship or aerial surveys listed in table 4.1. The boundary of the survey area in the northeastern Atlantic off Norway is not available, so its rough location is indicated by a dashed line. (From Whitehead 2002.)

1. Scaling by area. I assume that the ocean areas (>1 km deep) that have not been covered by the surveys have the same average density of sperm whales as those that have.

2. Scaling by catches marked in Townsend's (1935; see fig. 2.8) charts of the positions in which Yankee whalers made their kills. This method assumes that the current density of sperm whales in any area is proportional to the number of days on which catches were made in any area as represented on these charts.*

3. Scaling by primary productivity. As found by Jaquet et al. (1996) and noted in section 2.2, sperm whale abundance seems related, over broad scales, to primary productivity, so this can potentially be used to scale estimates upward.

These three techniques produce fairly similar, and not significantly different, results. The result obtained by scaling using primary productivity is intermediate between the other two and has the greatest precision. It gives a current global sperm whale population estimate of 361,400 (CV = 0.36) animals (Whitehead 2002). While this estimate has more credibility than previous ones, there are still important sources of uncertainty, which mean that it is probably more imprecise than is suggested by the CV of 0.36 (Whitehead 2002).

* See Whitehead and Jaquet (1996) for a justification of this assumption.

4.3   Population Parameters

The global sperm whale population rises or falls depending on the balance between fecundity and mortality. Thus, to understand something of its dynamics, we must look at the population parameters that describe birth and death rates.

Fecundity

Female sperm whales first ovulate and usually conceive at about age 9, so they first give birth at roughly age 10 (Rice 1989). Once she is mature, a female sperm whale produces a single offspring only about once every 4–6 yr (Rice 1989). However, sperm whale fecundity varies greatly in a number of ways. There is variation with place; whereas most inter-birth intervals calculated from whaling data were in the range of 4–6 yr (Best et al. 1984), R. Clarke et al. (1980) estimated a mean of 3 yr in the southeastern Pacific. Off South Africa, pregnancy rates seemed to increase for older females during the course of exploitation (Best et al. 1984), whereas in the southeastern Pacific they seemed to decrease (Clarke et al. 1980). This latter trend seems to extend into more recent times, as we have found a particularly low birth rate (0.046/mature female/yr, which translates into a mean inter-birth interval of about 20 years) off the Galápagos Islands near the southeastern Pacific whaling grounds (Whitehead et al. 1997).

Both an increase and a decrease in fecundity during the course of whaling are plausible. The increase off South Africa can be explained by the standard density-dependent response: fewer whales leave more food for all, and thus a better average physical condition results in higher fecundity (see section 2.7). In the southeastern Pacific an extremely high exploitation rate of large males may have so reduced the numbers of mature males that females no longer became pregnant. This effect might linger long after the end of whaling (Whitehead et al. 1997; see section 9.1), as males seem to take an effective role in breeding only after they reach their late twenties (Best 1979). If it is real, this effect makes the sperm whale one of the few species in which male abundance has affected female fertility rates.

One of the most important sources of variation in fecundity is age. Sperm whale birth rates fall with age, with females in their forties becoming pregnant at about one-third the rate of those in their teens (Best et al. 1984; fig. 4.3). In sperm whales, in contrast to humans, killer whales (*Orcinus orca*), and short-finned pilot whales (*Globicephala macrorhynchus*), there is no clear evidence for menopause, in the sense of females routinely living for decades

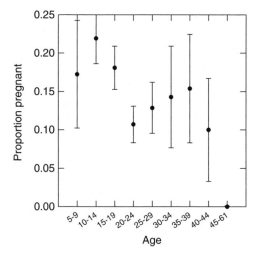

Figure 4.3 Age-specific pregnancy rates in sexually mature female sperm whales captured off Durban, South Africa. Only early pregnancies were counted; standard error bars were estimated assuming a binomial distribution. (Data from Best et al. 1984.)

after their last birth (Marsh and Kasuya 1986); however, this may simply be due to a lack of data. In the best available age-pregnancy data set (from whaling data collected at Durban, South Africa), the oldest pregnant female was 41 years old, and none of the twenty-two older animals (estimated ages 42 – 61) that were examined possessed a fetus or were ovulating, although six were lactating (Best et al. 1984; fig. 4.3). Thus it is entirely possible that female sperm whales, like humans, killer whales, and pilot whales, may stop reproducing during their forties and subsequently live for decades.

Mortality and Longevity

Information on sperm whale mortality is indirect and imprecise. When the Scientific Committee of the International Whaling Commission was trying to model sperm whale populations during the early 1980s, they used mortality rates of 0.055/yr for females and 0.066/yr for males (International Whaling Commission 1980a). These rates were obtained by looking at the age distribution of the catches, or at the length distribution using an age-length key, and observing how the proportion of animals changes with age, so that if, on average, there are x% fewer animals in the catch of age $t + 1$ than of age t, then an estimate of mortality is x. The major problem with this technique is that it includes whaling mortality as well as natural mortality, and so the estimates are an upper bound on natural mortality. Whaling mortality could not be assessed, as there was no accurate estimate of population size. These estimates have other problems, which include the assumption that mortality is constant at all ages beyond age 1, whereas the mortality rates of well-studied large mammals almost invariably show a characteristic "U-shaped" pattern with age (Ralls et al. 1980). The International Whaling Commission's estimate of mortality for juvenile sperm whales from birth to age 2, 0.0926/yr (International Whaling Commission 1980a), is even more problematic. It was obtained by plugging the other assumed population parameters for an unexploited sperm whale population (age at maturity of females, birth rate, and mortality) into a "balance equation" and seeing what juvenile mortality resulted in a population at equilibrium.

As far as I know, there is only one recent estimate of sperm whale mortality. This estimate, 0.021/yr (SE 0.066/yr), was obtained from our photoidentifications of female and immature sperm whales in the eastern tropical Pacific using a model that included movements (Whitehead 2001a). However, it is very imprecise and does not consider age-specific mortality, so it adds little, if anything, to the International Whaling Commission's attempts.

In the absence of realistic estimates of sperm whale mortality, I think that the best approach is to use estimates for another population that in other respects seems to have population parameters similar to those of sperm whales and has been well enough studied so that good mortality estimates are available. The population of resident, fish-eating killer whales studied off Vancouver Island by Olesiuk et al. (1990) meets these requirements. Its mortality rates are strongly "U-shaped," and, as expected for a sexually dimorphic species (Ralls et al. 1980), males have higher mortalities than females (fig. 4.4). These estimates were made before the rise in mortality in this population in the late 1990s, which may have an anthropogenic cause (Ford et al. 2000, 97). Sperm whales could have higher mortality rates than killer whales, as sperm whales are eaten by killer whales (see section 2.8) but killer whales probably have no natural predators. However, I believe that the killer whale mortality schedule is the best currently available for sperm whales.

How long do sperm whales live? I am afraid that, once again, our ability to answer a question about a population parameter is limited by aspects of the whaling industry and the science based on it, as well as by the constraints of modern studies. Many of the sperm whales that lived in the second half of the twentieth century were killed by humans, but scientists began estimating ages, using tooth layering, only toward the end of the catch, by which time most of the older animals had probably been killed. Also, the closure of

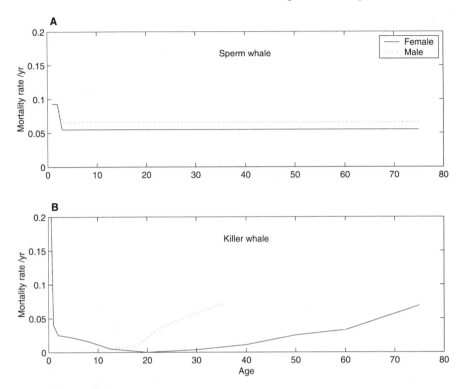

Figure 4.4 (A) Age-specific mortality rates assumed by International Whaling Commission for sperm whales. (B) Age-specific mortality rates calculated for resident killer whales off Vancouver Island (data from Olesiuk et al. 1990).

the pulp cavity within the tooth as animals age makes it hard to discern some layers, so that older animals cannot be accurately aged (Rice 1989). Thus, the oldest age estimates for sperm whales, 60–70 yr (Rice 1989), probably un-derestimate their longevity, perhaps substantially.

4.4   Population Dynamics

. . . whether Leviathan can so long endure so wide a chase and so remorse-less a havoc . . .
—Melville 1851, 571

The sperm whale's population dynamics are very uncertain, as we know so little of its critical population parameters, especially natural mortality, or its population numbers. It is clear, however, that sperm whale populations can increase only very slowly unless there is substantial immigration. If we accept the International Whaling Commission's population parameters, the maximum rate of increase of a sperm whale population with a stable age distribution is 0.9%/yr. Using the mortality schedule of killer whales and an age-specific pregnancy rate taken from the data presented by Best et al. (1984),* the annual rate of increase with a stable age distribution is 1.1%/yr. Both these numbers are very small.

Given the difficulty of assessing sperm whale populations, it is not surprising that there are only two assessments of any trend. In the Antarctic, the population of male sperm whales showed no substantial or significant trend in numbers between 1978 and 1992 (Branch and Butterworth 2001). The number of female sperm whales off the Galápagos Islands decreased dramatically between 1985 and 1999, seemingly mainly due to migration into waters off the South and Central American mainland (Whitehead et al. 1997; see box 4.2).

Box 4.2 Changes in Sperm Whale Populations off the Galápagos Islands, 1985–2002

The waters off the Galápagos Islands have been known to be favored by sperm whales for at least 300 years (Colnett 1798, 146; Shuster 1983). Their popularity is almost certainly related to the enhancement of productivity by the upwelling that is generated as the Equatorial Undercurrent, flowing westward under the equator, is forced to the surface by the islands, which lie astride the equator (see Houvenaghel 1978).

We selected the Galápagos area as a site for research on the social structure of female and immature sperm whales both because of the records of sperm whale abundance and because of the region's consistently calm seas, which are important for research efficiency. Initially, it seemed that we had chosen well. During studies in 1985, 1988, and 1989, groups of female and immature sperm whales were easily found and followed. In 1987, with warm-water El Niño conditions, the density was lower, but there was still a substantial number of animals near the islands (fig. 4.5A). During these years, large mature males constituted only about 1% of the population that we were studying (Hope

---

* The pregnancy rate for females older than age 10 is estimated to be $0.257 - 0.0038 \times$ age in years.

Box 4.2 continued

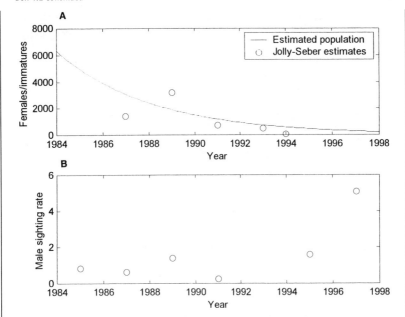

Figure 4.5 Changes in abundance of sperm whales using the Galápagos Islands, 1985–1997. (A) Estimated numbers of female and immature sperm whales using the waters off the Galápagos Islands (from Whitehead 2001a), as well as estimates for individual years using the Jolly-Seber method (from Whitehead et al. 1997). (B) Rate of sighting of mature males per day (from Christal 1998, 119).

and Whitehead 1991). But in 1991 we found groups of female sperm whales to be scarce around the islands, and the numbers grew fewer over the next decade. In 1998 and 1999, only one group of female and immature sperm whales was found—the nine members of "unit T" (see section 6.5)—and during 4 weeks of fieldwork in 2002, no females were sighted.

As we sailed around the Galápagos during the 1990s listening, so often fruitlessly, for the clicks of sperm whales through our hydrophones, we naturally wondered where all the animals had gone. The answer was supplied by photographs taken elsewhere. Eighty-eight animals first identified off the Galápagos were later identified from photographs taken off the mainland of the Americas between northern Chile and Panama, and a further seven in the Gulf of California (Whitehead et al. 1997; A. Coakes, unpublished data.; N. Jaquet et al., unpublished data; see fig. 3.12). Some of the animals might have moved westward out into the wider Pacific, but a survey of these wa-

Box 4.2 continued

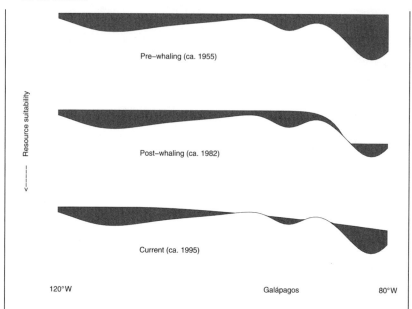

Figure 4.6 The dynamic geography of sperm whale populations of the eastern tropical Pacific from 1955 to 1995, visualized using basin theory (see MacCall 1990). In each of the three plots, the lower curve represents an approximation of the density of resources available to sperm whales (with higher resource values downward) at different longitudes (the *x*-axis), and the height of the shaded area above the curve indicates the density of sperm whales. Populations off the Galápagos Islands at 90° W decline following the end of whaling in Peru. (From Whitehead et al. 1997.)

ters in 1992 found few sperms there (Jaquet and Whitehead 1996). I have estimated that between 1985 and 1995, females and immatures were leaving the population that visits the Galápagos Islands at a rate of 24%/yr (SE 11%), and that animals were moving from the Galápagos Islands to mainland Ecuador at a substantially and significantly higher rate than in the reverse direction (Whitehead 2001a).

These findings indicated where at least some of the Galápagos animals had gone, and how quickly they went, but why? We do not know, but we have hypothesized the following scenario (Whitehead et al. 1997). Between 1957 and 1981 whalers based in Peru took several hundred sperm whales per year from the waters of the Humboldt Current (Ramirez 1989). I suspect that not too long into their operations they had removed most of the sperm whales that habitually used those waters, and thereafter were principally killing immigrants from farther west. When whaling ended in the early 1980s, the productive Hum-

Box 4.2 continued

boldt waters contained few sperm whales. We know that sperm whales move away from areas of poor feeding success (see section 3.4). Thus, during El Niño-induced, and perhaps other, periods of poor feeding, sperms moved away from the Galápagos area. Reaching the Humboldt, they found good conditions and relatively few other sperms, and stayed. In figure 4.6, I have illustrated this hypothesis (discussed in more detail in Whitehead et al. 1997) using the principle of "basin theory" (see MacCall 1990). This is, I think, the most plausible explanation for the rapid decline in female and immature sperm whale abundance off the Galápagos, but it is far from proven. Another potential explanation would involve changes in prey abundance (Whitehead et al. 1997).

As the females and immatures left the Galápagos during the 1990s, another unexpected change occurred: aggregations of large males appeared (fig. 4.5B). Prior to 1995, we had almost never found large mature males in that area apart from the groups of females and immatures, but from then on many encounters with sperm whales in these waters were with large males only, and in 2002 all encounters were with males. Their behavior, described in section 6.10 and by Christal and Whitehead (1997) and Christal (1998, 112–45), is quite similar to that of similar-sized, and presumably nonbreeding, males at high latitudes. That these nonbreeding males moved into Galápagos waters after most of the females had left suggests to me that large males are outcompeted for food by females and so, when not breeding, select areas where there are few females (see section 8.9). These areas are usually at high latitudes, but when suitable female-free habitat becomes available at low latitudes, then that habitat is also used.

Looking more widely, I have tried to produce a rough trajectory for global sperm whale populations from before the start of commercial whaling in 1700 to the present day (Whitehead 2002; fig. 4.7A). This attempt takes what we know of sperm whale population parameters (see section 4.3) and historical sperm whale catches (see fig. 1.9) and works backward from the present-day population estimate (see section 4.2) to produce the trajectory. There are a number of major uncertainties in the information input to this model, as well as in the nature of the model itself. I have tried to indicate their effects on the results by producing trajectories with input parameters chosen randomly from the distributions of those that I think might be rea-

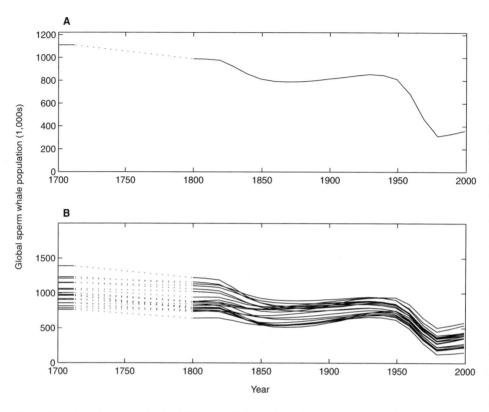

Figure 4.7 Estimated trajectory for the global sperm whale population from 1700 to 1999. (A) Trajectory calculated from my best estimates of the population and model parameters. (B) Twenty trajectories calculated using randomly chosen parameters within reasonable ranges. The period 1712–1800 is represented by dashed lines, as information is very limited. (From Whitehead 2002.)

sonable (Whitehead 2002; fig. 4.7B). Even when these uncertainties are taken into account, some patterns emerge.

First, none of the trajectories suggests that the open-boat hunt had a very substantial effect on the global population, which had been reduced by an estimate of only about 29% by 1880 (95% confidence interval 0–49%; see fig. 4.7). This finding is odd, because the open-boat whalers themselves noticed whales becoming scarcer (e.g., Melville 1851, 571–73), and their sighting rates decreased about 60% between 1830 and 1850 both off the Galápagos Islands and in the northwestern Pacific (Tillman and Breiwick 1983; Whitehead 1995b). The discrepancy between the severe drop in sighting rates over this period and the results of the model could be explained if the whales changed their schooling behavior or distributions as exploitation progressed, or if there were refuges, such as unavailable or undiscovered grounds, where the whales were fairly safe from the whalers.

The results suggest that the sperm whale population rebounded somewhat between 1870 and 1940 with diminished catches, but, except in a few of the trajectories, did not approach its unexploited level. The magnitude of modern whaling is clearly shown by all the trajectories: populations declined precipitously as the factory fleets and shore stations turned their attentions toward sperm whales after the Second World War. The results also suggest that the moratorium on sperm whaling of the 1980s came just in time to prevent severe overexploitation.

Finally, the model shows the overall effect of sperm whaling on sperm whale populations (Whitehead 2002). The initial population size seems to have been about 1,110,000 animals (95% confidence interval 659,000–1,460,000). My best guess at the parameters gives a total reduction between 1712 and 1999 of 68% (95% confidence interval 38–81%). Thus, the current population of sperm whales seems to form a fraction of their original population, but, unlike some of the baleen whale populations (Clapham et al. 1999), it is not a tiny fraction.

A final caution: While I have tried to indicate levels of uncertainty by running the model with randomly chosen parameters (see fig. 4.7B), there are many aspects of the biology and exploitation of sperm whales that are not incorporated (Whitehead 2000a). These aspects include the spatial distributions of sperm whales and whalers, age and sex structure, the effects of whaling on social structure and thus on population biology, temporal changes in resource availability, and modern anthropogenic influences, such as chemical pollution, climate change, fisheries, and noise. All of these factors add uncertainty beyond that suggested in figure 4.7.

4.5 Summary

1. Genetic studies have been unable to find significant differences in gene distributions among sperm whales of different areas within ocean basins. Differences between oceans have been found in the distribution of maternally inherited mitochondrial genes, but not biparentally inherited nuclear genes. These results suggest that movements of both sexes covering substantial parts of ocean basins are quite common, and that males, but not females, fairly frequently breed in a different ocean basin from the one in which they were born.

2. There are regional differences in the repertoires of coda vocalizations of female and immature sperm whales over scales of about 5,000 km, which result from the geographic distributions of clans with distinctive coda dialects.

These dialects constitute the only sperm whale character to show clear geographic variation over scales of less than an ocean basin.

3. Sperm whale density averages about 1.4 whales per thousand square
kilometers, but varies very considerably between ocean areas.

4. Extrapolating from the results of visual surveys, I conclude that there
are very approximately 360,000 sperm whales alive today.

5. Female sperm whales first give birth at about age 10. Inter-birth intervals are generally about 4–6 yr, but increase with age, so that pregnancy is
very rare after age 40.

6. Sperm whale mortality rates are poorly known. I suspect that the killer
whale's age-specific mortality schedule is closer to that of sperm whales than
are the constant mortality rates that were assumed by the Scientific Committee of the International Whaling Commission. Sperm whales can live at
least 60–70 years, and perhaps much longer.

7. Following whaling, there has been a decrease in the numbers of females
and immatures off the Galápagos Islands as animals moved eastward to waters off the South and Central American mainland, which had been heavily
whaled.

8. There seem to have been about 1,110,000 sperm whales alive before
whaling, but this number has now been reduced by very approximately 68%
by the two great sperm whale hunts.

5 Sperm Whale Behavior and Vocalizations

5.1 Measuring Behavior

We have two principal windows on the behavior of the sperm whale: what we can see from above the surface and what we can hear through hydrophones (underwater microphones). Although much of sperm whale behavior is expressed neither at the surface nor acoustically, together these two channels give us an insight into the daily lives of the whales. They also, to some extent, complement each other, as sperm whales are usually vocal at depth (see section 5.4) and probably perform most of their social behavior at or near the surface (see section 5.5). However, the description of sperm whale behavior in this chapter is based on far from perfect information, and many functionally important activities are undoubtedly missing. Some of these will be revealed in the coming decades as new approaches are applied to the study of sperm whale behavior (see section 9.3, Whitehead et al. 2000a).

For now, we have notes of what we can see the whales doing from above the surface and recordings of sounds—basically different patterns of clicks (see section 5.2). Thus, after describing the sounds produced by sperm whales, I will describe their visually observable behavior, mainly using "scan samples" collected off the Galápagos Islands during studies of female and immature whales in 1985 and 1987. There is no similar data set for mature and maturing males, but I will bring in what we know of them after discussing the results for females and immatures.

There are some important aspects of sperm whale behavior that occur fairly rarely, and so are not usually captured by routine scan samples. Some of these, such as defense against predators and stranding, are covered in the later sections of this chapter. Others, including aggressive behavior and mating, are included in chapter 6, on social structure.

As so much of the sperm whale's behavior is social in the broad sense, the division of material between this chapter and chapter 6 is somewhat arbitrary. Some of the constituent parts of sperm whale social structure that I consider in some detail in chapters 6 and 7 are also mentioned in this chapter. Here they are listed with brief definitions and the section of the book in which they are fully considered.

> *Cluster:* Animals swimming side by side in a coordinated manner a
>   few body lengths apart (6.8)
>
> *Group:* Animals moving in a coordinated fashion over hours or
>   longer (6.4)
>
> *Social unit:* Animals who stay together, and move together, over years
>   or more (6.5)
>
> *Aggregation:* Animals in a particular area, a few kilometers across, at a
>   particular time (6.3)
>
> *Clan:* Animals that use a similar repertoire of codas (7.2)

5.2   Vocalizations

Whales have some mysterious way of communicating with each other, although there may be miles of water between them.
—Hopkins 1922, 61

Patterns of Clicks

It is now clear that the vast nose of the sperm whale is principally a massive click producer (see section 1.4), and, not surprisingly, nearly all of the sounds heard from sperm whales under water are clicks. These vocalizations are not only the key to how sperm whales interact with one another and their environment, but also an indispensable tool with which we can study many elements of their behavioral and population biology (see appendix, section A.4).

Sperm whale clicks are sharp-onset, broadband, impulsive vocalizations with energy between 5 and 25 kHz (Madsen et al., in press). As noted in section 1.4, two or more pulses can often be heard, with the energy decreasing in consecutive pulses (see fig. 1.7). The inter-pulse interval is the time taken for a pulse to travel twice along the spermaceti organ (Norris and Harvey 1972; Gordon 1991a; Goold 1996). The clicks can be very powerful—up to 223 dB re 1 μ Pa @ 1 m (Møhl et al. 2000), the highest biologically produced source levels that have ever been recorded (Madsen et al. 2002)—and are very strongly directional (Møhl et al. 2000; Thode et al. 2002).

Sperm whales arrange their clicks in various patterns (usual clicks, slow clicks, creaks, codas, etc.), and use them in a variety of circumstances. The two broad classes of function ascribed to sperm whale clicks are echolocation and communication. Although it is likely that each pattern of clicks has one

Table 5.1. Approximate Characteristics of Principal Types of Sperm Whale Clicks

Click Type	Apparent Source Level (dB re 1μPa [Rms])	Directionality	Centroid Frequency (kHz)	Inter-click Interval (s)	Duration of Click (ms)	Duration of Pulse (ms)	Range Audible to Sperm Whale (km)	Inferred Primary Function
Usual	230	High	15	0.5–1.0	15–30	0.1	16	Searching echolocation
Creak	205	High	15	0.005–0.1	0.1–5	0.1	6	Homing echolocation
Coda	180	Low	5	0.1–0.5	35	0.5	~2[a]	Social communication
Slow	190	Low	0.5	5–8	30	5	60	Communication by males

SOURCE: Data from Madsen 2002; Madsen et al., in press, with additional material from section 5.2 in this volume.

[a]This value is inferred from values for other click types and subjective relative audibility of click types at sea.

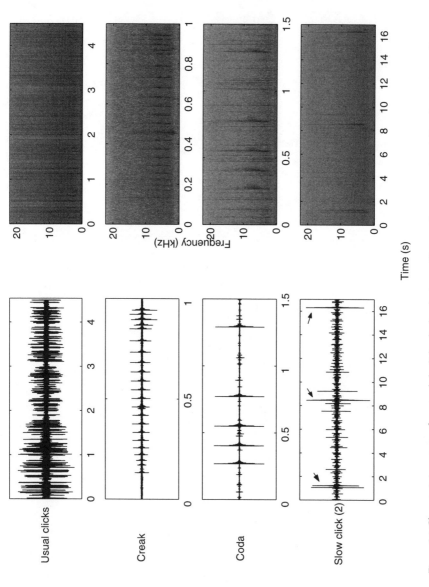

Figure 5.1 Characteristic patterns of sperm whale clicks, displayed using oscillograms (pressure vs. time; left) and spectrograms (frequency vs. time; right): usual clicks recorded from a group of foraging sperm whales (each whale makes about two clicks per second); a creak from a single whale; a five-click coda (this one is called "3 + 1 + + 1"); and a slow click sequence (arrows) over a background of usual clicks.

of these as its primary function, it is also possible that some patterns may have both broad functions (Backus and Schevill 1966; see below). Each pattern of clicks is usually heard under a characteristic set of circumstances, and thus sets of clicks are categorized by their temporal patterning. The clicks used in different patterns are generally structurally different (Madsen 2002; table 5.1; see below).

Scientists have categorized click patterns in different ways, and have given the same pattern different names. However, most of the patterns are quite distinct, and a majority of sperm whale scientists would probably agree with the following categorization (although maybe not with the names I use).

Usual Clicks

In the most common pattern, known as "usual clicks," sperm whales make a long train of regularly spaced clicks, often lasting for minutes. Inter-click intervals within usual click trains average about 0.5 s for females and immature males and 1.0 s for mature males (Whitehead and Weilgart 1990; Gordon 1991b; Goold and Jones 1995; Jaquet et al. 2001; Madsen et al. 2002). Usual click trains are almost universally made by sperm whales during deep dives and, especially with groups of females and immatures, are often heard as a jumble of many usual click trains superimposed on one another (fig. 5.1).

The great majority of sperm whale scientists (including Backus and Schevill 1966; Norris and Harvey 1972; Gordon 1987a, 221–24; Weilgart 1990, 136–39; Goold and Jones 1995; Møhl et al. 2000; Jaquet et al. 2001; Madsen et al. 2002) interpret usual clicks as a form of searching echolocation (i.e., sonar) with which the animal scans the waters in front of it for potential prey. In contrast, Watkins (1980) discounts echolocation as the primary function of sperm whale clicks, instead considering them more likely to be "contact calls" (see also Watkins et al. 1985). In box 5.1, I summarize the arguments against this theory. The patterns of use of usual clicks by foraging sperm whales are described and discussed in section 5.4.

Box 5.1 **Are Sperm Whale Clicks Echolocation?**

While nearly all sperm whale scientists agree that echolocation is the primary function of the sperm whale's clicks, William Watkins and his colleagues disagree with this conclusion (Watkins 1980; Watkins et al.

Box 5.1 continued

1985). Instead, they consider it more likely that the regularly spaced clicks that sperm whales make while under water (usual click trains) are social "contact calls." While noting that the frequency range and general form of sperm whale clicks suggest echolocation, Watkins (1980) has listed five features of sperm whale sounds that he believes make them unsuitable as echolocation signals. Jonathan Gordon (1987a, 221–26), a supporter of sperm whale echolocation, has replied to these arguments. With more information now available, an even stronger case for echolocation can be made. Here are Watkins's five features, each with a pro-echolocation reply (updated from points originally made by Gordon 1987a, 221–26):

1. "Sperm whale clicks do not appear to be highly directional." The recent work of Møhl et al. (2000), confirmed by Thode et al. (2002), shows the clicks to be highly directional, with perhaps 35 dB differences in source levels for the same click in different directions.

2. "Click repetition rates are generally very regular and have not varied with changing distances to an approaching obstacle." While click repetition rates during usual click trains are regular, I hypothesize that these clicks are made before the whale has found an interesting target. Once a target is found, it seems, the whale often begins creaks, which usually accelerate in a manner appropriate for an animal approaching a target (Gordon 1987a, 211; Jaquet et al. 2001; see section 5.4). Furthermore, inter-click intervals between usual clicks tended to decrease before creaks (Jaquet et al. 2001), which is also consistent with echolocation as an animal closes in on its target.

3. "Sperm whales are silent for long periods, especially when they are alone." Sperm whales are silent for long periods, particularly when they are resting or socializing near the surface (see section 5.5). However, solitary male sperm whales usually make clicks while foraging under water (Mullins et al. 1988). Sperm whale clicks are much more consistently associated with periods of foraging than with socializing (Whitehead and Weilgart 1991; see section 5.3).

4. "The level of their clicks appears to be generally greater than that required for echolocating prey or obstacles." Sperm whale clicks are loud, but the louder the echolocation signal, the longer the range, and the more effective the signal would be (Gordon 1987a, 226; see also calculations by Goold and Jones 1995, although these need to be updated using the results of Møhl et al. 2000). Fristrup and Harbison (2002) and others have pointed out that squid are poor targets

Box 5.1 continued

for echolocation, and so a strong signal is required to return a useful echo.

5. "Individual clicks are generally too long for convenient echolocation (a 30 ms click would obscure echo information from the first 22 m), and the duration of clicks usually remains constant throughout the click series." Sperm whale clicks are long, and long clicks have less discriminatory power. However, Jaquet et al. (2001) note that click length decreases from about 17 ms during usual clicks (which we interpret as long-range echolocation) to 3.6 ms within creaks (short-range echolocation). Furthermore, the duration of the pulse within the click, rather than the duration of the click itself, may set the level of discrimination. A 0.1 ms pulse (Madsen 2002) should be able to distinguish about 7 cm. Thus the length of the clicks changes adaptively as the whale draws closer to its prey, and, during the approach, the clicks shorten to give much better discrimination than indicated by Watkins's (1980) calculation.

Thus, Watkins's (1980) arguments against sperm whale echolocation do not hold up in the light of the evidence now available.

Creaks

On occasion, clicks can be much more closely spaced than the 0.5–1.0 s of usual click trains. Patterns of closely spaced clicks are usually called creaks (see fig. 5.1), and they may sound like an opening rusty hinge. Inter-click intervals within creaks range from 5 to 100 ms, and creaks last from 0.1 to 45 s. Jaquet et al. (2001) found that clicks within creaks were substantially shorter than those within usual click trains (~4 ms vs. 17 ms), and Madsen et al. (in press) have shown that they are less powerful but much shorter (table 5.1).

Creaks are produced by sperm whales both at depth and at the surface. When creaks are made at depth, the click rate often accelerates over the course of the creak, which can be interpreted as the sperm whale homing in on its prey (see section 5.4). Surface creaks tend to be shorter and to have more constant inter-click intervals, which might befit a sperm whale scanning its nearby social partners or our boat (see section 5.5). These surface creaks are sometimes designated differently from the creaks made at depth and may be subdivided into "coda-creaks" (Weilgart 1990, 85), "rapid clicks," or "chirrups" (Gordon 1987a, 213; Goold 1999).

Codas

The most unusual, and in many ways the most interesting, click pattern of the sperm whale is the "coda," a pattern of three to about twenty clicks lasting about 0.2–5 s (see fig. 5.1). Codas are sometimes made at the end of a usual click train, which is why Watkins and Schevill (1977b), following musical terminology, called them "codas." However, codas are more often made in other circumstances, particularly in exchanges with other whales (Watkins and Schevill 1977b), so the notation is not as apt as it initially seemed. One whale may make "click-click-click-pause-click" and another may appear to reply "click-pause-click-click-click-click." Codas can also be heard by themselves, with no apparent exchange involved, as well as in very complicated, multiply overlapping sequences in which several animals seem to be vocalizing at once. In the duet-like "echo-coda" two whales make almost exactly the same coda nearly simultaneously (Weilgart 1990, 109–11).

Structurally, the clicks within codas seem to differ from those that make up usual click trains, with codas generally having more pronounced secondary pulses (see fig. 1.7), less directionality, and reduced power (table 5.1; Madsen et al. 2002). Madsen et al. (2002) suggest that this may be achieved by changing the shape of the distal air sac to allow for more reflections within the spermaceti organ (see fig. 1.6), and thus a less directional click that is better suited for communication than for echolocation.

Codas containing a particular number of clicks, such as six-click codas, can be subdivided into quite well defined and distinct "types" depending on their relative inter-click intervals (Weilgart and Whitehead 1993; fig. 5.2, table 5.2; see box 7.1). Thus, in this respect, they are substantially discrete signals. We, and others (e.g., Moore et al. 1993), have largely categorized codas in this manner. Other features of codas, such as length, absolute inter-click intervals, and characteristics of the constituent clicks may also contain important information for the whales themselves, but so far have not been analyzed in much detail.

During, say, a 5 min recording of fifty codas, perhaps five different coda types (patterns of relative inter-click intervals) might be heard, although this varies considerably. Some coda types are very common, such as the "regular-five" coda with five equally spaced clicks, while others, especially those with very irregular inter-click intervals, may have been heard only once or twice (see table 5.2). Codas are used nonrandomly within exchanges; some codas, at least off the Galápagos Islands, tend to initiate exchanges, and some types tend to follow other coda types (Weilgart and Whitehead 1993).

Watkins and Schevill (1977b), in the first report on codas, noticed that

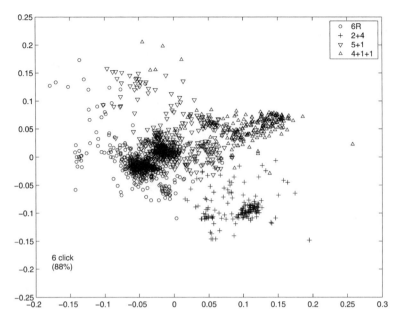

Figure 5.2 Six-click codas displayed using the first two principal components of the relative inter-click intervals, which together represent 88% of the variance among the relative inter-click intervals of the codas. Each coda is represented by one plot point. Codas were clustered into four types using *k*-means cluster analysis. The four types (the regularly spaced 6R type plus three others that were lumped into 6Var by the earlier analysis summarized in table 6.1) are represented by different symbols.

when they recorded codas on an array of hydrophones, it seemed that, during an exchange between two whales over a minute or two, each animal would prefer one coda type. They therefore proposed that coda types identified individuals, like the signature whistles of bottlenose dolphins (see Caldwell et al. 1990), and this suggestion has propagated through the literature (e.g., Tyack 1999). Our evidence, and that of others, strongly indicates that this is not the case, as individuals make several coda types, different individuals make the same coda type, and in any population there are many fewer coda types than individuals (Gordon 1987a, 218–20; Weilgart and Whitehead 1993; L. Rendell, unpublished data). In the extreme case of the sperm whales in the western Mediterranean Sea, 134 of 138 codas recorded by Pavan et al. (2000) were of just one type, "click-click-click-pause-click." However, recent unpublished studies off Crete in the eastern Mediterranean Sea, where social groups of females and immatures are more common than in the west, have found a much greater range of coda types (A. Frantzis and P. Alexiadou, unpublished data). In the West Indies and off the Galápagos, about twenty coda types are commonly recorded, although only a few are

Table 5.2. Classification (Using *K*-Means Cluster Analysis) and Frequencies of Codas Recorded in the South Pacific

Name[a]	Representation[b]	No. of Codas
"3R"		625
"3a"		131
"3b"		272
"1+2"		90
"2+1"		20
"4R"		209
"3+1"		51
"3++1"		68
"4L"		69
"4Various"		72
"4+1"		227
"4++1"		77
"5R"		269
"2+1+1+1"		53
"5Various"		55
"5+1"		160
"4+1++++1"		99
"6R"		149
"6Various"		98
"7R"		131
"5+1++1"		56
"6+1"		61
"7Various"		150
"8R"		109
"8L"		54
"8"		98
"9"		79
"10"		63
"11"		31
"12"		18

SOURCE: Weilgart and Whitehead 1997.

NOTE: A much larger sample has now been analyzed and recategorized into 70 coda types, see section 7.2).

[a] "*x* Various" categories are an amalgamation of all categories of codas with *x* clicks that had fewer than fifty codas allocated.

[b] Shown for each category is an approximate representation of the modal click pattern and mean duration (so "| | | |" is "click-click-click-pause-click"), except for "*x* Various" categories, where the representation is for a randomly chosen coda.

common to both areas (Moore et al. 1993; Weilgart and Whitehead 1993). Thus, there is regional geographic variation in coda types and other coda characteristics (see fig. 4.1; Weilgart and Whitehead 1997). This variation results from the distinctive coda dialects of particular vocal clans, and perhaps social units, which have preferred geographic ranges (see section 7.2). There may be individually distinctive aspects of codas, but it seems that these are in the characteristics of the clicks themselves, or in subtle aspects of their temporal patterning, rather than in the types of codas used.

A young calf kept briefly in captivity during rehabilitation made codas (Watkins et al. 1988), but large males rarely do (Gordon et al. 1992; Goold 1999). As young calves are highly social animals (see section 6.9) but large males are not (see section 6.10), this observation reinforces the linkages between coda production, communication, and sociality (see section 5.5).

Slow Clicks

Codas may be the most perplexing of the click patterns of the sperm whale, but the slow click, called the "clang" by Gordon (1987a, 215–16), is the most arresting. It poses a number of so far unsolved puzzles, but seems to be a key to elements of sperm whale behavioral biology, perhaps especially mating systems. The slow click is a loud, ringing click repeated about every 5–8 s (fig. 5.3). It is distinguishable from the clicks that make up usual click

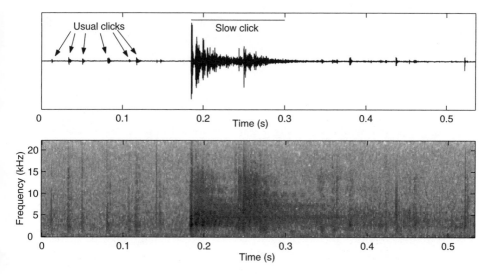

Figure 5.3 Structure of the slow click, or clang, made by mature males, displayed by oscillogram (top; pressure vs. time), and spectrogram (bottom; frequency vs. time). Compared with usual clicks on the same recording, the slow click is long, loud, and ringing.

trains, creaks, and codas not only by its much lower repetition rate, but also by its duration and its frequency structure, which includes emphasized "ringing" frequencies (Gordon 1987a, 215; Weilgart and Whitehead 1988; Goold 1999), emphasis of low frequencies, and much lower directionality (table 5.1; Madsen et al., in press). Generally, slow clicks seem louder than usual clicks or creaks (Gordon 1987a, 215; Weilgart and Whitehead 1988; Goold 1999), but this is probably at least partially a result of their lower directionality and frequencies. Whereas boat-based hydrophone systems can pick up usual clicks at ranges of about 4.5 km, slow clicks can be heard at an average of 20 km and sometimes more (Barlow and Taylor 1998; H. Whitehead, personal observation). Madsen et al. (in press) estimate that sperm whales may be able to hear slow clicks at ranges of about 60 km, as opposed to 6 km for creaks and 16 km for usual clicks, when listening off the axis of directionality (table 5.1).

Slow clicks are heard only in the presence of mature or maturing males (Weilgart and Whitehead 1988), and the inference is that only males make them, although it is possible that very occasionally a female emits slow clicks. I have estimated that, on the breeding grounds, mature males make slow clicks about 74% of the time (Whitehead 1993), and they do so both at the surface and at depth, when accompanying groups of females and immatures and when alone. On the feeding grounds at higher latitudes, slow clicks are made more rarely, and primarily when the male is at or near the surface (Mullins et al. 1988; Jaquet et al. 2001).

So, what is the function of this strange sound—my conception of a jailhouse door being slammed every 7 seconds? Almost all the circumstantial evidence indicates that it has something to do with the mating system, perhaps attracting females or repelling other males, like so many other male-specific vocalizations (see Bradbury and Vehrencamp 1998, 743–82). It would be an especially useful signal if it contained "honest," unfakeable elements (perhaps the combination of ringing frequencies) that indicated the size or other attributes of the male. Such a sound would allow females or other males to react appropriately to his presence, and thus could constitute an integral part of a system of female choice or male-male competition (see section 6.11).

For the moment, there is no concrete evidence for this function, and other possibilities should be considered. For instance, a slow click might be used to echolocate off other sperm whales or other large objects, such as boats or the ocean floor (Goold 1999). Its longer inter-click interval and different frequency structure could reflect the greater acoustic reflectivity of a large mammal containing air cavities and bone as compared with a relatively small, aqueous squid. In this case, however, one might expect that the characteristics of the slow click would differ between occasions when the male was alone and when he was surrounded by females, but it does not seem to. There is also the puzzle of the slow clicks made during surface periods by males on high-latitude feeding grounds, circumstances in which mating does not seem to be an immediate possibility. Jaquet et al. (2001) call these "surface clicks" and suggest that, although they may sound like slow clicks, they could serve a completely different function. Goold (1999) heard numerous slow clicks (but with an inter-click interval of 2–3 s rather than the 5–8 s that is normal) from a group of entrapped males who were being shepherded from the enclosed waters of Scapa Flow in the Orkney Islands, a circumstance in which mating is unlikely to have been on their minds. However, I think that the balance of evidence indicates that most slow clicks are some kind of mating signal, communicating information on the presence, location, identity, and perhaps size of the vocalizer.

Gunshots

There are two reports of "gunshots" being made by sperm whales. These are loud, impulsive, broadband sounds with a long duration (ca. 400 ms). Gordon (1987a, 216–17) heard them occasionally off Sri Lanka, and Goold (1999) heard them frequently from the entrapped males in Scapa Flow. Gordon could not be certain that the sounds he recorded were actually made by sperm whales, but this seems clear in Goold's case. In the spectrograms presented by Goold (1999), the gunshots look somewhat similar to slow clicks, and they seem to have comparable repetition rates. Could they be variants of slow clicks, or slow clicks heard in unusual circumstances? An alternative explanation, discussed by Gordon (1987a, 217), is that gunshots are used for stunning prey, following the hypothesis of Norris and Møhl (1983). However, if this is the case, prey are stunned very rarely.

Non-click Vocalizations

While almost all the vocalizations of sperm whales are made up of impulsive click-type sounds, there is credible evidence that they do make other kinds of sounds, although with much lower intensity. Gordon (1987a, 217) describes a narrow-band "trumpet" with harmonics, sounding like the "muffled trumpeting call of an elephant," that was heard following fluke-ups at the start of dives, and Goold (1999) recorded "short trumpets," "squeals," and series of "pips" from the entrapped whales in Scapa Flow. The functions of these relatively quiet non-click vocalizations are unclear, and some of them may be artifacts of other aspects of the whales' behavior (for example, Gordon [1987a, 217] suggested that trumpets were made "as the vocal apparatus was being readied for use, equivalent to a human clearing a throat").

5.3   Modes of Behavior

Every 5 minutes during daylight during our 1985 and 1987 Galápagos studies, we noted the position relative to our boat, size, composition (females/immatures, mature males, calves*), and visible behavior of all the clusters of whales, including single animals, that we could see. We also recorded the sounds of the whales through a hydrophone for 4 minutes every hour. After concatenating the visual records and quantitatively analyzing the acoustic tapes, we had a visual-acoustic record of behavior for each of the 932 hours we spent following whales. During the night we continued regular recordings, which gave us 24-hour coverage of the acoustic channel. The twelve measures we extracted from this analysis are listed in box 5.2.

Box 5.2 **Measures of Sperm Whale Behavior**

The following visible activities, shown in figure 5.4, were noted during our studies:

Fluke-up: Flukes raised above the water surface, usually at the start of a deep dive

Breach: A leap from the water, showing at least half the body

Lobtail: A thrash of the flukes onto the water surface

Spyhop: A slow raising of part of the whale's head above the water surface (indicating that the whale is oriented nearly vertically in the water)

Sidefluke: A portion of the flukes is visible above the water surface, but oriented vertically (indicating that the whale is turning sharply).

The following measures were calculated for each hour spent following female and immature sperm whales off the Galápagos Islands in 1985 and 1987 in which the appropriate data were collected, and are

---

* Throughout most of our fieldwork we separated the sperm whales sighted into three clearly distinct classes: large mature males, small first-year calves, and females/immatures (all other whales).

Box 5.2 continued

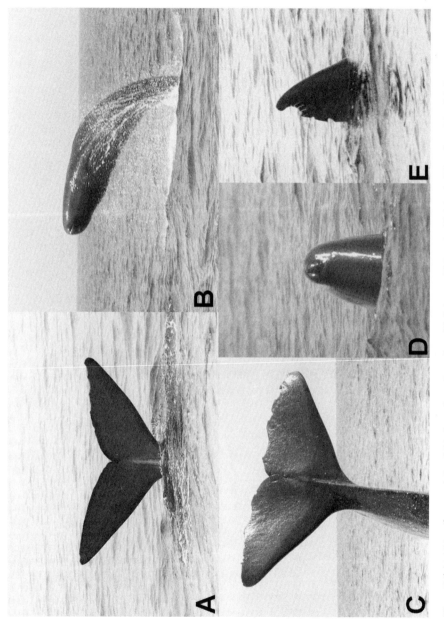

Figure 5.4 Activities of sperm whales visible from above the surface: (A) fluke-up, (B) breach, (C) lobtail, (D) spyhop, (E) sidefluke.

Box 5.2 continued

used in our analysis of behavioral patterns (for details, see Whitehead and Weilgart 1991):

Mean cluster size (see section 6.8 for more about clusters)
Maximum cluster size
Mean speed of whales near the boat (in km/hr)
Consistency of heading: a measure of the consistency among the headings of the whales observed (e.g., if all were heading northwest, then *Consistency* = 1.0; if half the clusters were heading east and half west, *Consistency* = 0)
Fluke-up rate per whale seen at the surface per minute
Breach rate per whale seen at the surface per minute
Lobtail rate per whale seen at the surface per minute
Spyhop rate per whale seen at the surface per minute
Sidefluke rate per whale seen at the surface per minute
Coda rate per minute of recording (see section 5.2)
Creak rate per minute of recording (see section 5.2)
High click rate: Highc = 1 if the measured usual click rate from a group or aggregation is greater than 20 clicks/s; *Highc* = 0 if the measured usual click rate is less than 20 clicks/s. As sperm whales normally click at a rate of about 2 clicks/s off the Galápagos Islands when making trains of usual clicks (see section 5.2), *Highc* indicates whether or not roughly ten or more animals were vocalizing in this way.

The summary statistics for these measures given in table 5.3 provide a basic overview of the behavior of female and immature sperm whales off the Galápagos Islands. The whales were often clustered, but while cluster size ranged up to fifty-two animals, more usually just two or three swam together at the surface (see section 6.8 for more about clusters). The headings of clusters containing different animals observed within the same hour were usually quite consistent, indicating coordinated movement and a broader social structure (the "group"; see section 6.4). Apart from fluke-ups, which were performed at a relatively consistent rate (CV = 0.81 among hours), the visually observable activities, as well as codas and creaks, were extremely variable in their occurrence, with CVs greater than 2.0 among hours.

In an attempt to make some sense of all this variation, we carried out a principal components analysis of these data (Whitehead and Weilgart 1991).

Table 5.3. Summary Statistics for Visually Observable and Acoustic Variables (from Whitehead and Weilgart 1991), Plus the Means when Foraging and when Socializing

Measure[a]	n (hr)	Minimum	Maximum	Mean	SD	Foraging[b]	Socializing[b]	Unit
Mean cluster size	879	1	24.8	2.55	2.53	1.57	5.81	whales
Maximum cluster size	932	1	52.00	5.69	5.83	3.63	13.28	whales
Mean speed	739	0.93	8.95	3.69	1.20	3.80	3.40	km/hr
Heading consistency	440	0.14	1.00	0.81	0.18	0.84	0.75	
Fluke-up rate	932	0	0.300	0.042	0.034	0.050	0.015	/whale/ min at surface
Breach rate	932	0	0.340	0.007	0.022	0.006	0.009	/whale/ min at surface
Lobtail rate	932	0	0.874	0.009	0.045	0.007	0.017	/whale/ min at surface
Spyhop rate	932	0	0.200	0.004	0.014	0.003	0.010	/whale/ min at surface
Sidefluke rate	932	0	0.200	0.007	0.015	0.006	0.013	/whale/ min at surface
Coda rate	1572	0	28.75	1.01	2.56	0.75	4.04	codas/ min
Creak rate	1572	0	34.50	0.63	1.96	0.45	2.07	creaks/ min
High click rate	1751	0	1.00	0.80	0.40	0.92	0.54	

SOURCE: Summary statistics from Whitehead and Weilgart 1991.

[a]See box 5.2 for definitions of these measures.

[b]Foraging and socializing were discriminated by a value of 0.5 on the first principal component of the behavioral and acoustic measures; see table 5.4, Whitehead and Weilgart 1991.

The results are summarized in table 5.4. One strong component ("principal components" are linear combinations of the original measures) explains 31% of the variance in the twelve measures. It is highly correlated with cluster size, and quite strongly related to the rates of spyhops, sideflukes, codas, and creaks, but is negatively correlated with speed, heading consistency, and fluke-ups. Thus, it contrasts what I call "foraging" and "socializing" behavior. When the whales are foraging, members of groups move steadily and

Table 5.4.   Results of Principal Components Analysis of Relationships between Variables

	Principal Components			
	1	2	3	4
Percentage of variance accounted for	31.09	13.41	12.08	10.52
Loadings[a]				
Mean cluster size	0.82	0.35	0.01	−0.14
Maximum cluster size	0.83	0.24	0.17	−0.12
Mean speed	−0.38	0.3	0.44	−0.09
Heading consistency	−0.48	0.39	0.19	−0.04
Fluke-up rate	−0.65	−0.19	0.30	0.25
Breach rate	0.24	0.24	−0.13	0.74
Lobtail rate	0.29	0.30	−0.09	0.71
Spyhop rate	0.46	−0.60	−0.23	0.01
Sidefluke rate	0.49	−0.57	−0.2	0.07
Coda rate	0.68	−0.08	0.53	0.03
Creak rate	0.57	−0.11	0.70	0.02
High click rate	−0.41	−0.55	0.47	0.31
	"Socializing/ foraging"	"Directed movement"	"Vocal"	"Aerial"

SOURCE: Whitehead and Weilgart 1991.

[a]Loadings are the correlations between the original measures and the components.

usually in much the same direction. They are alone or form small clusters, and they fluke up at the start of deep dives, during which usual clicks are often produced. When they are socializing, clusters of females and immatures are much larger, slower, and less consistent in their movements, and dives (indicated by fluke-ups and usual clicks) are rare, while maneuvering (as indicated by spyhops and sideflukes), as well as codas and creaks, is much more common (see table 5.3). The contrast between the behavior of the whales in these two modes is quantified on the right side of table 5.3, in which all measures except breaching rates show substantial differences between foraging and socializing, despite the rather crude way in which I have divided the behavioral modes.

Some elements of this contrast are illustrated in figures 5.5 and 5.6, each of which represents 4 selected hours of surface observations of sperm whale groups. During the "foraging" hours represented in figure 5.5, cluster sizes were small, movement was quite fast and consistent, and fluke-ups were frequently seen. During "socializing" hours (fig. 5.6), clusters were much

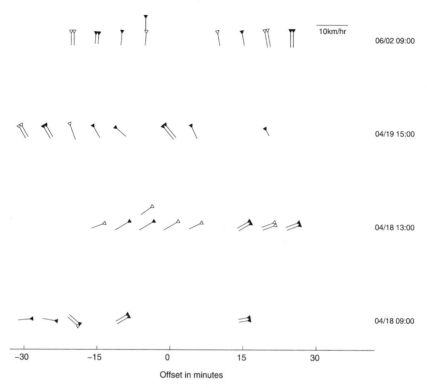

Figure 5.5 Representation of foraging female and immature sperm whales observed off the Galápagos Islands during four hour-long periods in 1987. For each cluster closely observed during a 5 min interval, the diagram illustrates the number of whales in the cluster (represented by close parallel lines), the heading (north is up the page), the speed (proportional to the length of the line), and whether the animal fluked up (represented by a solid triangle at the rear of the representation of the whale). An individual whale may be represented more than once if it was observed in more than one 5 min interval. These whales moved quite fast in a consistent direction, and were in small clusters that frequently fluked up.

larger, and the whales moved more slowly and less consistently, rarely fluking up. However, the female and immature sperm whales often showed intermediate behavioral patterns of various types, as indicated by the representations in figure 5.7, when some whales seemed to be foraging and others did not.

So the females spend some time socializing, some time foraging, and some time in rather intermediate behavior. How is this time arranged? Because of the presence of intermediate states (see fig. 5.7), the division between socializing and foraging has to be rather arbitrary. However, it seems that the females and immatures spend about 75% of their daylight time for-

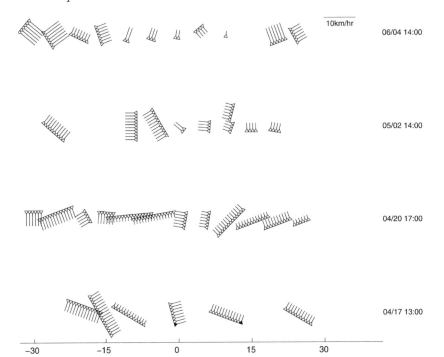

Figure 5.6 Representation of socializing female and immature sperm whales observed off the Galápagos Islands during four hour-long periods in 1987. For each cluster closely observed during a 5 min interval, the diagram illustrates the number of whales in the cluster (represented by close parallel lines), the heading (north is up the page), the speed (proportional to the length of the line), and whether the animal fluked up (represented by a solid triangle at the rear of the whale). An individual whale may be represented more than once if it was observed in more than one 5 min interval. These whales generally moved slowly with little consistency, were in large clusters, and rarely fluked up.

aging and 25% socializing off the Galápagos Islands (Whitehead and Weilgart 1991), Azores (Gordon and Steiner 1992), and northern Chile.* However, the proportion of time spent socializing varies considerably between days. Many days included no socializing, but on others the whales spent virtually all the daylight hours at the surface. Off the Galápagos Islands, the lengths of bouts of socializing averaged about 2–3 hr. Measures of visually observable (but not acoustic) behavior suggested that these social bouts tended to last either about 1–2 hr or 5–6 hr (Whitehead and Weilgart 1991). Socializing peaked in the afternoon (fig. 5.8), becoming known among British crew members as "tea-time." By about 16:00 the whales were

---

* During our 9-month study off Chile in 2000, we noted the general behavior of the sperms being followed for 776 hourly intervals in daylight. Of these intervals, in 18% the whales were socializing, in 68% they were foraging, and in 14% their behavior was intermediate.

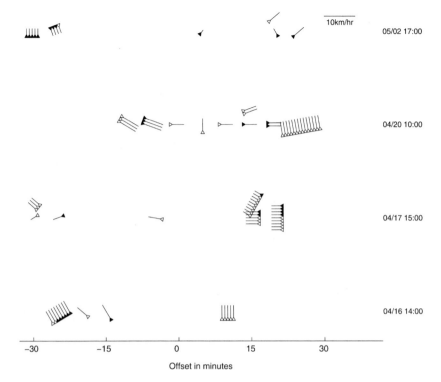

Figure 5.7 Representation of female and immature sperm whales observed off the Galápagos Islands during four hour-long periods in 1987 when behavior was intermediate between socializing and foraging. For each cluster closely observed during a 5 min interval, the diagram illustrates the number of whales in the cluster (represented by close parallel lines), the heading (north is up the page), the speed (proportional to the length of the line), and whether the animal fluked up (represented by a solid triangle at the rear of the whale). An individual whale may be represented more than once if it was observed in more than one 5 min interval. These whales showed variable behavior.

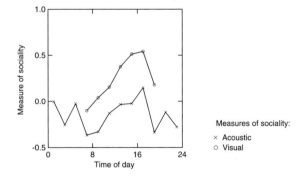

Figure 5.8 Diurnal variation in sociality among female and immature sperm whales off the Galápagos Islands, based on visual and acoustic measures. (Redrawn from Whitehead and Weilgart 1991.)

Figure 5.9 A large, quiet cluster during a period of socializing.

Figure 5.10 Sperm whales, including a large male, fluking up at dusk at the start of deep dives.

often lying quietly at the surface in large clusters (fig. 5.9). These social periods frequently ended at dusk, with large numbers of animals fluking up and
beginning deep dives just as the sun set (fig. 5.10). However, "tea-time" was
not compulsory; groups quite frequently foraged through the afternoon
hours and/or were social in the morning. Acoustic data show that the whales
usually foraged at night (fig. 5.8), but on occasion they would stop their
usual clicks and gather quietly at the surface during darkness (also described
by Barlow and Taylor 1998)—behavior that made tracking them extremely
difficult and sometimes caused us to lose the group.

In addition to "foraging/socializing," three other components (see table
5.4) emerged from the principal components analysis with eigenvalues
greater than 1.0 (indicating a greater contribution toward the total variance
of the original data set than any of the original measures). As shown by the
fan-shaped plots of component scores in figure 5.11, there was little variation in any of these components when the whales were foraging; rather, com-

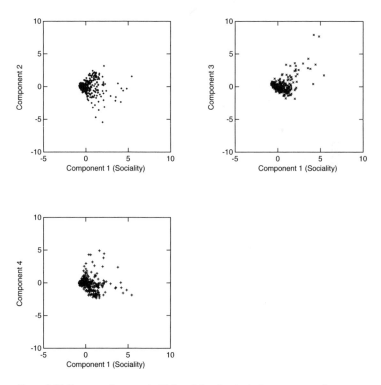

Figure 5.11 Scores on the second, third, and fourth principal components of measures
of sperm whale behavior plotted against the first, the foraging-sociality axis. Each plot
point represents behavior during 1 hr spent following female and immature sperm
whales. (Adapted from Whitehead and Weilgart 1991.)

ponents 2–4 principally distinguish between types of socializing. Component 2 contrasts larger, quieter clusters on fairly consistent headings with smaller, louder clusters within which there is considerable maneuvering, a situation that is often observed at the start of socializing periods. Component 3 relates to acoustic activity, indicating that there are some social situations with much vocal activity and some with little. Component 4 loads heavily with breaching and lobtailing. The rates of these aerial activities are not closely correlated with any of the other measures; they will be discussed in section 5.6.

Thus, as indicated by the measures listed in box 5.2, the behavior of female and immature sperm whales is principally stereotyped foraging, with the whales making repeated dives. However, for periods of hours at a time, and often in the afternoon, they aggregate at the surface, showing us a much wider behavioral repertoire. Foraging and socializing are described in more detail in the next two sections.

5.4   Foraging

The foraging behavior of sperm whales is made up of two distinct phases with different principal functions. During their dives, the whales search for food; while at the surface, they breathe. We think we have a reasonable understanding of their behavior during surface periods, but what happens at depth is probably more significant, and certainly more mysterious.

At the Surface between Dives

All the cetacea, as is well known, . . . possess lungs . . . and require consequently a frequent intercourse with atmospheric air, and for this purpose it is of course necessary that they should rise to the surface of the water at certain intervals.
—Beale 1839, 41

Usually, a sperm whale ascending from a deep dive breaks the surface suddenly at a fairly steep angle, blows loudly and forcibly (Caldwell et al. 1966), and arches its back in the process of settling on a horizontal axis. This fairly dramatic initial surfacing is unpredictable to human observers at the surface, and so is rarely seen closely or photographed clearly. However, such active surfacings are not universal. Tagged males off Dominica sometimes did not blow immediately after returning from the depths, remaining at, or just beneath, the surface for several minutes before blowing (Watkins et al. 1999).

During the remainder of the surface period, the whale's behavior is much easier to see, and usually appears rather lethargic. The animals generally move steadily (ca. 4 km/hr; see section 3.4) in a fairly straight line with the body almost horizontal. Either the ridge of the back and the dorsal fin remain just above the water—so that the whale looks like a log—or the whale "porpoises" gently to rise from just beneath the surface to blow (fig. 5.12). Sometimes whales lie almost still during this breathing period; at other times they move rapidly and determinedly, arching their bodies between blows (fig. 5.12). I suspect that some of this variation may be due to whales at the surface trying to stay abreast of their group members beneath while coping with variation in foraging speed and layering of currents. If a surface current is running counter to the one at foraging depth and the deep whales are moving down-current, the surface animals may need to work hard to keep up. In other situations, the surface current may assist the breathing animals, so that they need to do little swimming to stay above their group members. We face the same range of circumstances when tracking the sperm whales from our boat.

The sperm whale's blow comes from the end of the snout and is projected forward and to the left (see fig. 5.12). It is relatively low and "bushy" compared with that of similar-sized baleen whales, presumably because of a lower exhalation speed, which is probably related to the more convoluted nasal passage of the sperm. The blow itself is mainly formed by water, which gathers in the entrance of the nasal passage between respirations (Berta and Sumich 1999, 234). Little of the blow is visible if the blowhole stays above water between exhalations, as is sometimes the case with sperms in calm weather. Inhalation quickly follows the blow, and then the blowhole closes. For females and immatures, exhalation lasts about 0.2–1.2 s and inhalation 0.6–2 s, with the blowhole remaining open for between 0.5 and 2.5 s at a time (Gordon 1987a, 98).

Blow rates have been measured in a few studies (table 5.5). Inter-blow intervals during surface periods terminated by a fluke-up are typically about 12.5 s for females and immatures and 17.5 s for large males. Inter-blow intervals are generally shortest at the beginning of a surface period, rising gradually by about 20% over the time at the surface, but the last two intervals before the fluke-up are approximately 10% shorter than their predecessors (Gordon and Steiner 1992; Gordon et al. 1992, 32). Gordon and Steiner (1992) could find no significant variation in blow rates with time of day, sea-

Figure 5.12 Sperm whales blowing at the surface, moving (A) fast and (B) slow.

Table 5.5.   Mean Intervals between Blows of Sperm Whales during Surface Periods Terminated by a Fluke-Up

Area[a]	Females and Immatures	Large Males	Reference
Azores	12.7 s (SD = 8.3 s; n = 5,683 intervals)	19.3 s (SD = 7.8 s[b]; n = 910 intervals)	Gordon and Steiner 1992
Sri Lanka	12.7 s (n = 46 encounters)	—	Gordon 1987a, 97
New Zealand		16.8 s (SD = 3.3 s; n = 177 encounters)	Gordon et al. 1992, 30
New Zealand		16.5 s (SD = 6.0 s; n = 1,175 encounters)	Richter 2002, 65 (largest sample)
Nova Scotia		18.9 s (SD = 4.4 s; n = 34 encounters)	Recalculated from Whitehead et al. 1992

[a]In the Azores study each interval was considered a unit for the analysis, whereas in the other studies, each encounter with a whale at the surface was considered a unit.

[b]The reported SD of 37.8 s in the text of this paper appears to be a typographic error.

son, or cluster size, nor does there appear to be much difference in blow rates between geographic areas (table 5.5).

The periods of breathing at the surface last about 8 min (table 5.6), although there is some variation. For instance, off Kaikoura, New Zealand, male surface times had a standard deviation of 4 min around a mean of 9.5 min (Gordon et al. 1992, 30). One might expect that the longer the dive, the longer the recuperation period at the surface. Such a correlation was found for males studied off Kaikoura, (Gordon et al. 1992, 31), but not for females and immatures off Sri Lanka (Gordon 1987a, 177).

Although breathing is the most important, and most often observed, activity during these inter-dive periods, the whales occasionally spyhop, sidefluke, breach, or lobtail, but much less often than when they are socializing (see table 5.3). Surface periods between deep dives are sometimes interrupted by shallow, non-fluking dives of a few minutes' duration. If whales are disturbed during the surface period between dives, they may sidefluke, spyhop, or make such shallow dives. While at the surface, they often actively cluster with one or more other whales, moving toward one another to gain proximity and then, frequently, synchronizing their dives (see section 6.8).

At the surface between dives, the sperms are usually silent. The occasional coda or creak may be heard—often, it seems, when clusters are forming (although we have not confirmed this quantitatively). However, large males quite frequently make slow clicks while at or near the surface, both on the breeding grounds (H. Whitehead, personal observation—the slow clicks of

Table 5.6. Surface Times and Dive Times from Recent Studies of Living Sperm Whales

Area	Surface Time	Dive Time[a]	Source
Canary Islands (females and immatures)	Mean 8.5 min	Mean 34.9 min	André 1997, 125
Azores (mainly females and immatures)	Mode 7.5 min	Mode 45.0 min	Gordon and Steiner 1992
Sri Lanka (females and immatures)	Median 7.5 min	Median 39.5 min	Gordon 1987a, 177
New Zealand (males)	Mean 9.5 min	Mean 45.5 min	Gordon et al. 1992, 30
New Zealand (males)	Mean 9.1 min	Mean 41.3 min	Jaquet et al. 2000
Galápagos (males alone)	Mean 9.7 min	Mean 43.3 min	Christal 1998, 138–139
Galápagos (females and immatures)	Median 9.5 min	Mode 45.5 min	Papastavrou et al. 1989
Dominica (males)	Mean 8.3 min[b]	Mean 29.3 min	Watkins et al. 1999
Dominica (1 male)		Mean 44.4 min (deep dives only)	Watkins et al. 2002
Nova Scotia (males)	Median 8.0 min	Mode 37 min	Whitehead et al. 1992
New York (males)	—	Median 10 min	Scott and Sadove 1997

[a]In most cases, dive times were obtained by subtracting the published surface time from the fluke-to-fluke or "cycle" time.

[b]Given are the means of the means of all surfacings for two radio-tracked males.

a nearby male can sometimes be heard through the surface when he is within 50 m and the boat and sea conditions are quiet) and at higher latitudes (Mullins et al. 1988; Gordon et al. 1992, 41; Jaquet et al. 2001; Madsen et al., in press). Making slow clicks at the surface on the breeding grounds makes sense if they are signals used in mating tactics (Whitehead 1993; see section 6.11), but why do males make slow clicks at the surface at high latitudes? Jaquet et al. (2001) have considered this question using data from Kaikoura, New Zealand, and have suggested that the short series of clicks with low repetition rates heard from large nonbreeding males near or at the surface may be qualitatively different from slow clicks in function, even though they sound similar. They called such clicks "surface clicks," but were unable to suggest a likely function for them. In contrast, Madsen et al. (in press) have suggested that slow clicks by males at high latitudes may be used in competition for food aggregations.

As the surface period draws to a close, the whale speeds up somewhat, raising its rostrum during breaths and beginning to arch its back between them. At the final exhalation and deep inhalation, the snout is raised well above the surface, and is then plunged into the water, leading an arched, rolling body

Figure 5.13 Stages in the fluke-up: the arched back (left), the flukes emerging from the water (right), and nearly vertical (center).

in an elegant movement that usually ends in a fluke-up (fig. 5.13). The fluke-up presumably increases the effectiveness of the transfer of horizontal momentum into vertical momentum as the weight of the flukes above the surface helps push the whale downward. However, a small proportion of deep dives, perhaps especially those of younger animals (Whitehead 2001b), are not preceded by fluke-ups.

At Depth

Like other marine mammals, but more so than most,* the sperm whale can be thought of as a "surfacer" (A. Pinder, personal communication) rather than a diver—it spends most of its life deep under water. The movements of sperm whales under water are described in section 3.2; here I will concentrate on what is done during their dives.

At depth the sperm whale is, unfortunately, largely hidden from us. We

---

* But, in terms of time under water, the sperm whale is eclipsed by the elephant seal (*Mirounga* spp.), which when at sea makes deep dives almost continuously, including, seemingly, to rest (Le Boeuf et al. 1988). Hochachka (1992) calls elephant seals "mesopelagic mammals."

currently have two principal sources of information on the behavior of sperm whales at depth: records from sonar traces and acoustic recordings. The latter are easily obtained, but in the case of females, usually contain the sounds of too many animals at any time to yield information on the behavioral patterns of individuals at depth; for this reason, we are mainly restricted to recording of males at high latitudes. In the near future, however, our conception of the sperm whale's behavior at depth may change dramatically, as tags that can sample the whales' visual and acoustic environment at high resolution, as well as aspects of their physiology, start to tell us what it looks like, sounds like, and feels like to be a sperm whale at depth (see section 9.3).

While the trace (plot of depth against time) of a sperm whale's dive is usually "U-shaped," consisting of a descent, a period at depth, and then an ascent, the "U" is not necessarily smooth (see section 3.2). The whales make sudden ascents, descents, and probably sideways movements as well, of up to 50 m or more (fig. 5.14; Gordon 1987a, 186–87). We imagine that these movements are excursions to chase or capture prey, but we cannot be sure yet.

The vocal output of the sperm whale changes during its dive. Usually silent as they fluke up, the whales start making usual clicks at depths of about 150–300 m off the Galápagos Islands (Papastavrou et al. 1989), 160–425 m off Sri Lanka (Gordon 1987a, 209), and 50–250 m for males off northern Norway (Madsen et al., in press), but only 25–50 m for males at Kaikoura, New Zealand (Jaquet et al. 2001). Interestingly, off Kaikoura, in 24% of the dives, the first vocalization was a loud slow click ("surface click"; Jaquet et al. 2001), a pattern not reported from any other study area. The reasons for the initial slow click, and the much more rapid start of usual

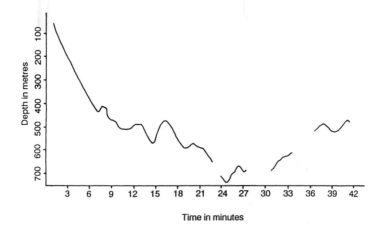

Figure 5.14 Depth-sounder trace of a sperm whale dive off Sri Lanka. (From Gordon 1991b, 234.)

clicking, off Kaikoura are unknown, but may relate to differing oceanographic conditions. As it seems very likely that usual clicks are used to echolocate for food (see section 5.2), perhaps potential food is more likely to be found at depths of 50–300 m off Kaikoura than in the other locations. The first few clicks of the sperm whale's usual click train after its fluke-up seem sometimes be used to echolocate off the bottom (Gordon 1987a, 210; Goold and Jones 1995; Jaquet et al. 2001; Thode et al. 2002).

While usual clicks are "usual" during sperm whale dives, they are not continuous. Trains of usual clicks are typically interspersed with periods of silence as well as creaks and other vocalization patterns. The consensus interpretation of these patterns is that usual clicks function as general echolocation to find prey, creaks are used for homing in on acoustically located prey, and silences indicate times when prey are visually observable or in the process of being eaten (Gordon 1987a, 224–26; Jaquet et al. 2001; Thode et al. 2002; but see Watkins 1980 and box 5.1 for a contrary view). Creaks frequently coincide with changes in direction and are often followed by about 2–10 s of silence (Gordon 1987a, 211; Jaquet et al. 2001), bolstering this interpretation. However, some creaks are not followed by silence (Gordon 1987a, 209–10), indicating either that some prey that are chased are not caught or that usual clicks can be made during ingestion.

The proportion of time spent making usual clicks, creaks, and in silence varies considerably between study areas, whales, and dives, as do the durations of bouts of the different types of vocal activity (table 5.7). This variation probably reflects different expected and actual prey concentrations and types. For instance, when feeding on small prey such as histioteuthids, which

Table 5.7.   Vocal Behavior during Deep Dives Compared between Studies

Between First and Last Clicks of Dive	Galápagos	Azores	Nova Scotia	Kaikoura, New Zealand	Bismarck Sea	Andenes, Norway
Time spent making usual clicks	—	85%	81%	—	66%	—
Time spent creaking	—	9%	1%	—	—	—
Time silent	—	6%	18%	7%	—	—
Duration of usual click trains	—	—	160 s	—	—	—
Duration of creaks	5 s	—	44 s	16 s	—	10–30 s
Duration of silences	—	—	24 s	—	5–30 s	5–20 s
	Weilgart 1990, 83	Gordon 1987a, 210	Mullins et al. 1988	Jaquet et al. 2001	Madsen et al. 2002	Madsen et al., in press

can be detected using echolocation only at short ranges but are easily caught, creaks and silences might be shorter than with larger, more muscular prey, such as *Dosidicus,* which might require a prolonged chase and more processing (Clarke et al. 1993). Similarly, at times when prey are numerous, usual click trains might be shorter and creaks and silences more frequent than when food is scarce.

Creaks have the potential to tell us a fair amount about the prey capture process if we make the assumption that the whales make each echolocation click when the echo of the previous one returns from its target, so that the inter-click interval within creaks is directly proportional to the target range. This has been shown to be the case with other echolocating species (e.g., Au 1993, 115–18), and there is some evidence, but no proof, for it in sperm whales (Gordon 1987a, 210). However, if it is a tenable assumption, inter-click intervals at the start of creaks, about 40 ms (Gordon 1987a, 211–12) for females and immatures off Sri Lanka and about 50–60 ms for males at Kaikoura (Jaquet et al. 2001), indicate that homing in may start at ranges of around 30 m for females and immatures and 35–40 m for males. Rates of change in inter-click intervals suggest closing speeds of about 1–10 km/hr (Gordon 1987a, 211; Jaquet et al. 2001). These speeds are similar to the ones measured for sperm whales at the surface (see section 3.4) and during dives (see section 3.2), a correspondence that bolsters the interpretative method. Using this rationale, maximum click rates within creaks suggest that creaking stops when the whales are a few meters or less from their prey (Gordon 1987a, 211; Jaquet et al. 2001). However, when making this calculation, the time to process and make a click is crucial, and so the estimated range when creaking stops is only an upper limit that assumes instantaneous processing.

If we make another similar assumption, that a sperm whale produces a usual click when there is little hope of receiving a useful echo from the previous one, inter-click intervals within trains of usual clicks can indicate search ranges (Goold and Jones 1995). Thus, the 0.5 s median inter-click interval for females and immatures off the Galápagos, Azores, and Sri Lanka (Whitehead and Weilgart 1990; Gordon 1991b; Goold and Jones 1995) suggests a useful echolocation range of about 375 m, while the 1.0 s interval for large males off the Azores and Kaikoura indicates that they may be looking ahead about twice as far (Goold and Jones 1995; Jaquet et al. 2001). This difference may be a consequence of the males making louder clicks with their larger spermaceti organs, or of their usually searching for larger prey, which produce better echoes, or some combination of the two factors.

During his studies off Sri Lanka, Gordon (1987a, 216–17) several times heard and recorded very loud impulsive sounds, which he called "gunshots" (see section 5.2). These sounds could be the "big bangs" that Norris and Møhl (1983) suggested that sperm whales might make to stun squid. However, as Gordon notes, they were not heard after creaks, as might be predicted by the stunning hypothesis, and, to my knowledge, have not been reported from other studies, with one exception: Goold (1999) heard "gunshots too numerous to quantify" from sperm whales trapped in the shallow enclosed waters of Scapa Flow. It seems unlikely these animals were feeding much. Thus the possibility that sperm whales use sound to stun prey remains open, although it seems unlikely that this is a common feature of sperm whale foraging (see Madsen 2002 for a more detailed discussion of this hypothesis).

Sperm whales may also use vision to discover and approach their prey. Many squids are bioluminescent (see table 2.2), and it is possible that the whales use their bioluminescence to locate them. Fristrup and Harbison (2002) have made the interesting additional suggestions that sperm whales may either swim beneath the layers containing potential prey and look upward for silhouettes against the surface, or use bioluminescence produced by other organisms as the prey move through them. The bioluminescent prey method would clearly work only with squid species that possess photophores, while, although Fristrup and Harbison (2002) argue to the contrary, silhouettes would probably not be visible at night (when much sperm whale foraging takes place; see section 5.3), and this method would not be possible with benthic prey. The bioluminescent trail technique would require not only a reasonable density of other bioluminescent organisms, but also that they be bioluminescing. However, Widder et al. (1989) found no spontaneous luminescence in mesopelagic waters with a high density of bioluminescent organisms. Beale (1839, 36) describes an obviously blind, but otherwise healthy, sperm whale. All this argues against Fristrup and Harbison's (2002) visual predation hypothesis. However, Southall et al. (2002) found that sperm whales, like other deep-diving animals, have blue-shifted eye pigments, suggesting that vision is important at depth for this species. I conclude that vision may be an important way in which sperm whales find some of their food, but suspect that it is generally a supplement to echolocation.

So, from sonar and acoustic records, we have a reasonable, but unverified, rough model of how sperm whales find and approach their prey. But what about the final stages of the encounter? How does the sperm whale convert a rapidly approaching image on its sonar screen, or a visual image, into food? If not stunned, many squids (but perhaps not the histioteuthids) can accelerate rapidly for short distances and so potentially evade the momentum of

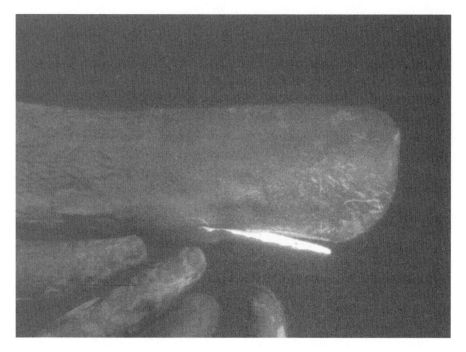

Figure 5.15 The white mouth of the sperm whale—a lure for squid, or for other whales? (Photograph by Flip Nicklin, courtesy of Minden Pictures.)

an onrushing sperm whale. The whale's long and narrow jaw may help, as a sperm whale can, by turning its head, cover 2 or more meters of ocean (laterally as well as dorsoventrally) without changing the orientation of the rest of its body. However, this ability cannot be crucial, as sperm whales with twisted or missing lower jaws have been found healthy and with recently caught food in their stomachs (Berzin 1972, 266).

Beale (1839, 35) believed that the sperm whale, after reaching foraging depth, would stop and open its remarkable white mouth (fig. 5.15), and the squid, which are known to be attracted to light, would be lured in. Gaskin (1967) elaborated on this hypothesis by suggesting that luminescent mucus smeared on the mouth from previously eaten squid might attract new meals. However, according to Fristrup and Harbison (2002), living squid do not possess such luminous mucus—the slime Gaskin reported was that of luminous bacteria growing on dead squid. Additionally, we now know that sperm whales are not usually motionless at depth. However, Fristrup and Harbison (2002) have suggested an alternative, and more plausible, version of the luring hypothesis: bioluminescence is produced by smaller organisms disturbed by the movement of the sperm whale's head, and this, together with the white lining, attracts squid to the vicinity of the sperm whale's mouth. The final step, ingestion, seems to be achieved by sucking the squid in, as indicated by the small ventral grooves on the whale's throat (Caldwell et al. 1966; Berzin 1972, 266; Rice 1989; see fig. 1.1). When the target is a large squid, ingestion is probably not always clean, with the squid sometimes being bitten into parts, only some of which are ingested (Clarke et al. 1988).

The information available on the spatial arrangement of foraging sperm whales at depth is incomplete and somewhat contradictory (see section 6.8), so we can only speculate on the social dimension of sperm whale foraging. It probably varies depending on the overall social situation and the types and concentrations of prey present.

The sperm whales' foraging dives usually last about 30–45 min (see table 5.6), but there is much variation. In the shallow (40–70 m) waters off Long Island, New York, most dive times were 10–15 min, with a maximum reported submergence of 21 min (Scott and Sadove 1997). Using most methodologies, it is hard to be sure that a whale has not surfaced during an apparent dive, so determining maximum dive times is difficult. However, the maximum recorded dive times are 76 min for tagged males off Dominica, West Indies (Watkins et al. 1999; see also Watkins et al. 1993), 83 min for carefully watched males off South Africa (Clarke 1976), and 60 min for carefully watched females off Sri Lanka (Gordon 1987a, 178). The older literature from nineteenth-century whaling suggests maximum dive times of 60–90 min, with the largest whales making the longest dives (Caldwell et al. 1966).

We know less about the ends of sperm whale dives than the beginnings, as it is hard to stay in touch with a whale throughout its dive. However, Jaquet et al. (2001) found that males off Kaikoura, New Zealand, made their final vocalizations an average of 3.6 min before surfacing. As the same whales started their click trains about 25 s after diving, this finding suggests that while the animals were surveying for prey during their descent, they, like the males off Norway (Madsen et al., in press), usually did not bother to do so while ascending. In 57% of the dives in Jaquet et al.'s (2001) study, the final clicks made before surfacing were the slow, metallic "surface clicks."

So, from the available information, I try to envisage a foraging sperm whale cruising the deep waters, clicking. Sometimes it is just above the bottom and mainly "looking" down with its clicks; sometimes it is a few tens of meters beneath a "scattering layer" where mesopelagic animals are concentrated, "looking" up with its clicks and, in daylight, its eyes. When it finds something interesting, the whale turns and moves in, often starting a creak. Some of the potential prey are small and fairly immobile, and are easily caught and swallowed. Others lead the sperm on a chase, but mammalian physiology is generally superior over longer distances, so the sperm eventually outruns its prey and closes in. The squid's best chance of avoiding capture may come from last-second maneuvers, but these will often be thwarted by the length and mobility of the whale's head, or by the inrush of water into the gullet as the whale sucks in. The prey may also be confused by, or attracted to, the whiteness of the whale's mouth, or stunned by loud clicks, and so caught. Protruding parts of larger squid may fight back as the animal is eaten, leaving scars on the whale's face or pieces of squid bitten off. After each capture, the whale wastes little time before resuming the hunt—a female must, on average, consume about 37 squid per dive.*

5.5   Socializing

> *In the sea, the deeper you go, the darker and colder it gets, and it is down there, in the dark and cold, that dangerous things live—the squid and the Sea Serpent and the Kraken. The valleys are the wild, unfriendly places. The sea-people feel about their valleys as we do about mountains, and feel about their mountains as we feel about valleys. It is on the heights (or, as we would say, "in the shallows") that there is warmth and peace. The reckless hunters and brave knights of the sea go down into the depths on quests and adventures, but return home to the heights for rest and peace, courtesy and council, the sports, the dances and the songs.*
> —Lewis 1955, 189

When they are not foraging, and are clustered at or near the surface, the behavior of sperm whales is diverse, ranging from highly energetic breaches and lobtails to almost complete immobility. The whales may be totally silent, or they may emit codas, creaks, or other types of sounds. However, they rarely raise their flukes or make the trains of usual clicks associated with echolocation, usually move slowly, and are clustered together, sometimes very tightly. Hence, I call this class of behavior "socializing," although it probably has other functions, especially resting (see below).

Some of the variety within this behavioral mode is illustrated in figures 5.6 and 5.16, which illustrate diagrammatically and photographically some examples of socializing behavior as seen from above the surface. However, we

---

* If a female consumes about 750 squid per day (see section 2.3) during 18 hours of foraging (see section 5.3) and dives once every 53 min (see table 5.6), the number of squid consumed per dive is about $750 \times 53/(18 \times 60) = 37$.

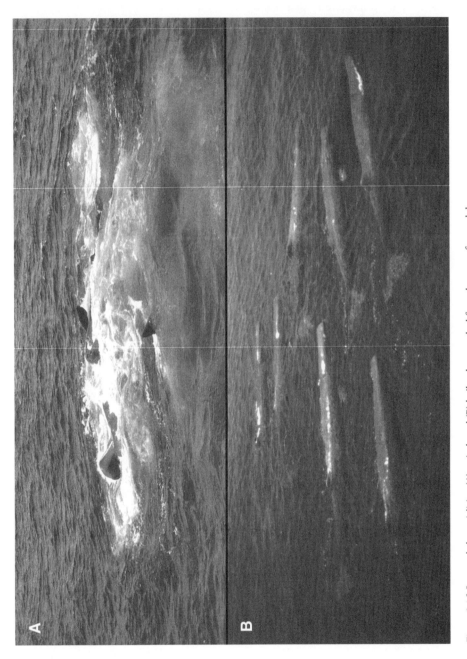

Figure 5.16 Sperm whales socializing (A) actively and (B) lazily, photographed from the mast of research boat.

see only a little from above the surface. Underwater observations by snorkelers have been few and biased (toward times when the whales are swimming slowly, seas are calm, and there is good visibility). However, underwater photographs (fig. 5.17) and quotes from Jonathan Gordon, who has swum with sperm whales as much as any other scientist, illustrate what can be observed of socializing sperms from beneath the surface:

> Sperm whales were revealed to be highly manoeuverable and graceful swimmers, belying their clumsy log-like appearance at the surface. They would readily assume a variety of orientations when swimming underwater. They sometimes swam on their sides and more commonly on their backs, ventral surface uppermost. It often seemed that sperm whales would turn their ventral surface towards objects which they wished to observe visually. It was also common to see sperm whales hanging more or less vertically in the water . . .
>
> Sperm whales often touched and rubbed their bodies against each other as they milled together near the surface. Underwater, groups were often seen to assume a three-dimensional "star-like" formation, with all heads pointing towards the centre. . . . On two occasions whales were observed swimming belly to belly with their jaws slightly agape and touching. (Gordon 1991b, 235)

> They seem to love to touch each other, often rolling along each other's bodies in what can seem like an underwater dance. . . . Their jaws seem to be particularly important areas, and it is not unusual to see animals gently clasping jaws. (Gordon 1998, 18)

The vocalizations of socializing sperm whales are also varied and representative of the intimate interactions among members of the group. At the most basic level, we hear few usual click trains, indicative of foraging, and more creaks and codas from sperm whales gathered at the surface (see table 5.3). Sometimes groups of socializing whales are silent for an hour or more, generally when they are most inactive. At other times, a hydrophone may pick up a cacophony of creaks and codas (although André and Kamminga [2000] heard few codas from socializing sperm whales off the Canary Islands).

Creaks are heard both during foraging, when, as noted above, they are interpreted as a whale homing in on its prey, and during socializing. The creaks heard during socializing, called "coda-creaks" by Weilgart (1990, 85) and "rapid clicks" and "chirrups" by Gordon (1987a, 213), are usually shorter (0.1–4 s; Weilgart 1990, 85) and have longer inter-click intervals (50–100

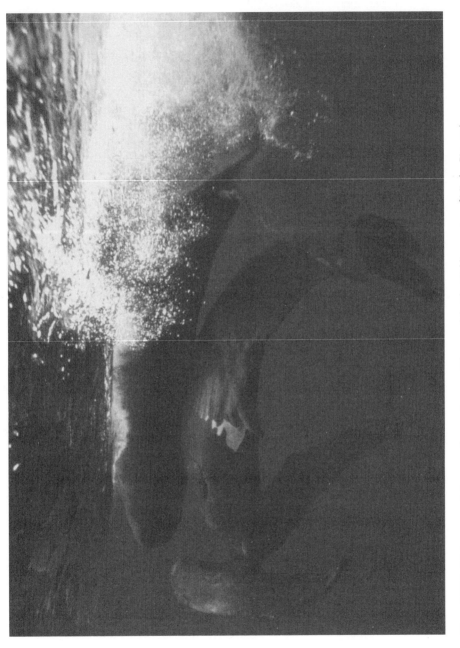

Figure 5.17 Socializing sperm whales as seen from under water. (Photograph by Flip Nicklin, courtesy of Minden Pictures.)

ms; Gordon 1987a, 213 for "rapid clicks") than the foraging creaks (lengths ca. 5–45 s, see table 5.7; inter-click intervals ca. 5–40 ms; Gordon 1987a, 211). Thus (applying the methodology used in section 5.4), these socializing creaks seem to be suitable for a brief sonar assessment of an object at ranges of up to about 30–80 m. They certainly seem to be used in this way when we hear, and feel, them apparently directed at us as we swim with socializing whales (H. Whitehead, personal observation; see also Gordon 1987a, 213). They may be used similarly on group members, perhaps to assess their orientations, movements, or internal organs. They may also intentionally communicate information (Gordon 1987a, 218; Weilgart 1990, 145), as has been suggested for the click series of other odontocetes (Watkins and Wartzok 1985).

The most distinctive communicative vocalizations of sperm whales are codas, patterned series of clicks (see section 5.2). These vocalizations are frequently heard as exchanges between socializing animals (fig. 5.18), as well as together with the shorter socializing creaks (fig. 5.19). Most remarkable in

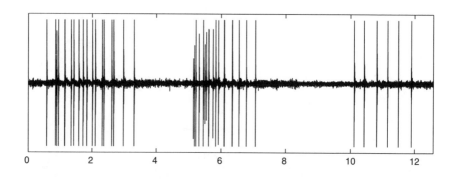

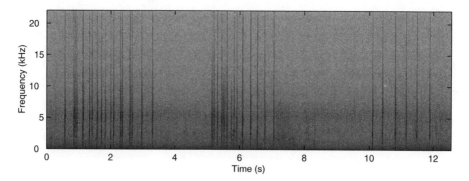

Figure 5.18 An exchange of three codas among socializing sperm whales, displayed by oscillogram (pressure vs. time; above) and spectrogram (frequency vs. time; below).

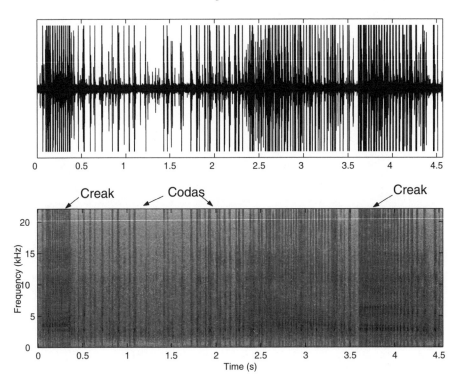

Figure 5.19 Codas and creaks produced by socializing sperm whales, displayed by oscillogram (pressure vs. time; above) and spectrogram (frequency vs. time; below).

some ways are "echo-codas," in which two animals make precisely (within a few ms) the same coda pattern, offset by about 50–100 ms (Weilgart 1990, 109–11; fig. 5.20), in a form of duet (Weilgart 1990, 151). The ordering of codas within exchanges is to some extent nonrandom, suggesting "conversations" (Weilgart and Whitehead 1993), but we do not know what information is being transmitted.

How important is socializing? Is it just a filler for time when the animals do not need or want to forage? Some indication can be obtained by looking at variation in the amount of time spent socializing. The chimpanzees (*Pan troglodytes*) at Gombe socialize less when they have poor feeding conditions (Wrangham 1977). However, for sperm whales off the Galápagos Islands, there was no significant difference in the number of hours spent socializing between month-long time periods or groups, and virtually no correlation, positive or negative, between the amount of time spent socializing and feeding success as indicated by defecation rates (Whitehead 1999c). Thus, while groups socialize on some days but not on others, and for longer on some days

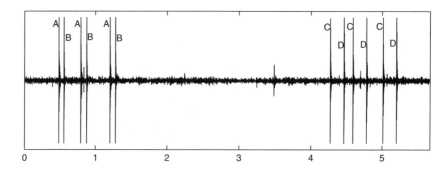

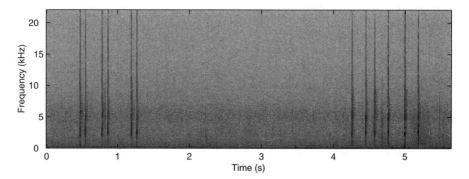

Figure 5.20 Echo-codas, displayed by oscillogram (pressure vs. time; above) and spectrogram (frequency vs. time; below). In each of these three-click codas, each click is repeated precisely by a second whale after a very short delay,

than on others, it seems that, in any area, all groups socialize at roughly the same overall rate in good times and bad—and for sperm whales, there can be very prolonged periods of poor feeding success (see section 2.6).

Thus far in this section, I have described the socializing of groups of females and immatures. What about the larger mature males? When they are with groups of females on the breeding grounds, they take part in socializing sessions much as the females they are visiting do, often becoming the focus of social behavior (see section 6.11; fig. 5.21). When they are away from the females, the situation is less clear. In the Gully, Nova Scotia, mature males sometimes remained at or near the surface for at least 30–100 min, moving slowly and not fluking up, but this behavior accounted for only 5.6% of the 66.7 hr spent following them (Whitehead et al. 1992), as compared with 25% of the time of the females and immatures (see section 5.3). Off Kaikoura, New Zealand, most of the males spend most of their time foraging, with prolonged periods at the surface seemingly rare (Gordon et al. 1992, 23). However, it is possible that the Kaikoura males socialize or rest

Figure 5.21 A large male socializing with a group of females and immatures.

farther offshore, where they are less accessible to researchers (Gordon et al. 1992, 49). Large males, then, may sometimes spend considerable periods at the surface without diving, although they do this less often than the females, and the concomitant social behavior, including large clusters, codas, spy-hops, and sideflukes, is much rarer (Whitehead et al. 1992; Gordon et al. 1992, 41; Christal 1998, 137).

I have called this behavioral mode "socializing," and some of the activities that are characteristic of it, such as large clusters, touching, and coda ex-changes, clearly are social in nature. However, this does not mean that so-cializing is the only function of prolonged periods at or near the surface. The most likely alternative function, especially for solitary large males, is resting, and the whales seem to be resting when they lie very still. Perhaps the close clustering of females sometimes functions to reduce predation risk for rest-ing whales, rather than the clustering itself being the primary behavior. However, the apparently large proportion of time spent by females in this mode—a proportion maintained even when foraging is difficult—suggests that close proximity and communication during these periods is important to the animals, perhaps in the maintenance of important social bonds after dispersion during foraging (Whitehead and Weilgart 1991).

5.6   Aerial Behavior (and Other Peculiar Activities)

> *Rising with his utmost velocity from the furthest depths, the Sperm Whale*
> *thus booms his entire bulk into the pure element of air, and piling up a*
> *mountain of dazzling foam, shows his place to a distance of seven miles and*
> *more. In those moments, the torn, enraged waves he shakes off, seem his*
> *mane; in some cases, this breaching is his act of defiance.*
> —Melville 1851, 667

For humans, visual creatures of the air, the breach (fig. 5.22) is the most
spectacular of all sperm whale activities. I suspect that it may be impressive
to other sperm whales as well. However, breaching, and its fellow aerial behavior, lobtailing (fig. 5.23), remain mysterious in many respects (Whitehead 1985b). Compared with our other measures of visually observable and
acoustic behavior, rates of breaching and lobtailing are not as well correlated
with the foraging-socializing axis (see table 5.4). They do occur somewhat
more often during socializing (see table 5.3), but the difference is not as dramatic as with the other activities that can be observed visually from the surface: fluke-ups, spyhops, and sideflukes. To try and shed a little light on what
these activities mean to the whales performing them, as well as the whales
nearby, I will describe variation in the performance of breaches and lobtails,
look at the circumstances in which aerial behavior takes place, and consider
their energetics.

Breaching and lobtailing are defined in box 5.2. The body of a lobtailing
sperm whale may be vertical, horizontal, or at any intermediate angle. The
whale raises its peduncle and flukes above the water surface and then slams
them down, creating a splash and a noise. Most lobtails are performed with
the flukes ventral side down, but it can also be the dorsal surface that slams
the water.

Breaches are also variable. However, the whale usually makes a steep and
rapid dive, frequently, but not always, preceded by a fluke-up (fluke-ups during social periods are often a sign of an impending breach). It then descends
perhaps 70–110 m before turning and emerging from the water, roughly
25–40 s later (Waters and Whitehead 1990a).* The sperm whale's vertical
approach to the breach (Beale 1839, 48; Gaskin 1964; Waters and Whitehead 1990a) contrasts with that of breaching baleen whales, which usually

---

* The approximate depth of descent before a breach was estimated assuming a mean speed underwater prior to the breach of 20 km/hr, a little less than that estimated for the speed at emergence (see
later in this section), and a 25–40 s delay between the fluke-up (at which time the whale has already
descended about 9 m) and the breach (Waters and Whitehead 1990a).

Figure 5.22 Breaches of sperm whales.

Figure 5.23 Lobtails of sperm whales.

make a horizontal approach, then raise their heads and use their large flippers to convert horizontal motion into the leap (Whitehead 1985a; Würsig et al. 1985). The baleen whales are often constrained to the horizontal technique by shallow water depths, which is very rarely an issue with sperm whales. Conversely, the relatively smaller flippers and narrower head of the sperm whale may make rotating about a lateral axis less efficient (Waters and Whitehead 1990a). Once clear of the water, the sperm whale usually emerges at an angle of 20°–60° to the horizon, often twisting in the air (figs. 5.22 and 5.24). At peak emergence, with 50–100% of the body above the surface, the whale is usually on its side, but there are a few "back breaches" and "belly breaches" (fig. 5.24). The whale occasionally clears the surface (Gaskin 1964), but this is rare (e.g., Gordon 1987a, 81). Following the breach, the whale reenters the water with a large splash. The reentry can be heard easily in calm conditions from above the surface at ranges of more than 1 km, "with a sound that reverberates like a broadside of guns" (Darwin 1882, 223, cited in Caldwell et al. 1966), and beneath the surface through hydrophones (H. Whitehead, personal observation for sperm whales, and Würsig et al. 1989 for bowhead whales, *Balaena mysticetus*).

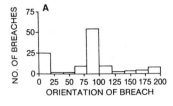

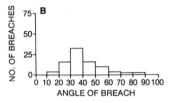

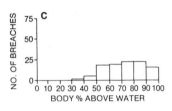

Figure 5.24 Form of the sperm whale breach, as indicated by photographs taken off the Galápagos Islands in 1985 and 1987. (A) The orientation of the breaching whale (in degrees; 0° is a bellyflop; 90° is a whale on its side; and 180° a whale on its back). (B) The angle between the breaching whale and the water surface at full emergence (in degrees). (C) The percentage of the body out of the water at full emergence. (From Waters and Whitehead 1990a, fig. 5.)

Breaching and lobtailing are rarely isolated events. Often a sperm whale breaches several times in succession at about 20 s intervals (Gordon 1987a, 82), and, even more commonly, may lobtail continuously (once every 3–7 s or so) over a minute or more. For instance, on 8 March 1983 off Sri Lanka, a female or immature was observed to lobtail 10 times in 35 s (Gordon 1987a, 82), and on 3 June 1995 off the Galápagos Islands, another lobtailed 12 times in 72 s. Generally, breaches and lobtails decrease in intensity through a sequence (Gordon 1987a, 81–83; Waters and Whitehead 1990a). However, among female and immature sperm whales, breaching and lobtailing are rarely independent events. Many members of the group join in, breaches accompany lobtails, and bouts of aerial behavior occasionally last for hours at a time. On 22 March 1985 we observed a 6.5 hr bout of aerial behavior off the Galápagos Islands, with most 30 min intervals containing at least 10 lobtails and 2 breaches (Waters and Whitehead 1990a). On 11 March 1985, 40 breaches were observed from a group within 30 min, and 284 lobtails were observed in 30 min on 9 January 1987 (Waters and Whitehead 1990a). The most spectacular instance that I have observed was on 9 April 2000 off northern Chile, when, during the height of a 4 hr bout of aerial activity, a group of between fifteen and thirty whales was producing an average of 7 breaches and 12 lobtails per minute.

In trying to work out the functions of aerial behavior, we can get useful hints from the circumstances in which they are observed. Breaching and lobtailing seem to be expensive in terms of time and energy (see below), and so, if their functions are not crucial, might be expected to be curtailed during times when resources are scarce. Defecation rates, our measure of feeding success (see section 2.6), were fairly strongly, but not significantly ($r = .53$, $P = .13$, one-sided test), correlated with breaching rates for the seven calendar months of our 1985 and 1987 Galápagos studies for which there are sufficient data to estimate defecation rates with reasonable accuracy (at least 100 records of the presence of defecations with fluke-ups). There was no relationship between defecation rates and lobtails ($r = -.10$, $P = .58$). Thus, there is a suggestion, but no proof, that rates of the more expensive of the aerial activities, breaching, decrease when feeding conditions are poor.

Aerial activity rates also seem to be influenced by other aspects of the environment. For instance, off the Galápagos, both breaching and lobtailing rates peak in the afternoon (Waters and Whitehead 1990a). However, this is not unexpected, given that both activities are somewhat more prevalent when whales are socializing (see table 5.3) and that socializing itself peaks in the afternoon (see fig. 5.8). Breaching and lobtailing are quite often heard, and very occasionally seen (by moonlight), at night, but we have no quantitative data on nocturnal aerial activities (Waters and Whitehead 1990a).

In other species of whales, breaching rates rise with wind speed (Whitehead 1985b). This correlation could not be examined with the Galápagos data, as the weather was so frequently calm, but off northern Chile in 2000 we experienced a wider range of conditions. For fifteen bouts of breaching, I compared the wind speed during the bout with that at the same time on either the following day (if feasible) or the previous day (if whales were breaching or not followed on the following day) when we were with whales, but no breaches were being made. Wind speeds were slightly higher during the breaching bouts (mean of Beaufort 3.0) than during the control periods (mean of Beaufort 2.6), but the difference was not significant (one-tailed paired t test, $P = .19$).

Although rates of aerial activity are not strongly correlated with the broad foraging/socializing continuum (see table 5.4), these activities do seem to be coincident with more specific social events. For instance, bouts of aerial activity by females and immatures off the Galápagos were longer and contained more breaches when members of more than one social unit were photographically identified (means of 14 breaches in 276 min for multiple units versus 6 breaches in 177 min for single units; Waters and Whitehead 1990a). High rates of aerial activity often preceded the breakup of large clusters, and

Table 5.8.   Rates of Breaching and Lobtailing

	Breaches/ Fluke-Up	Lobtails/ Fluke-Up
Large males: Galápagos Islands	0.018	0.063
Large males: Scotian Shelf	0.027	0
Females and immatures: Galápagos Islands	0.177	0.22

SOURCE: Christal 1998, 137.

rates increased, although not statistically significantly, when large males were present (Waters and Whitehead 1990a). However, the large males themselves almost never breach, and rarely lobtail, at either low or high latitudes (table 5.8). In contrast, small calves quite frequently breach and lobtail, but we have no quantitative data on their aerial behavior.

How much energy does it take to breach? Using a method that I originally developed for humpback whales (*Megaptera novaeangliae*) (Whitehead 1985a), I calculated the speed of an 11 m sperm whale as it breaks the surface at the start of a breach during which 80% of the body emerges from the water at 35° (see fig. 5.24) to be about 22 km/hr, similar to that for humpback whales. This seems to be close to the maximum speed that whales can sustain over a minute or two (see section 3.4), so, when performing a breach, sperm whales, like humpback whales (Whitehead 1985a), typically put all, or most, of their available power into it. But how much energy is involved? The locomotive energy of a breach in which a physically mature female weighing 13.5 t (Lockyer 1981) emerges from the water at 22 km/hr, assumed to be 90% of her maximum swimming speed, is about 617 kcal (using methodology in Whitehead 1985a)—about 0.075% of her estimated daily active metabolic rate (Lockyer 1981). Thus, as in the case of humpback whales (Whitehead 1985a), one breach does not represent a substantial part of a sperm whale's daily energy budget, although a sequence of 20 breaches begins to become significant. A lobtail is almost certainly less of an energy drain than a breach.

Having marshaled what evidence there is on the circumstances, acoustics, and energetics of aerial behavior in sperm whales, what can be concluded about the function of these activities, in which animals spend time and effort to purposefully leave their natural medium? Breaching, the most dramatic observable activity of the whales, has attracted both lay and scientific speculation. Some have suggested that breaching may be a nonsocial part of the daily life of an aquatic mammal and have a relatively prosaic function, such as stretching, parasite removal, looking at things above the surface, a reaction to excitation, or ascent from a deep dive, or be a method or by-product of feeding (see references in Gaskin 1964; Whitehead 1985a). However, modern scientific writing, as well as some of the older literature, generally favors a communicative function (Herman and Tavolga 1980; Whitehead 1985b). While not providing conclusive proof, several of our results support this hypothesis: the increase in such behavior when clusters are splitting up, when more than one social unit is present, and when males attend groups, as well as the lower rates of aerial behavior from the less social males.

If breaching and lobtailing are often signals to other whales (or maybe sometimes to humans), what information is being transmitted, and why choose this method of conveying it? There is an evolutionary puzzle here. A breach may be the best visual signal a sperm whale can make for an observer above the surface (being visible at longer ranges than anything else the whale can do), but it is not so spectacular under water, where most intended recipients are likely to be swimming. Although breaches and lobtails produce conspicuous sounds under water, sperm whales can almost certainly make signals of louder amplitude, at lower cost, and with better transmission characteristics (the surface is not a good place to make a signal with good propagation) vocally (e.g., Würsig et al. 1989). So why breach or lobtail to convey information, when another whale will hear a click, or perhaps some other type of vocalization, at greater ranges and at a greater amplitude? I think the answer probably lies in the theory of honest signaling (Bradbury and Vehrencamp 1998, 649–76): breaches and lobtails make good signals precisely because they are energetically expensive, and thus are indicative of the importance of the message and the physical status of the signaler. A breach is made at close to an animal's maximum power output, as indicated both by the energetic analysis and by the reduction in magnitude of breaches through a sequence. The sound, and perhaps also the bubble pattern, of a large breach cannot be produced in any other way than by a whale that is physically fit traveling at nearly its maximum speed through the surface. The breach therefore signals to other whales that the breacher considers the message important, as well as her physical state, in a way that the standard vocalizations may be unable to match, as their received character depends greatly on directionality, range, and oceanographic conditions. Breaches and lobtails may be used to accentuate other, vocal, signals (Whitehead 1985b).

So, an important message is being conveyed. But what? Given my ignorance of the fine-scale social behavior of sperm whales, I can only speculate. Breaches and lobtails do not seem to be courtship signals, at least by males. However, they do seem to be involved in group and perhaps unit dynamics

Figure 5.25 A jaw-clap. (Photograph by T. Arnbom.)

(perhaps "Please leave!" or "May I join?") and the arrangement of group activity (perhaps "Let's get going!"). Lobtailing is often suggested to be an agonistic signal (Pryor 1986; Gordon 1987a, 83), and sometimes, but far from always, the context suggests this. It may also be a more cooperative social signal, as suggested by the observations reported by Gambell (1968). Another sperm whale activity that is often perceived as aggressive is the "jaw-clap" (fig. 5.25), but this is rarely seen.

Jaw-claps, as well as breaches and lobtails, even if they function primarily as communicative signals, probably have other purposes at different times. As suggested by Beale (1839, 48) and Scammon (1874, 77) breaches can usefully remove ectoparasites.

Some breaches and lobtails, especially those by young animals, may fall into that difficult category of behavior, play. Play can be defined as activities that, while having no immediate function, may benefit the animal later in life—for instance, by improving its physical abilities or social skills (see Beckoff and Byers 1998). The category of play, however, tends to become a garbage-can for any behavior whose function is not obvious. For instance, D. K. Caldwell et al. (1966) devote a whole section of their review of sperm whale behavior to play, and within it they include some activities, such as breaches and lobtails, for which (as they note) there *are* plausible immediate functions. However, there are accounts of other behaviors in their review, such as sperm whales surfing on large waves and diving back and forth beneath a plank (Caldwell et al. 1966), that are probably play. We have also seen behavior that seems to be play in sperm whales, including an animal pushing a stick along with its head (Whitehead 1989b, 104). Nishiwaki (1962) shows an aerial photograph of a group of sperm whales playing with a piece of timber. Gordon (1987a, 91) describes sperms pushing small trees and branches and suggests that sperm whales may play socially by touching, nudging, and showing sexual behavior in nonreproductive situations.

Play behavior is usually perceived as pleasurable, at least by humans, and is performed at times and in ways that increase pleasure. Perhaps the increase in aerial behavior with wind speed is a result of whales finding it more pleasurable or satisfying, or less painful, to slam the body on rough, rather than smooth, water. Perhaps not. Aerial behavior in sperm and other whales will remain something of a mystery for some time to come.

5.7   Birthing

The birth of a sperm whale is a rare event, since each sexually mature female gives birth only about once every 4–20 years (see section 4.3). However, if groups of females and immatures, each containing perhaps fifteen mature females (see section 6.4), are followed for a total of a year or so, as in our studies, there is a chance of being present at a birth, and this has been our good fortune (see the prologue). Here I will summarize some of the most pertinent aspects of our observation, which is described in more detail in our scientific and popular notes (Weilgart 1985; Weilgart and Whitehead 1986).

At 08:52 on 21 October 1983, off Sri Lanka, we observed a sperm whale "making unusual movements,"* raising her flukes and head simultaneously and then arching her back, and thus flexing her body strongly in the vertical plane with little forward movement. After 3 min of this behavior, the whale rolled on her side, and a rush of blood and a dark object were expelled from her genital area at the surface. Then a wrinkled 3–4 m calf with a bent dorsal fin was seen at the surface beside the mother's head. The calf "swam awkwardly in a bobbing, rocking fashion." At 09:00 "the mother and calf were joined by one other adult, which jostled and pushed the calf." By 09:10 there were four adults maneuvering closely around it, jostling it and some-

---

* All quotes in this paragraph are from Weilgart and Whitehead (1986).

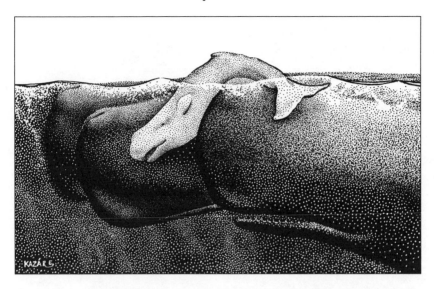

Figure 5.26 A newborn sperm whale receives a rough introduction to social life. (Illustration based on observations of Weilgart and Whitehead 1986; Copyright Emese Kazár.)

times squeezing it between them (fig. 5.26). After 10 min of this behavior, the mother and calf were briefly left alone. Soon the calf swam toward a human swimmer in the water, but was intercepted and gently nudged away by the mother, and at 9:30 we observed a possible attempt to suckle. Over the next half hour the calf made another, closer, approach to the human swimmer (fig. 5.27), but was also seen in close contact with other adults, as well as by itself with no adults visible nearby.

To my knowledge, no other scientists have described the actual moment of birth of a sperm whale, but there are some accounts of some very recently born calves and apparently imminent parturitions (Pervushin 1966; Gambell 1968; Gambell et al. 1973; Best et al. 1984). These accounts describe close and physical interactions between the newborn calf and several adults, in agreement with our observation. Newborn calves may have tooth-rake marks, apparently from adult females (Best et al. 1984), further indicating a socially boisterous arrival. However, in several of the other reports, sharks, and in one case killer whales (*Orcinus orca*), were observed close to the newborn and its adult companions. In contrast, during our observation, only dolphins were seen in the vicinity.

From the perspective of sperm whale societies, I think that there are two important, and probably linked, messages that come from these accounts: that birth is a social event, and that the newborn may be in substantial danger during its first hours outside its mother.

Figure 5.27 A sperm whale calf a few minutes old. (Photograph by Linda Weilgart.)

5.8 Behavior of the Calf

The young calf has three important activities to perform: breathing, swimming, and suckling. The breathing behavior of calves appears to be much like that of adults in miniature, although calves do not make the prolonged dives of adults (see section 3.2). Although newborns are clumsy (Gambell et al. 1973; Weilgart and Whitehead 1986), their flukes soon uncurl, and they quickly become adept swimmers. They have to—the calf whose birth we observed was tracked (we think—we have no positive photographic matching identifications between the beginning and ends of the tracking) moving 290 km during its first 4 days of life (Weilgart and Whitehead 1986). Calves swim along at or near the surface, sometimes apparently alone and sometimes with one or more adult escorts (Gordon 1987b; Whitehead 1996a). They seem to stay above the foraging group beneath, which could easily be achieved by listening for its usual click trains (Gordon 1987a, 90). If the foragers are moving fast relative to the surface waters, perhaps because of adverse current layering (see section 5.4), the young sperm seems to charge along determinedly, flinging its head above the surface to breathe in its attempt to keep up (fig. 5.28). At other times, perhaps when the layered

Figure 5.28 A sperm whale calf swimming fast (in the center of the photograph).

currents are favorable to the calf at the surface, the calf can dawdle and meander while staying above its elders.

Suckling is rather more of a puzzle (Beale 1839, 38). However, both the observations of nineteenth-century whalers and modern observations by snorkelers suggest that the nipple is usually held in the gape of the calf's jaw (Beale 1839, 38; Best et al. 1984; Gordon 1991b). Bennett (1840, cited in Best et al. 1984) describes females turning on their sides so that calves can suckle while lying at the surface. However, we have not seen this, and Gordon (1987a, 86–87) saw suckling taking place in a variety of orientations relative to the surface, but often with the mother dorsal side uppermost at the surface. He also describes observations of two calves suckling from the same female at once. However, he rightly calls for caution in interpreting such observations, noting that the object of one apparent suckling attempt turned out to be the penis of a male!

As the calf grows and develops, its behavior starts to resemble that of the older members of its group: its dives become longer, fluke-ups are seen occasionally, it breaches and lobtails, and it seems to take a more active role in the socializing of the group. The sperm whale calf is born into a highly social milieu. It probably depends on animals other than its mother, and who may not even be genetically related to it, for protection from predators, and

Figure 5.29 Young sperm whales, like this newborn with our boat, may approach, and sometimes follow, mother-sized objects in their environment. (Photograph by Phillip Gilligan.)

may gain sustenance from them as well (see section 6.9). Therefore, the development of social skills is likely to be an important part of its upbringing, although we do not know the details of this process. Young calves will sometimes swim alongside a sailboat (fig. 5.29) or whaling boat (Best et al. 1984), or even a Bryde's whale (*Balaenoptera edeni*) (Whitehead 1989b, 112–13)—behavior that can be interpreted as a "following response" toward any object that might be an adult sperm (Best et al. 1984). However, this behavior seems to wane with age (Best et al. 1984).

5.9   Defense against Predators

The instant the biggest whale was hit all the individuals of the herd made a circle like a marguerite flower centering around the biggest whale. These radially gathered whales put their heads together and made many splashes with their tail flukes. They did not swim away or dive. The gunner, therefore took the whales very easily, starting with the largest one.
—Nishiwaki 1962

The sperm whale has enemies, and largely tries to thwart them behaviorally. Of its natural enemies, the most serious is almost certainly the killer whale (see section 2.8). Six of the most detailed observations of interactions between killer whales and sperms come from observations off the Galápagos Islands by myself and others. To give a flavor for the range of behavior that takes place when the two great odontocete species interact, these Galápagos observations are summarized in box 5.3, together with the crucial observations of Pitman et al. (2001), who observed three encounters in the eastern North Pacific, including an attack off California during which at least one sperm whale was killed (see the prologue).

Box 5.3 **Interactions between Sperm Whales and Killer Whales**

This box summarizes some of the clearest accounts of encounters between sperm whales and killer whales. The majority of these accounts come from research off the Galápagos Islands. This is not as surprising as it might seem, as more research on living sperm whales has probably been carried out there than in any other location. Also summarized are three encounters in the eastern North Pacific described by Pitman et al. (2001). Within each group I list the encounters in order of increasing severity. For other accounts, see Jefferson et al. (1991).

The Galápagos Encounters

On 2 June 1987, about four killer whales passed close to the foraging sperm whales (including a calf) that we were following, but there was no noticeable change in behavior by either the sperm whales or the killer whales.

On 5 April 1991, a group of eight killer whales moved through a dispersed foraging formation of sperm whales that we were following. After the killer whales had passed, the sperm whales ceased their foraging behavior, came to the surface as a tight cluster of about twenty whales (without a visible calf), and produced codas. They followed the track of the killer whales for about 5 min and then turned away.

On 7 March 1991, during the early morning, we were following a well-spread group of foraging sperms in clusters of one or two animals, moving steadily northeast. At 08:00 we suddenly saw a cluster of eight animals moving south determinedly, and a male killer whale apparently fleeing from them. There were two other killer whales visible.

Box 5.3 continued

The sperms seemed to chase them for a number of minutes, then slowed, and turned back to the north. The sperm whales then gathered to socialize at the surface.

On 28 April 1996, G. Merlen (personal communication) was tracking a group of twelve sperm whales, with no calves, from a 13 m vessel. Here is his report:

The group, which had been diving asynchronously and moving steadily westward on a front about 2 km wide, made a sudden change of direction to the east, and the animals were observed to be stationary at the surface and clustered close together, blowing strongly. Almost at the same time a killer whale was seen moving eastward approximately 400 meters to the south. Three minutes later, at 17:36, at least four killer whales, including a large male, turned toward the closely grouped sperm whales, which were facing southward in total silence. This was the last sighting of the killer whales. Of the original 12 sperm whales, 11 were in the cluster together. The twelfth was seen at 17:40, surging through the surface waters toward the main cluster. Its back was arched strongly by the force of its swimming motion and a large bow wave was formed around the head. Three minutes later it joined the others. Now the whole cluster moved very slowly northward, still in silence. At 18:10 they changed direction to the east, possibly due to the fact that they were closing with the coast (although only 7 km from the land, they did not cross the 1000 m isobath). They were arranged line abreast, with each one being no more than a few meters from its neighbour. From time to time, an individual would make a shallow dive of a few hundred meters and a number of side flukes were observed. Several non-continuous trains of clicks were heard. A breach was seen and heard at 18:17 and lobtailing began at 18:24. Between 18:17 and 19:07 a total of 17 breaches and 35 lobtails were counted! At 19:07, the cluster began to travel a little faster, and all the animals fluked-up together, with the usual clicks associated with foraging beginning 3 min later.

On 26 October 1993, while following a dispersed group of twelve sperm whales from a 28 m ketch, Brennan and Rodriguez (1994) saw a killer whale, in a cluster of three, porpoise from the water and bite a sperm whale behind its dorsal fin. A large patch of blood was then seen. Some minutes later, five killer whales, including an adult male and a

Box 5.3 continued

calf, circled the patch of blood at high speed and then left. The sperm whales stopped making long dives and came to the surface, where they formed a tight cluster and were mostly silent. Apparently only one sperm whale was attacked.

On 18 April 1985, we observed killer whales making a prolonged attack on a group of sperm whales (fully described by Arnbom et al. 1987). The sperm whales, who were spread out, diving, clicking, and presumably foraging, suddenly became silent and were seen in a very tight (animals less than 3 m apart), slow-moving cluster containing over thirty whales. A first-year calf stayed in the middle of the cluster, and a large male on its periphery. In the vicinity were fifteen to twenty-five killer whales, including two large males, one smaller male, and two juveniles, arranged in clusters of two to seven individuals. The killer whales made fairly discrete attacks on the sperm whales, with two to seven killers approaching the sperm whale cluster from the flank or rear. They dived when close to the sperms, staying under water for about 3 min before retreating. The sperm whales, by turning and facing downward, appeared to be trying to keep their heads toward the killer whales (fig. 5.30). Lobtails were seen and intense clicking was heard from the sperm whales during the attacks. Bloody cuts, consisting of a single line or three to five parallel lines, were seen on the heads of at least three sperm whales, but there was no sign of more serious injury. After 2.8 hr, the killer whales moved away to the west, and the sperm whales started swimming northeastward at 9–11 km/hr. They maintained this course and speed for at least the next 5.5 hr, and, except for occasional trains of slow clicks from the large male, were silent. The attack seems to have been unsuccessful.

Three Encounters from the Eastern North Pacific

On 21 August 1989, approximately 1,100 km SW of Acapulco, Mexico, Pitman et al. (2001) simultaneously observed clusters of sperm whales, killer whales, short-finned pilot whales, and bottlenose dolphins (*Tursiops truncatus*). The sperm whales adopted a Marguerite formation when the pilot whales approached and swam among them, and the killer whales were never seen within 1 km or so of the sperm whales. The behavior of the sperm whales could be interpreted as a reaction either to the approach of the pilot whales or to the presence of the killer whales, with the pilot whales perhaps "seeking safety" among the sperms (Weller et al. 1996; Pitman et al. 2001).

Box 5.3 continued

On 26 October 1997 off California, Pitman et al. (2001) saw several clusters of sperm whales, including one large male, coalesce rapidly from ranges of up to 7 km in the presence of at least five killer whales. One adult female killer whale approached the sperm whales closely and, "after several arching dives" by the killer whale, an oily slick formed, suggesting that a sperm whale had been bitten. The sperm whales coalesced into a tight cluster, usually in a "spindle formation" about one to five animals wide and twelve to fifteen animals long, with almost all animals facing the same way. About 20 min after the female killer whale "lost interest," the tight aggregation of about fifty sperm whales broke up into clusters, which dispersed.

The most dramatic and important of these observed encounters between killer and sperm whales was observed by Pitman et al. (2001) 5 days earlier, between 07:05 and 13:00 on 21 October 1997, also off California, and perhaps involving the same killer whales.* When the whales were first sighted, the attack was already taking place, with nine female or immature sperm whales, but no calves, in a Marguerite formation (fig. 5.31) surrounded by a slick of oil and blood. During the course of the incident, the number of killer whales within the 3 km visual range of the observers varied between twelve and thirty-five, but they were scattered in clusters of one to six, and only a few were in the vicinity of the sperms at any time. The killer whales attacked in clusters of four or five, moving fast toward the sperm whale formation and then biting the sperm whales below the water surface, often at about mid-body. Each attack lasted a few minutes, after which fresh blood was usually observed. The killer whales seemed to have some difficulty in biting through the tough hide of the sperms.

As the incident progressed, the attacks became more "frenzied," involving generally more killer whales and with shorter time intervals between them. It became increasingly apparent that the killer whales were attempting to detach sperm whales from the Marguerite formation. Animals isolated from the main formation were preferentially attacked, and those animals attempted to regain the formation. All the sperms in the Marguerite seemed to take similar roles and be attacked indiscriminately. However, the sperm whales made little apparent attempt to defend themselves with their flukes, jaws, or in other ways, apart from forming the Marguerite and actively maintaining it through almost all of the encounter. At about 09:00, when the killer whales had separated one sperm from the Marguerite and were biting

Box 5.3 continued

it, another sperm left the Marguerite and swam over to the detached animal, which it led slowly back to the formation, as both were being attacked by the killers. There were several similar instances of apparent altruism in which animals left the relative safety of the Marguerite to accompany detached colleagues.

By 11:00, a number of the sperm whales were showing dramatic injuries, including large chunks of flesh missing, protruding intestines, and a broken jaw. At this time, two sperm whales had been isolated from the Marguerite. Up to this time, the adult male killers had taken little part in the attack, but now one slammed into one of the detached sperm whales, taking it into his jaws and shaking it violently "in an immense display of power not shown by any of the females." Other killer whales joined in the assault; soon a sperm whale was dead and being carried off by the male killer whale and others. The killer whales dragged the carcass about 1 km away and started to feed on it. The other sperm whales were not sighted after this time. It is likely, given the nature of the injuries observed, that additional sperm whales, and perhaps the entire group, died as a result of the attack.

* Pitman and Chivers (1999) give a vivid "popular" account of this attack, a section of which is reproduced in the prologue.

As illustrated by the accounts in box 5.3, as well as in the summary of pre-1990 accounts by Jefferson et al. (1991), there is great variation in these interactions. Some lasted little more than a few minutes, while others went on for hours. On some occasions the killer whales seemed to ignore the sperms, while on other occasions they made repeated attacks. In response, the sperms sometimes showed no change in behavior as killer whales passed, but at other times they adopted distinctive defensive measures. A variety of responses to the presence of killer whales is characteristic of their marine mammal prey species (Jefferson et al. 1991). Jefferson et al. (1991) make the reasonable suggestion that this variation may at least partially result from the prey's ability to distinguish between killer whales with and without predatory intent.

If sperm whales do react to the presence of killer whales, they show some characteristic responses. In particular, they cease foraging dives, come to the surface, swim fast toward each other, sometimes from ranges of several kilometers, and cluster actively and tightly (see box 5.3). The speed with which foraging ceases and the whales actively cluster at the surface was remarked by the observers in several cases. For instance, Brennan and Rodriguez (1994) noted that "these responses . . . differ from anything that has been observed in Galápagos during the past years' work." In no instance did the sperms attempt flight (at least until all killer whales had left the area) or deep dives as a method of escape. These strategies would probably not have been effective, in the first case because of the greater speed of killer whales, and in the second, on account of their probable ability to track sperms under water using echolocation, and thus to attack them during breathing periods at the surface. Instead, the defensive behavior of the sperms resembled that of muskox (*Ovibos moschatus*) being attacked by wolves (*Canis lupus*) (Klein 1999): they stood up to their attackers in numbers, adopting defensive formations within which small calves, if present, were securely positioned. Particularly noteworthy is the way some of the California sperm whales left the defensive formation to assist each other at considerable personal risk—"heroic" acts by almost any definition (Pitman and Chivers 1999).

The clearest accounts of attacks by killer whales (see box 5.3) describe two principal defensive formations characterized by the orientation of the sperms relative to their attackers. The contrast is exemplified by the two longest encounters. The sperm whales that successfully defended themselves off the Galápagos Islands in April 1985 tried to keep their heads toward their attackers (Arnbom et al. 1987; fig. 5.30; see also fig. 2.20), whereas the doomed California group (Pitman et al. 2001) adopted the "Marguerite formation"* (Nishiwaki 1962), with their tails outward, their heads together, and their bodies radiating outward like the spokes of a wheel (figs. 5.31 and 5.32). Why do sperm whales sometimes keep their heads away from killer whales and sometimes face them? Pitman et al. (2001) suggest that the Marguerite is more effective with smaller ($< 10-15$) numbers of sperms, and heads-out with larger clusters. There are other possibilities: the defensive formation may be adapted to the behavior of the attackers, or sperm whale social units or clans may have characteristic defensive methods (Whitehead et al. 1998). However, such attacks occur sufficiently rarely that it is likely to be a long time before we have sufficient data to distinguish between these hypotheses.

In both the Marguerite and heads-out formation, the sperm whales remain primarily at the surface. This behavior makes sense for animals that stand their ground when attacked, as it allows them to breathe whenever they wish and halves the avenues open to attackers.

All the attacks by killer whales on sperm whales described in box 5.3, as

---

* Named by Nishiwaki (1962) after the marguerite flower. Others have called this the "wagon-wheel" or "rosette" formation.

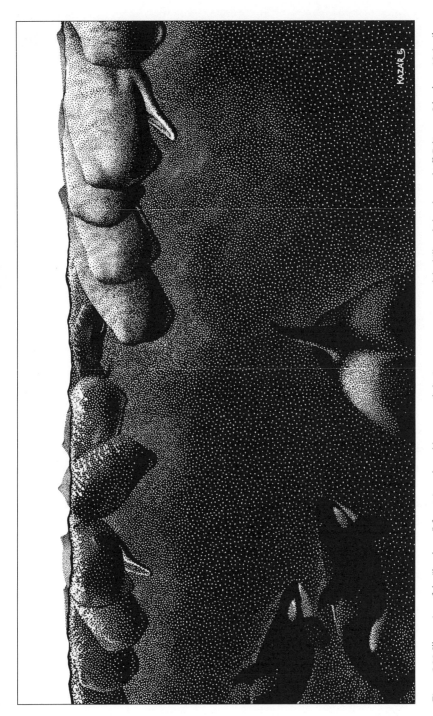

Figure 5.30 Illustration of the "heads-out" formation adopted by sperm whales in response to an attack by killer whales observed off Galápagos Islands on 18 April 1985. A calf is situated behind the main defensive ranks. (Copyright Emese Kazár.)

Figure 5.31 Marguerite formation adopted by sperm whales in response to an attack by killer whales on 21 October 1997. (Photograph by Robert Pitman, courtesy of Robert Pitman.)

well as those in the accounts compiled by Jefferson et al. (1991), involved females and immatures, with two exceptions. Large males were part of the defensive formations of the females and immatures in the incident described by Arnbom et al. (1987)—in this case generally in a rear, safe, part of the formation—and in that of 26 October 1997 described by Pitman et al. (2001). We have only limited information on whether killer whales attack large males when they are alone or in "bachelor" groups and what defensive measures are taken.

The most detailed account of an encounter between large male sperm whales and killer whales is probably that of Nolan et al. (2000), who made observations on several occasions of 12–15 m male sperm and killer whales feeding in the vicinity of a deep-water longline vessel off the Falkland Islands, perhaps taking Patagonian toothfish (*Dissostichus eleginoides*) from the hooks, perhaps not. Usually, at the appearance of killer whales near the fishing vessel, the sperm whales moved off. But on 18 November 1995, in the face of five killer whales, the six sperms stood their ground. The two species were seen in close proximity, with physical contact initiated by the killer whales, but there was no sign of injuries. The sperms struck the killers with their flukes. After a few minutes the sperms moved off about 300 m, formed

Figure 5.32 Illustration of the Marguerite defensive formation adopted by sperm whales in response to an attack by killer whales. (Copyright Emese Kazár.)

a "line abreast," and accelerated, "creating a foaming bow wave," back toward the preferred position where the fishing line came aboard the ship. They swept the killer whales away, and the killers moved off 1,000 m. On another day, four sperms moved in toward the same group of five killers who were in the preferred position. This time the killers formed a line abreast and drove the sperms off. In these observations all was not clear, but it seemed that the two species were contesting the preferred position beside the ship, and would tolerate members of their own species, but not the other. The killers and sperms seemed fairly evenly matched, with the killers usually, but not always, being dominant.

It may be that large male sperms are generally too formidable for killers to consider potential prey and so are left alone, except when a valuable resource is at stake. Thus, the male sperms and killers are more occasional competitors than predator and prey. In the ocean environment, contest competition such as that observed by Nolan et al. (2000) is likely to be rare. These observations might, in turn, relate to the less social nature of the large males, if social structure largely functions in defense against predators (see section 6.10).

There are other species that seem to harass, and maybe even attack, sperms, such as the false killer whale (*Pseudorca crassidens*), pilot whales (*Globicephala* spp.), and large sharks (see section 2.8). It seems that sperm whales show reactions to these species somewhat similar to those they use to defend themselves against killers: they bunch up, slow down, and, especially in the case of approach by pilot whales, may adopt the Marguerite formation (Palacios and Mate 1996; Weller et al. 1996). However, in these cases, dives were sometimes used to end the encounters.

Since 1712, the sperm whale has faced another, and often much more dangerous, predator: the human whaler. As described in section 1.6, whalers have come in two principal varieties. In the eighteenth and nineteenth centuries, they searched from slow wooden sailing vessels, rowed or sailed toward their prey in slim open whaleboats, attacked them using hand-held harpoons, and killed with a lance. Compared with killer whales, they were slower, less good at detecting and tracking sperm whales (being restricted to the surface and having no sonar or even any effective means of listening under water), had poorer coordination, and were probably more vulnerable to sperm whale counterattacks (see fig. 1.8). Only in his killing apparatus would Captain Ahab seem to have possessed any advantages over the orca when preying on sperms, the harpoon and lance being perhaps more effective than a killer whale's teeth. However, killer whales are clearly able to kill sperm whales, and in all other respects, they would seem to have the advantage over the open-boat whaler.

Why, then, with the exception of the attack off California witnessed by Pitman et al. (2001), does all evidence suggest that the killer whale is not a "serious enemy" (Berzin 1972, 273) of the sperm whale, in the sense of causing significant mortality, while Ahab so clearly was? Part of the answer lies in the behavior of killer whales. For reasons both economic and cultural, they generally forgo preying on sperms. The sperm whale is a difficult (Pitman et al. 2001) and perhaps dangerous prey, and so may be left in peace when easier foraging is available. Additionally, killer whale groups have culturally distinctive diets and foraging behavior, consistently passing by food that is not in their group's repertoire (Boran and Heimlich 1999). For most groups of killer whales, sperm whales seem not to be on the menu. In contrast, open-boat whaling for sperm whales was both economically lucrative and culturally significant for eighteenth- and nineteenth-century Yankees and Europeans.

But the defensive behavior of the sperms may also be a factor. Whereas their cooperative vigilance and communal defense may usually thwart killer whales, some of the same behavior is useful to human whalers. Staggering dives so that calves are always accompanied (see section 6.9) makes a group easier for visual predators like the open-boat whalers to follow. Gathering at the surface and facing the attackers, rather than fleeing, is also a maladaptive defense against humans.

So did sperm whales change their defensive behavior in response to the new predator? Caldwell et al. (1966) reviewed the whaling literature and concluded that "flight is the usual reaction of sperm whales to danger." Whales either swam fast away from the whalers, particularly upwind, dived deep, or dived shallow ("horizontal settling"). All of these responses are more appropriate reactions to open-boat whalers than the standard "surface and face-up" response to killer whales. However, sometimes the whales seemed to behave less adaptively; for instance, surfacing after a long dive almost exactly where they had gone down, or swimming downwind (Caldwell et al. 1966). When confronted, they would sometimes attack the whalers (Caldwell et al. 1966), with varying success (see fig. 1.8). So, overall, the sperm whales did react differently to whalers than to killer whales and their other natural predators in ways that make adaptive sense, but they were not as effective as their natural abilities might suggest. Sperm whales can easily swim 3 km during a 35 min dive, and if they kept track of the whaleboats using sonar while they were under water and maneuvered so as to increase their distance from the boats, they should almost always have been able to escape. They often did escape, but were captured frequently enough to indicate that their response to the whalers was not always optimal. Furthermore, their social nature was exploited by the whalers, who harpooned, but did not kill, calves, and then could easily pick off the other whales who "stood-by" (Berzin 1972, 256–57). Other members of the group would remain beside a wounded female and then be killed themselves.

In the twentieth century, the human whaler became a much more formidable predator. With his steel mechanized catcher vessels, harpoon guns, and sonar, the whaler could now move fast in any direction, locate the sperms under water, kill them more easily, and was never threatened by their defensive behavior. The sperms separated, dived, swam fast, and maneuvered under water (Gambell 1968; Mano 1986), but were usually doomed. In contrast to their strong aggregative reaction when faced by killer whales, sperms tended to scatter in the presence of modern whalers (Lockyer 1977), although there are exceptions, such as the animals described by Nishiwaki (1962) in the quote at the beginning of this section that assumed the Marguerite formation around a wounded colleague in front of a catcher boat. Perhaps they were trying out all their defensive maneuvers in the face of this new danger, to no avail.

Observers, from the open-boat whalers to modern scientists, have remarked upon the skittishness of sperm whales. Thomas Beale put it well, describing the sperm whale as "a most timid and inoffensive animal . . . readily endeavouring to escape from the slightest thing which bears an unusual appearance" (Beale 1839, 6). They can react strongly to the presence of dolphins or to sounds as quiet as the click of an underwater camera (Caldwell et al. 1966). I have seen a group of twenty or so resting sperm whales dive suddenly when a relatively tiny fur seal (*Arctocephalus australis*) appeared in front of them. The sperm whale seems to rightly consider the ocean a potentially dangerous habitat.

5.10 Stranding

Of all the actions of the sperm whale, the one that has most perplexed humans is the stranding, or beaching, of single sperms or groups of sperms. By running up on a beach, a deep-ocean animal becomes accessible to a much larger number of people than ever see it in its natural environment—an estimated 200,000 people converged on Koksijde, Belgium, in November 1994 after four sperm whales stranded there (Jaques 1996). And they can see the whole animal, touch it, dissect it, smell it, and perhaps even taste it. Sperm whale strandings, although rare, occur often enough so that most countries with a few hundred kilometers of coastline can expect a sperm whale stranding at least every few decades or so (e.g., Berzin 1972, 270; Rice 1989). Sperm whale strandings have puzzled humans for at least centuries, and still do (Smeenk 1997). In this section I will summarize some of the available information and speculations about their causes.

An important dichotomy is between strandings of single animals and those of two or more whales, called mass strandings (Simmonds 1997). Single strandings often involve animals that are already dead (e.g., from disease or old age), morbid, or newborn (Rice 1989), and thus the stranding involves an animal without the ability to survive at sea, and so may be explained fairly simply (although such events still present puzzles; see Christensen 1990). In contrast, the majority of sperm whales that come ashore multiply seem healthy, although there are exceptions (Jauniaux et al. 1998). For example, only one of the fifty-nine animals that stranded near Gisborne, New Zealand, in 1970 appeared seriously injured or diseased, an old female with peeling skin that appeared to be in such poor condition that it "gave the impression that it would not survive very long" (Robson and van Bree 1971). Mass strandings can include only males, only females and their young, or a combination of the sexes. The number of animals involved can be large, especially when females are present. Rice (1989) calculated a mean of 32 animals (range 8–72) for eighteen mass strandings including females, and a mean of 13 (range 3–37) for all-male strandings. The 1979 stranding of 41 animals described by Rice et al. (1986) appears to have involved a fusion of one group of females with a set of males 14–21 years old. All age and sex classes of sperm whales mass strand in proportions that are not obviously different from those in the living population at sea. Single strandings, and to a lesser extent, mass strandings, may be clustered along coastlines hundreds or thousands of kilometers long over periods of months, such as the strandings or deaths at sea of 36 sperms over 12 months off Norway in 1988–1989 (Christensen 1990) and three mass strandings in Tasmania involving 112 whales during February 1998 (Evans et al. 2002) (fig. 5.33).

Scientists have looked for spatial or temporal correlates of mass strandings as a clue to their causes. Usually sperm whale mass strandings occur on beaches with high wave action and a gentle slope (Brabyn and McLean 1992), suggesting that beach topography may play a role in these events. Klinowska (1985a,b), who analyzed British stranding records, suggested that strandings happen at times of geomagnetic disturbances and that they may occur in places of magnetic anomalies, thus indicating that magnetic sensing

Figure 5.33 A mass stranding of sperm whales in February 1998 in Tasmania. (Photograph by Karen Evans, courtesy of Karen Evans, Antarctic Wildlife Research Unit, University of Tasmania.)

may be important in navigation. However, this hypothesis is not universally supported (Brabyn and Frew 1994; Berta and Sumich 1999, 441). Robson and van Bree (1971) note that the Gisborne, New Zealand, mass stranding occurred just after the most violent electrical storm in 50 years, and suggest that unusual weather patterns may be involved in mass strandings. It is also possible that anthropogenic effects, such as chemical pollution or loud sounds, may contribute to (Geraci et al. 1999) or cause (e.g., Frantzis 1998; Balcomb and Claridge 2001) mass strandings.

Further clues come the from the behavior of whales during mass strandings. Sperm whales have been described as swimming actively into the shallow water, in clusters of a few animals at a time, with some minutes between the arrival of the clusters at the beach (Robson and van Bree 1971; Lucas and Hooker 2000; Evans et al. 2002). Often other sperm whales can be seen swimming just offshore during the stranding of a cluster (Robson and van Bree 1971; Lucas and Hooker 2000; Evans et al. 2002). Some of these offshore animals have gone on to strand later (e.g., Robson and van Bree 1971). The stranded clusters of sperms may be separated by a few meters, tens of meters, or hundreds of meters along the beach (Robson and van Bree 1971;

Evans et al. 2002). If returned to the water, a living sperm whale will often restrand (Evans et al. 2002).

Sperm whales are not the only species to mass strand. Both long-finned and short-finned pilot whales mass strand frequently, and some other highly social odontocetes, such as Atlantic white-sided dolphins (*Lagenorhynchus acutus*) and false killer whales (*Pseudorca crassidens*), are also notable for this behavior (Sergeant 1982). In contrast, the more solitary cetacean species, such as the baleen whales, very rarely mass strand alive.

Where do these observations leave theories about the causes of this most maladaptive behavior of running up on a beach to die? External factors, such as weather, ocean conditions, bottom topography, magnetic fields, noise, or chemical pollution may have a role in confusing the whales, or in disrupting their sonar or other sensory systems. However, as noted by many authors (including Robson and van Bree 1971; Sergeant 1982; Evans 1987, 242; Rice 1989; Stevick et al. 2002), mass stranding has a strong social aspect. The whales are not on the beach together because they were brought there independently by some common external factor; they actively behaved as a group, coming ashore together, and dying together. Why?

A general, but not universal (see Sergeant 1982), consensus suggests that mass strandings are usually caused by some combination of a failure of the sonar system to give clear information; a "sensitive nervous system which may cause panic and a blind response" to the misleading information, perhaps intensified by external events such as poor weather; "non-adaption to shallow water"; and strong social cohesion (e.g., Gilmore 1959; Robson and van Bree 1971). Animals may follow a sick or confused colleague onto the beach and refuse to leave her/him.

So much for the proximate causes of mass stranding, but if, as seems to be the case, this behavior causes significant mortality—often among groups of relatives—in a slowly reproducing, long-lived species, why is it not strongly selected against? One would think that any individuals that genetically or culturally evolved the means to avoid stranding would be advantaged, and so the behavior would disappear. It has not. This observation suggests that the internal factors that contribute to mass stranding have benefits that outweigh their contribution to mass stranding mortality. I suggest that the "sensitive nervous system" may be useful in responding quickly and appropriately to predators (see section 5.9), "non-adaption to shallow water" a consequence of the animals usually and adaptively staying at depths suitable for their body sizes and foraging techniques, and strong social cohesion a result of the advantages conferred by sociality in combating predation and obtaining prey (see section 6.7). Additionally, sociality can be reinforced by cultural processes, so that "doing as the group does" becomes imperative for the individual (see section 7.3). Collective action can be advantageous in many ways, and so favored by evolution, but on occasion it can lead animals—including humans—into highly maladaptive collective behavior.

5.11 Summary

1. The sperm whale's vocalizations consist principally of loud, broadband, directional clicks. These sounds are used for echolocation (both searching and tracking) and communication. Communicative vocalizations include codas, patterned series of 3–20 or more clicks made in social circumstances, and slow clicks, loud, ringing clicks repeated every 6–8 s and heard from large males, particularly on the breeding grounds.

2. The behavior of groups of female and immature sperm whales has two principal modes: "foraging," when animals seen at the surface are in small clusters of one to three animals, moving relatively fast and consistently, fluking up, and making usual clicks; and "socializing," when clusters seen at the surface are larger, movements are slower and less directed, and the whales show a wide range of aerial behavior as well as emitting codas and creaks.

3. Although there is much variation, groups of females and immatures spend approximately 75% of their time foraging and 25% socializing. Socializing often, but not always, takes place in the afternoon.

4. Sperm whales spend about 8 min at the surface between foraging dives. During these periods they breathe about every 12.5 s (females and immatures) or 17.5 s (large males). The surface period ends with the raising of the flukes.

5. Foraging dives last about 30–45 min. During most of their dives the whales make trains of regularly spaced usual clicks, which are probably a form of searching echolocation. However, usual click trains are interrupted by creaks and silences, which may indicate squid being chased or consumed. With their diverse prey, sperm whales probably use a wide variety of techniques to find, close in on, and consume them.

6. When female and immature sperm whales are socializing, their behavior is very varied. They may lie still, apparently resting, or be extremely active; they may be very vocal, especially in the production of codas, or silent. They are often in physical contact with one another. The amount of time spent socializing varies considerably from day to day, but shows no consistent variation with environmental conditions. Large males on the breeding

grounds socialize with females and immatures, but when they are at higher latitudes their social/resting periods seem reduced.

7. Sperm whales show two principal types of aerial behavior: breaching, in which a whale leaps into the air, and lobtailing, a slam of the flukes on the water surface. The functions of these activities are not clear, but it is likely that they are often used as communicative signals. These signals are energetically expensive, but have relatively poor propagation characteristics for receivers under water. They may have evolved and persist because they are honest indicators of the whale's physical condition and the importance attached to the message. However, these and other puzzling activities of sperm whales may sometimes be types of play or have more prosaic functions, such as ectoparasite removal.

8. The birth of a sperm whale is a social event. The newborn may be in substantial danger during its first hours outside its mother.

9. An infant sperm whale is precocial, accompanying its mother and her group in their substantial movements. Young calves seem to have a following response to nearby adult sperm whales (and similar-sized elements of the marine environment). They lead very social lives.

10. When attacked by killer whales, sperm whales gather quickly and closely at the surface. They adopt two principal defensive formations, the "Marguerite" formation, in which their heads are together and their tails radiate outward, and the "heads-out" formation, in which the sperms face their attackers as a concerted mass. Defensive behavior against human whalers is different in functional ways (less clustering, long dives, and movement upwind), but does not always seem optimal for reducing the risk of mortality.

11. From time to time, sperm whales, of all age and sex classes, mass strand on beaches. The stranded animals usually appear to be healthy. These strandings are probably generally caused by some combination of a failure of the sonar system to give clear information, a "sensitive nervous system which may cause panic and a blind response" to the misleading information, external events such as poor weather, "non-adaption to shallow water," and strong social cohesion, perhaps culturally reinforced.

6 Sperm Whale Societies

6.1 Social Scales

I hope that it is clear from the preceding chapter that sperm whales are social both in their routine daily lives and during crucial events. At least for females, other sperm whales are generally the most significant elements of their environment, and the evidence suggests that these animals have important and diverse relationships with one another. A rough calculation, assuming an even distribution of sperm whales at large scales and based on the results described in chapters 3 and 4, suggests that each animal may share at least part of its range with about 7,510 other female and immature sperm whales,* although this figure will be an underestimate if, as seems to be the case, there is large-scale variation in sperm whale density (see table 4.1). It will also be an underestimate for the wider-ranging males. So, each animal has several thousand potential relationships with other sperm whales. In this chapter, in many ways the core of the book, I will try to describe the major types of relationships that make up sperm whale society. As noted in section 1.7, because of the difficulty of observing interactions—the foundations of relationships—I am forced to use patterns of association to infer a model of social structure.

An "association" between animals should refer to circumstances in which interactions are likely to occur (see section 1.7). Virtually all interactions between sperm whales are mediated by sound, touch, or vision. Of these, sound has by far the greatest range in the sperm whale's underwater habitat. Sperm whales can probably hear each other at ranges of tens of kilometers (Madsen et al., in press) and may respond to each other's sounds over these ranges. However, I suspect that at such ranges, except possibly in the case of large males, the identity of the vocalizer may be unknown or unused, so that the interaction is not truly social (in at least one sense of the word), and that

---

* The estimated global population of sperm whales, 360,000 (see section 4.2), is spread over 316,620,000 km² of ocean (see table 4.1), giving a mean average density of 0.0011 whales/km². The ranges of female and immature sperm whales typically seem to span about 1,450 km (see section 3.4). Thus, an animal shares part of its range with all others whose range centroids lie within 1,450 km of its own range centroid, or roughly $0.0011 \times \pi \times 1,450^2 = 7,510$ animals.

most social interactions between sperm whales occur at ranges of a few kilometers or less. This suspicion is based on the principal types of spatial and social structure we observe when studying female and immature sperm whales in tropical and subtropical waters:

Concentrations: During large-scale surveys, sperm whales are characteristically, but not universally, encountered in patches a few hundred kilometers across (e.g., Berzin 1978; Jaquet 1996a, 118).

Aggregations: Within a concentration, a hydrophone generally picks up either no sperm whales or more than seem to be present visually, even after correction for the animals under water at any time (Whitehead and Kahn 1992). Such aggregations often span about 10–20 km (Jaquet 1996a, 122; see section 6.3).

Groups: When we are with female and immature sperm whales for more than a few minutes, it is usually apparent that a group of whales is moving in a coordinated fashion, although often spread over hundreds or thousands of meters of ocean (see the definition of "school" in Best 1979).

Clusters: At the smallest scale, animals may swim in the same direction, at the same speed, and within a body length or two of one another (Whitehead and Arnbom 1987).

Groups and clusters show coordinated behavior, and thus involve behavioral interactions among animals. Hence, groups and clusters seem to be more "social" than aggregations and concentrations, for which coordinated behavior has not been found. In the following sections I consider each of these structures in some detail, examining their spatial and temporal scales, social significance, and function.

In our analysis of our photoidentification data, a further social entity appeared: the social unit. These units are not apparent in real time at sea, emerging only from the analysis of long-term data, but they may be the most significant of all sperm whale social structures. Concentrations, aggregations, groups, units, and clusters all have characteristic spatial and temporal scales (fig. 6.1), and I will consider them from largest to smallest.

In the analysis of our recordings of coda vocalizations, an additional population structure emerged: the vocal clan. Clans seem to be larger, both in spatial extent and number of animals, than concentrations, but unlike concentrations, they probably persist over generations and are maintained through cultural transmission. Consequently, they would qualify as social structures under most definitions of the term. Clan membership seems to structure relationships among units, and clans are distinct in at least some

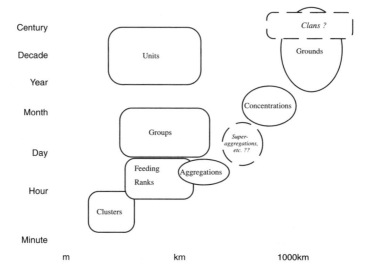

Figure 6.1 Approximate temporal and spatial scales of the spatial and social structures of female and immature sperm whales. The rounded boxes represent primarily social structures ("mutualistic groups"), and the ellipses indicate animals aggregated because of extrinsic factors, primarily food distribution ("nonmutualistic groups"). (Adapted from Jaquet 1996a, 132; Whitehead and Weilgart 2000.)

behavioral features in addition to vocal dialect. Thus, clans may be highly significant elements of the female sperm whale's social structure. I discuss clans in more detail in sections 7.2 and 7.3.

After a consideration of non-clan social structures, I will move on to other segments of the sperm whale population—first the calves, which, although members of the units of females and immatures, are particularly interesting as potential drivers of much of sperm whale sociality. The chapter closes with a description of the apparently simple social life of the adult male and his breeding interactions with females.

6.2   Concentrations

As described in section 2.2, sperm whales are not evenly distributed over the ocean, instead being more common over "grounds," areas a few thousand kilometers across that are usually characterized by higher primary productivity than surrounding regions. However, as noted repeatedly by frustrated whalers as well as whale scientists, the presence of a ground does not guarantee the presence of sperm whales.

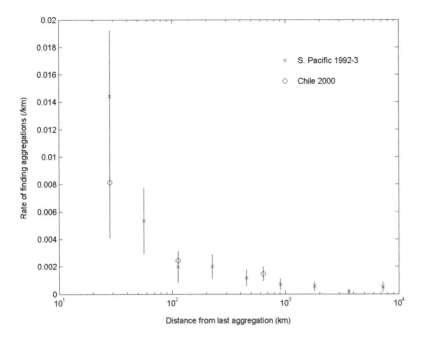

Figure 6.2 Spatial organization of aggregations of sperm whales, plotted from data collected on a survey around the South Pacific in 1992–1993 and off northern Chile in 2000. The *y*-axis gives the rates of encountering new aggregations (excluding single whales, assumed to be males) at different ranges from the previous aggregation. Range intervals were 20–40 km; 40–80 km; 80–160 km; 160–320 km; 320–640 km; 640–1,280 km; and 1,280–2,560 km. Intervals with fewer than four encounters were combined. Bars indicate approximate standard errors, calculated assuming independence of encounters (see Jaquet 1996a, 116).

Day after day passes and we see nothing . . . it seems to me that it is quite useless for me to go to places where I have seen whales with the expectation of finding them again for I am almost sure to meet with disappointment. (Captain Samuel Braley, 15 June 1851, quoted in Wray and Martin 1983)

Grounds, as represented so clearly on Townsend's (1935) charts (see fig. 2.8), are time-averaged geographic structures. From the perspective of a whale, it is perhaps more pertinent to ask about the distribution of other whales around it at any time. Jaquet (1996a, 112–26) has used data from our 1992–1993 survey of the South Pacific to examine whale distributions from the whale's perspective. I have reanalyzed these data, as well as data from an extensive study off northern Chile in 2000 (fig. 6.2).*

* Jaquet (1996a, 120) also analyzed encounter rates from our study off the Galápagos Islands and mainland Ecuador in 1991, but because there were few data, the results are not clear, and so I have not used them here.

It is clear in both Jaquet's original presentation and in figure 6.2 that the rates of finding new aggregations of sperm whales fell substantially, and significantly (*G* tests, *P*<.0001 for South Pacific, *P* = .042 for Chile), as we moved farther from the last aggregation. Thus, there is empirical justification for the growing frustration experienced by whalers and whale scientists who are experiencing "dry cruising." Figure 6.2 also shows that whales are aggregated over a wide range of spatial scales, from tens of kilometers to at least several hundred kilometers.

Jaquet (1996a, 118) noted that there seems to be a plateau in these plots at scales of a few hundred kilometers, suggesting that there are areas of ocean about 300 km across that contain concentrations of whales. This interpretation depends greatly on the low encounter rate at ranges of a thousand kilometers and more in the South Pacific survey (see fig. 6.2). However, there were 15,250 km of survey data (and just six encounters) at ranges of over 1,280 km from the last encounter. Therefore, I think that, while recognizing that the density of animals generally falls with range from an arbitrarily chosen individual, sperm whales do seem often to form concentrations a few hundred kilometers across.

A few hundred kilometers is also the scale at which sperm whale density in our 1992–1993 South Pacific survey data was best correlated with the distribution of subsurface biomass (as indicated by depth sounder records) and a contour index of the steepness of underwater topography (Jaquet and Whitehead 1996). Thus, it seems reasonable to infer that these concentrations reflect regions of the ocean with increased abundance of sperm whale food. As the scales of movement of sperm whales over periods of weeks to months (see section 3.4) approximately coincide with the sizes of concentrations, the concentration may be considered the context within which scramble competition for food resources takes place.

The number of sperm whales within a concentration undoubtedly varies very substantially, but, using the approximate sizes of concentrations and the rates at which we encounter aggregations within them, I come up with an approximate estimate of 750 sperm whales per concentration.* It should be

---

* A very approximate estimate of the number of sperm whales within a concentration can be obtained from the following equation:

$$\frac{\text{Rate of encounter within concentration} \times \text{Whales in aggregation} \times \text{Size of concentration}}{\text{Strip width of search}}$$

The encounter rate with aggregations within concentrations is about 0.0015 encounters/km (see fig. 6.2); there are approximately thirty whales in an aggregation (see section 6.3); the size of a concentration, if circular and spanning 300 km, is roughly 270,000 km²; and the acoustic search strip width for aggregations was about 16 km (2 × 8 km; see fig. 6.3). This gives a population within a concentration of about 750 whales.

stressed that this is a *very* approximate estimate that depends crucially on the assumption that a concentration spans 300 km; using the same methodology, a concentration assumed to be 100 km across would contain an estimated 85 animals, while one 500 km across would have about 22,000.

Jaquet (1996a, 123) noted another feature in her plots of the rate of encountering aggregations against distance from the previous aggregation: "super-aggregations," which are sets of groups close together over scales of 9–72 km. However, these structures are not clearly apparent in my reanalysis of the South Pacific survey data or in the Chile data (see fig. 6.2). It may be more accurate simply to recognize that sperm whales aggregate over a range of spatial scales from tens to hundreds of kilometers (see fig. 6.1).

This spatial pattern, together with the movements of animals (see section 3.4), means that individuals may have access to anywhere between ten and several thousand social partners over time scales of a few days. For instance, Paterson (1986; see the prologue) describes dense concentrations of over a thousand sperm whales. At the other extreme, on our South Pacific survey, in February 1993 we encountered one small group of sperm whales (probably containing fewer than fifteen animals) in the unproductive waters near Easter Island, 2,000 km from any other encounter.

6.3   Aggregations

Moving to smaller scales, the tendency of sperm whales to aggregate becomes even more apparent. We see them together and, using a hydrophone, hear them together. Over ranges of kilometers, sound is usually a more effective way to detect sperm whales than vision, so acoustic analysis gives the best perspective on spatial organization at these scales.

By sailing away from a group of female and immature sperm whales on a course approximately perpendicular to their direction of movement and listening at regular intervals, we have charted the spatial extent of sperm whale aggregations. As a sperm whale can be heard at a range of about 5 km using our equipment, the results, shown in figure 6.3, indicate that sperm whales are usually aggregated over about 2–20 km.

From the overall rate at which sperm whales produce clicks throughout the day (about 1.22 clicks/s for the Galápagos whales; Whitehead and Weilgart 1990), we can convert click rates heard on standard hydrophone recordings to estimates of the number of whales in the vicinity. This method may underestimate the number of whales in an aggregation if it is spread widely, but gives us some indication of how many animals are present much of the

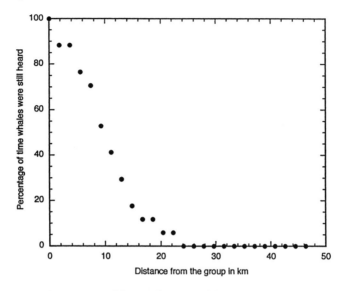

Figure 6.3 Percentage of the time that sperm whales were heard through hydrophones while sailing away from a group of sperm whales on a course perpendicular to its direction of movement, plotted against distance from the group. (From Jaquet 1996a, 122.)

time. The results from the four studies in which regular recordings were made are summarized in table 6.1. Aggregation sizes clearly varied substantially between places and with time, but may have a mode of about thirty whales. There is also great variation in the number of whales within a particular aggregation over time scales as short as a few hours, but this variation is confounded with variance in the click rates of individual whales with time and so is not presented quantitatively here. However, it is apparent when watching, and listening to, female and immature sperm whales at sea that aggregations have dynamic membership over these brief intervals.

Aggregations themselves, as well as differences in aggregation size, are likely to be related to aspects of the whales' food supply, although there is no clear relationship between aggregation size in studies of several months' duration and our measure of feeding success, the defecation rate (see table 6.1). There is probably scramble competition among members of an aggregation for food, but it may be at a low level if the food is in large patches or the animals are sufficiently spread out that they do not interfere much with one another. From the social perspective, aggregations over scales of tens of kilometers may be important because these are the ranges at which sperm whales can probably hear one another. Sperm whale vocalizations contain a great

Table 6.1. Estimates of Mean Aggregation Sizes and Feeding Success from Four Studies

Study	No. of Recordings[a]	Aggregation Size (Whales)[b]	Typical Aggregation Size (Whales)[c]	Feeding Success (Defecations/ Fluke-Up)
Galápagos 1985	20	60.1	77.4	0.062
Galápagos 1987	35	32.8	47.3	0.021
Seychelles 1990	39	9.4	15.7	0.156
Ecuador 1991	21	30.7	49.2	0.058

SOURCES: Whitehead and Weilgart 1990; Whitehead and Kahn 1992; Kahn et al. 1993; Whitehead et al. 1997.

[a]These are calculated from the 07:00 recordings for the Galápagos studies, and from recordings at least 14 hr or more apart for the Seychelles and Ecuador studies.

[b]The aggregation size as experienced by an outside observer.

[c]The aggregation size as experienced by a whale within the aggregation.

deal of information about the activity of an animal, including whether it is foraging and (probably) its feeding success (see section 5.2). This information may assist other members of an aggregation in moving and foraging efficiently. Thus, members of an aggregation may affect each other's feeding success, either negatively or positively, but I suspect that, because the aggregation is usually spread out over a few kilometers, these effects are usually quite small.

6.4 Groups

What Is a Group?

We are now reaching the social scale at which terminology becomes more than a little confusing. There are descriptions of "schools," "pods," "groups," "herds," and "bodies" of sperm whales (e.g., Caldwell et al. 1966; Best 1979; table 6.2). All of these terms suggest spatial proximity at some, usually undeclared, scale, but may, or may not, also imply behavioral coordination or long-term relationships among members. In our research we have tried to standardize and define terms for describing social structures, of which the "group" is one of the most fundamental.

In our definition, a sperm whale group is *a set of animals moving together in a coordinated fashion over periods of at least hours.* This coordination implies active maintenance of the formation, so that the group is a social entity.

Table 6.2. Estimates of the Mean Size of Groups of Female and Immature Sperm Whales from Various Studies

Mean Group Size	Term Used	Region	Reference
27.1	Nursery school	Off Japan	Ohsumi 1971
29.8	School	?	Mazaki et al. (unpubl. cited in Best 1979)
22.9	Mixed school	South Africa	Gambell 1972 (see Best 1979)
28	Mixed group	South Africa	Best 1979
~20	School	Sri Lanka	Gordon 1987b
~25	School/group	Global	Rice 1989
22.1	Group (typical)	Galápagos	Whitehead and Kahn 1992
18	Group (typical)	Seychelles	Whitehead and Kahn 1992
27.2	Group (typical)	Off mainland Ecuador	Whitehead and Kahn 1992

NOTE: Groups are defined as animals moving together in a coordinated fashion over periods of at least hours. All group sizes are as experienced by an outside observer, except those marked "typical," which refer to group sizes as experienced by members of the population.

However, this use of the term is different from that of many primatologists, for whom "group" refers to a nearly closed interacting community of animals. Cetologists use the term inconsistently, sometimes referring to short-term spatial aggregation and coordination of animals (e.g., Clapham 2000; cf. "cluster" below), and sometimes to population subsets with strong long-term affiliations (e.g., Connor et al. 1998). Our definition corresponds approximately to the entity that was once pursued by whalers ("we fell in with a school of sperm whales," Beale 1839, 233) and is followed by today's sperm whale scientists ("the team has been up for the last four hours tracking these whales through the night with hydrophones"; Gordon 1998, 53).

Female sperm whales and their attendant young nearly always—perhaps always—seem to be members of groups. In hundreds of days spent with sperms, we have never seen a female without others within a kilometer or two. For males, the situation is different (see section 6.10).

Group Size

Assessing the size of a sperm whale group is not straightforward, as some animals are usually invisible under water at any time (see box 6.1). However, if the whales are followed for several hours, and especially if the group coalesces at the surface to socialize (see section 5.5; fig. 6.4), observers can make a rea-

Figure 6.4 A group of sperm whales socializing/resting at the surface.

sonable estimate of group size, although it may be biased. There have been a number of different estimates of mean group size for female sperm whales. Some of the variation may be due to different conceptions of a "group" and some to variation with time and place (Whitehead and Kahn 1992; see table 6.2 and below). However, there is general agreement that the mean group size for females and immatures is about twenty to thirty animals.

Most of these mean group sizes are as experienced by an outside observer, such as a whaler, scientist, or possibly a roving male sperm whale. For the females and immatures themselves, the mean group size will be different, as there are more animals in larger groups. The mean group size as experienced by a randomly chosen member of the population has been called the "typical group size" by Jarman (1974), and will never be smaller than the group size as experienced by an observer.* The mean group sizes off the Galápagos Islands, Seychelles, and mainland Ecuador at the bottom of table 6.2 refer to analytically estimated typical group sizes (see Whitehead and Kahn 1992 for the methodology, and box 6.1 for a summary).

---

* $\Sigma(g^2 \times n_g)/\Sigma(g \times n_g)$, the typical group size, is always greater than or equal to $\Sigma(g \times n_g)/\Sigma n_g$, the group size as experienced by an observer (where there are n_g groups of size g).

Box 6.1 **The Problems of Estimating Group Size**

Estimating the sizes of sperm whale groups is not easy (e.g., Ohsumi 1971; Gordon 1987b; Christal et al. 1998). This box contains a summary of the principal methods used.

Observations of Animals Together

As sperm whales can dive for long periods (see section 3.2), and groups can spread out over several kilometers, a count of animals visible at the surface may well not represent the number of animals traveling in a co-ordinated manner. When the whales are socializing at the surface (see section 5.5 and fig. 6.4), then a full count is more likely, but rarely can an observer or hunter be certain that none are missed. Occasionally, when a group is followed over several days in an area of low whale density, so that other groups are not encountered and counts are consistent over days, then the group size can be estimated accurately (e.g., Christal and Whitehead 2000; see box 6.3). However, this is more likely to be the case with small groups, and the groups found in areas of low density may not be representative.

Estimates of Mean Typical Group Size
Using Lagged Association Rates

Given that two identified animals, A and B, are associated at any time, we can estimate the probability that, after any time lag τ, a randomly chosen associate of A will be B (Whitehead 1995a, appendix B). This probability is the "standardized lagged association rate," and its expected value is the inverse of the mean number of companions of a randomly chosen individual over that time scale (i.e., the typical group size minus one, if τ is less than the time interval used to define groups—in our case, a day or two). Therefore, an approximately unbiased estimate of the mean typical group size is 1 + 1/(lagged association rate at small τ). This is the approach taken by Whitehead et al. (1991) and Whitehead and Kahn (1992) and which produced the estimates at the bottom of table 6.2. However, only mean typical group sizes for a whole data set can be produced by this method; we do not have an estimate of the size of any particular group.

Estimates Using a Mark-Recapture Method

The mark-recapture method also uses photographic identifications, but divides each day's identifications into two sets (either before and af-

Box 6.1 continued

ter midday, or the first and second half of the identifications) and assumes, reasonably in most cases, that we were with the same group all day. If x_1 whales are identified in the first set, x_2 in the second, and x_{12} in both, then a "Petersen" estimator of group size is

$$g = \frac{(x_1 + 1)(x_2 + 1)}{(x_{12} + 1)} - 1.$$

Using this method, I was able to estimate group sizes, and coefficients of variation (see Seber 1982, 60), on 251 days spent tracking sperm whales in the South Pacific. Unfortunately, because x_{12} is sometimes small (<5), the estimate of group size, g, has low precision and a high CV. When presenting results on the distribution of group sizes, I wished to remove low-precision estimates. Unfortunately, doing so leads to a bias, as larger group sizes are harder to estimate with the same precision as smaller groups. Thus, in table 6.3 and figure 6.5, there may be a bias against the largest groups when only the more accurate data are used.

With identifications of individual animals, we can make objective estimates of the sizes of particular groups (see box 6.1). These estimates can be compiled to give estimates of mean group size and variation in group size as experienced by an observer, as well as the typical group size experienced by an animal (table 6.3). However, there is a trade-off: if only high-precision estimates are used, there is a bias against larger groups, as their sizes are harder to estimate with precision, whereas if lower-precision estimates are included, measures of group size variation may be inflated. Hence, in table 6.3 and figure 6.5, I present estimates both from the most precisely estimated group sizes (CV <0.25) and from a wider collection (CV <0.4). Despite these biases, both estimates agree quite well with subjective, and other analytical, estimates of a mean group size between about twenty and thirty animals (see table 6.2).

Group size varies quite substantially around the mean (see fig. 6.5). While some groups contain fewer than ten animals (for instance, the group of nine animals from unit T followed on 18 days off the Galápagos Islands in 1999), in other situations more than forty female and immature sperm whales may swim as a coordinated group for a day or more. There also seems to be some variation with place. For instance, groups were generally larger off mainland Ecuador and Peru than near the Galápagos Islands (see tables 6.2 and 6.3).

Table 6.3. Summary Estimates of Group Size and Group Size Variation for Our South Pacific Data

Data	Estimates with CV < 0.25			Estimates with CV < 0.40		
	n Days	g (SD) Whales	g_t (SD) Whales	n Days	g (SD) Whales	g_t (SD) Whales
All	120	19.4 (9.8)	25.1 (10.4)	177	24.1 (15.0)	36.0 (19.1)
Galápagos-all	97	18.8 (10.0)	24.8 (11.0)	139	23.8 (15.5)	35.5 (19.6)
Ecuador-Peru	15	26.2 (7.4)	28.8 (6.4)	20	30.0 (13.5)	37.6 (18.3)
Southeastern Pacific	5	11.9 (5.4)	13.9 (2.7)	11	18.4 (9.6)	25.6 (5.5)
Western Pacific	3	18.3 (6.3)	20.4 (4.7)	7	21.6 (15.3)	33.8 (16.3)
Galápagos 1985	8	14.9 (6.7)	18.2 (4.7)	15	26.8 (17.7)	41.3 (16.7)
Galápagos 1987	28	18.1 (8.8)	22.7 (8.8)	42	22.2 (12.7)	31.6 (14.7)
Galápagos 1989	24	24.7 (11.4)	30.9 (9.8)	41	30.0 (17.2)	42.9 (21.1)
Galápagos 1995	12	18.6 (3.8)	19.6 (3.3)	12	18.6 (3.8)	19.6 (3.3)

NOTE: Results are subdivided by geographic area and by major study year for studies off the Galápagos Islands. Data were analyzed using the Petersen mark-recapture method (see box 6.1), with a day's individual identification data split in half (estimates were very similar when data were split by morning-afternoon). Results are given for two levels of precision in group size estimates: g is the estimated mean group size as experienced by an outside observer; g_t is the typical group size as experienced by a member of the group. Estimated standard deviations (SD) are corrected for imprecision in the estimated group sizes.

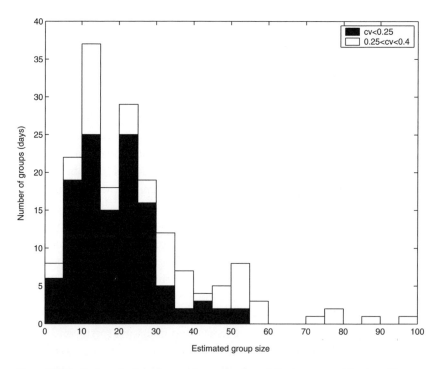

Figure 6.5 Distribution of estimated group sizes on days spent following groups of female and immature sperm whales in the South Pacific. The more accurate estimates are represented by solid bars.

Although much smaller group sizes for females and immatures are sometimes reported for particular regions, such as the northern Gulf of Mexico (Weller et al. 2000), in many of these reports the term "group" may actually refer to what I call a "cluster." There may also be some variation in group size between the years of our Galápagos studies (see table 6.3), although the differences between 1985, 1987, and 1989 are not statistically significant (Whitehead and Kahn 1992).

Formations of Groups

When sperm whales are socializing (see section 5.5), a group sometimes gathers at the surface in one large cluster, with each animal just a few meters from its neighbors (e.g., fig. 6.4). However, socializing occupies a minority of the sperm whale's day; when foraging, sperm whales are spread over much larger ranges. This was shown in Ohsumi's (1971) plot of the positions of sighting and capture of members of three sperm whale groups, as well as by Watkins and Schevill (1977a), who located sperm whales under water by listening for their clicks using a widely spaced hydrophone array, and concluded that sperm whales could be spread over several kilometers when foraging under water.

By making scan samples every 5 min during our 1985 and 1987 Galápagos studies, we amassed a large data set on the spatial distributions of clusters of sperm whales at the surface (Whitehead 1989a). A random sample from the patterns observed in 1985 is illustrated in figure 6.6. These patterns are varied, with the sperm whales sometimes occupying an area just a few hundred meters across (e.g., 26 April, 14:05), and at others being spread over more than 3 km (e.g., 19 April, 12:10). However, there is a tendency, apparent in nine of the fifteen patterns (as well as in one of Ohsumi's [1971] three groups), for the clusters to form a rank, aligned roughly perpendicular to their direction of travel.

Statistical analyses of the 1,884 5-min intervals in the 1985 and 1987 data confirm these impressions (Whitehead 1989a). The tendency of objects to form a line in two dimensions can be quantified by comparing the length of the major axis (the standard deviation of the points in the direction of greatest scatter; Sokal and Rohlf 1981, 594–601) to that of the minor axis (the standard deviation in a perpendicular direction). For both the 1985 and 1987 data, the median of the ratio of the minor axis to the major axis was 0.3, whereas for randomly positioned data it was 0.5 (Whitehead 1989a), indicating that the sperm whales spread out in a line. The angle between the direction of movement and the major axis was more likely to

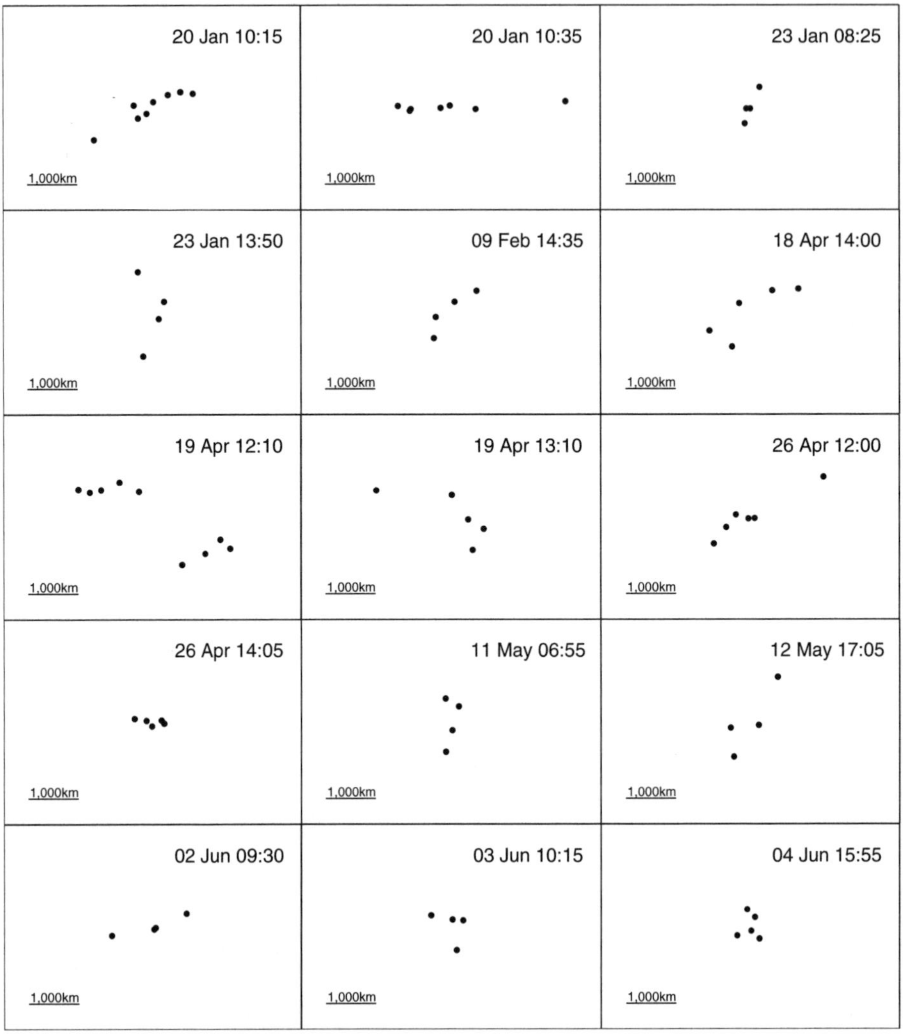

Figure 6.6 Spatial arrangement of clusters of sperm whales during fifteen randomly chosen 5 min intervals when at least four clusters were observed (off the Galápagos Islands, 1985). Each dot represents a cluster of one or more sperm whales. The general direction of movement of the group (as indicated from consecutive satellite navigator fixes) is up the page.

be 60°–90° (49% of 5 min intervals) than 30°–60° (31%) or 0°–30° (21%), so the line was more often a "rank" than a "file." The tendency of sperm whales to form a rank, "like soldiers on parade," was also noted in both the older whaling literature (see Caldwell et al. 1966) and by modern whalers (e.g., Gambell 1968).

Off the Galápagos Islands in both 1985 and 1987, clusters visible at the surface were separated by about 200 m along an axis perpendicular to the direction of movement, and there was a small but significant tendency for them to be spaced out rather than clumped along this axis (spacing was suggested for 56% of 5 min intervals compared with 45% for random data; Whitehead 1989a). The lengths of the ranks (i.e., the spread of visible whales along an axis perpendicular to the direction of movement) had a mean of 1,160 m and a standard deviation of 1,080 m in 1985 ($n = 587$ 5 min intervals with at least ten whales visible at the surface), but the ranks were rather shorter in 1987, with a mean length of 790 m (SD 670 m, $n = 323$).* Group sizes were not obviously larger in 1985 (see table 6.3), but aggregation sizes were (see table 6.1), suggesting that the generally longer ranks of 1985 are the result of members of several dispersed groups being seen more often together at the surface.

Although there are a number of errors and biases that affect these statistics on the formations adopted by sperm whale groups (see Whitehead 1989a, table 5), they are dwarfed by the variation that naturally exists (see fig. 6.6). The modal formation of a group of foraging sperm whales is represented in figure 6.7.

Within groups of lions (*Panthera leo*), individuals may possess preferred positions in hunting formations (Stander 1992): there are "left wing" specialists, "center" specialists, and so on. In contrast, there was no indication of such preferred positions within foraging formations of sperm whales, either within the large data sets collected in 1985 and 1987 (Whitehead 1989a) or in two groups studied intensively in 1995 (group A2B) and 1999 (group T), when GPS systems allowed us to obtain accurate locations of individual whales (Christal and Whitehead 2001). Similarly, there was no indication that either large mature males or calves were closer to, or farther from, the center of the rank than other animals (Whitehead 1989a). However, as explained below, positions within ranks of foraging sperm whales are distinguishable from random in one respect: the proximity of two animals within a rank indicates a long-term social bond.

---

* These calculations use only those 5 min intervals when more than ten whales were visible at the surface. However, the estimated mean lengths of the ranks were little changed whether this criterion was relaxed so that only five or more whales had to be visible (mean length = 1,133 m; SD = 973 m for 1985) or strengthened so that at least twenty whales had to be visible (mean length = 1,134 m; SD = 1,100 m for 1985). These calculations of mean rank length are less biased than those in Whitehead (1989a), which inappropriately assumed equal numbers of whales at the surface at all times.

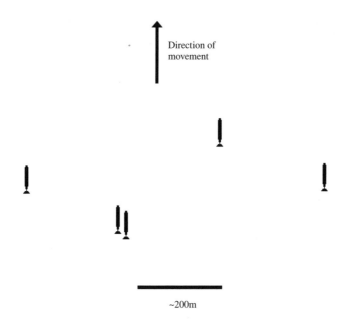

Figure 6.7 Representation of the approximate modal formation of a group of foraging sperm whales as seen at the surface at any time. (Redrawn from Whitehead 1989a.)

Social Structure of Groups

So, like many marine animals, female and immature sperm whales travel together as spatially coherent groups for periods of at least hours. However, even the early whalers could see that there was more to sperm whale social structure. Beale's (1839, 53) quote in the prologue is worth repeating: "The females . . . are also not less remarkable for their strong feeling of sociality or attachment to one another." Modern whalers collected more concrete data. In particular, Ohsumi (1971) reported on four pairs of female sperm whales, and one female–immature male pair, who were marked together in the same group (using metal "Discovery" tags; see appendix, section A.1) and killed 5–10 years later and hundreds of kilometers distant, still in the same group. As he concluded (Ohsumi 1971, 12), "From this viewpoint, it will be probable that sperm whale schools have a tight family union for long time."

That at least some female and immature sperm whales stayed together over periods of weeks and months was confirmed by the first photoidentification studies of living sperms off Sri Lanka and the Galápagos Islands (Gordon 1987b; Whitehead and Arnbom 1987). Our hypothesis at that time was that groups were essentially closed and had stable membership over the long term, although they might interact with other groups for periods of hours. However, as studies matured and photographic identifications were compiled over longer time periods, exceptions appeared. There were animals that clearly transferred between groups, which we initially called "transients" (Whitehead and Waters 1990). But when the large set of photoidentifications (843 high-quality photographs of 293 females and immatures) collected off the Galápagos Islands in 1989 was analyzed and matched to the catalogue built up during the previous study years (1985, 1987, and 1988), it became clear that the data were no longer consistent with a model of closed and stable groups, even allowing the presence of some "transients." There were many clear cases of individuals that spent days together, but then split up.

With the crumbling of our ad hoc model of social structure, I set out to develop techniques that would reveal the pattern of relationships contained in our large quantity of photoidentification data and allow models to be fitted to those data objectively. The problem was unusual in that we had sparse data on a large number of animals (over a thousand animals with just a few identifications of each) rather than extensive data on fewer individuals, as is more common in studies of social structure in terrestrial mammals, such as that of Goodall and her colleagues on the chimpanzees (*Pan troglodytes*) at Gombe Stream (Goodall 1986). The method that emerged was based on the work of Myers (1983) on shorebirds and Underwood (1981) on ungulates, and became known as the "lagged association rate" (Whitehead 1995a). This rate is a function of time lag ("time lag" is the interval between two instants in time), τ: given that two animals X and Y are associated at any time, it is simply an estimate of the probability that, after τ time units, X will still be an associate of Y (Whitehead 1995a). With the sperm whales, because we did not always photoidentify all the companions of a given animal, we had to use a variant of this method, called the "standardized lagged association rate": the probability that Y *is the animal in an identification* of an associate of X τ time units after their known association (see box 6.1).

Plots of the standardized lagged association rate against time lag (e.g., fig. 6.8) indicate temporal aspects of social structure. For instance, if the standardized lagged association rate is roughly constant at K for all τ, this suggests that animals form closed groups with a mean typical group size of about $1 + 1/K$. If the standardized lagged association rate falls to zero after lags of about τ, then associations do not last longer than this. Quantitative

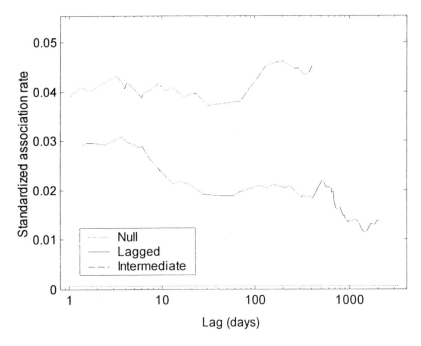

Figure 6.8 Standardized null, lagged, and intermediate association rates for data on female and immature sperm whales of the eastern tropical Pacific in 1985–1997, using only high-quality photographs. The plots show moving averages (100,000 pairs of potential associates for lagged rates; 10,000 for null and intermediate rates).

models of social structure can be fitted to these plots (Whitehead 1995a). I have found models containing exponential decays in the probability of association with time lag, so that individuals have a constant probability of disassociating after any time lag, to be particularly useful.

Using this methodology on our data for female and immature sperm whales in the eastern tropical Pacific (mainly from the Galápagos Islands, but also from the waters off Peru and mainland Ecuador) results in the plot of standardized lagged association rates shown in figure 6.8.* Here I define whales as associated if they were identified within 12 hr of each other. This definition is an operational stand-in for "in a group together" (see Christal et al. 1998 for a justification of this). Two other standardized association rates are also shown in figure 6.8. The "standardized null association rate" is the rate

* This presentation of lagged association rates for female and immature sperm whales from our eastern tropical Pacific data for 1995–1997 essentially follows that of Christal (1998, 20–37). Additional data from 1998 and 1999 have not been added, as they refer exclusively to one, rather atypical group with nine members from just one social unit (unit T).

expected if there were no preferred or avoided companionships, and is simply the inverse of the estimated population size minus one. The "standardized intermediate association rate" is calculated similarly to the standardized lagged association rate, but uses only associations, and potential associations, between pairs of animals during the interval between their first and last recorded association together (Whitehead 1995a). It is therefore useful for examining whether relationships remain constant over time or whether, alternatively, animals often associate, separate, and then reassociate (Christal 1998, 26).

What can we conclude from the plots in figure 6.8 showing standardized lagged, null, and intermediate association rates for the Galápagos data? Several significant features emerge from these plots:

- The lagged association rate is well above the null association rate over all time lags from a day to a decade. The low null association rate is a function of the large population size (in the thousands of animals): if association were completely random, the probability that two randomly selected associates (group members) of any given animal would be the same would be very small, whatever the time lag. This is not the case, confirming earlier conclusions that female and immature sperm whales often maintain relationships over periods from days to at least a decade.
- There is a fairly constant lagged association rate up to about 3 days, indicating that groups largely maintain their identity over these time scales.
- There is a drop in the lagged association rate over time scales from about 3 days to 2 weeks, indicating some breakup of groups over these time scales.
- There is little disassociation over scales of 2 weeks to 2 years.
- There is a drop in the lagged association rate over scales of several years, which could be caused by mortality, emigration from the study area, or transfers between nearly closed social units.
- There is a fairly constant intermediate association rate over time scales from about 1 day to 1 year, indicating that long-term associates do not separate and then regroup.
- There is a mean typical group size of about twenty-five to thirty-five animals (inverses of lagged association rates over short time lags, plus one).

This analysis thus confirms earlier discoveries about sperm whale groups: their approximate sizes, and the observation that they contain a mixture of long-term (over years) companions and short-term (over days) acquaintances. If measures of precision are added to the lagged association rates (fig. 6.9), it becomes clear that smaller features of the plots, such as the apparent fall in the lagged association rate at lags of about 2 years, should not be taken literally, especially as the patterns of social organization indicated by

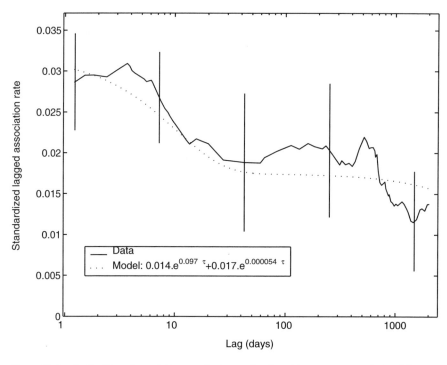

Figure 6.9 Standardized lagged association rates for data on female and immature sperm whales of the eastern tropical Pacific in 1985–1997 (as in fig. 6.8), with approximate standard error bars (using the conservative jackknife method; Efron and Gong 1983) and the best fit model of exponential decay with time lag, τ, in days (see Whitehead 1995a; the model is fit to the original data, not the moving averages displayed here).

the lagged association rates appear to change between the two decades of our study: the 1980s, when densities were high, and the 1990s, when they were low (Christal 1998, 31–34). However, additional useful information can be extracted from the lagged association rates by fitting models to the data.

Of the models containing exponential decays fitted to these data by Christal (1998, 30–31), the one with the best fit (shown in fig. 6.9) is

$$s(\tau) = 0.014.e^{-0.097\,\tau} + 0.017.e^{-0.000054\,\tau},$$

where $s(\tau)$ is the standardized lagged association rate over a lag of τ days. Thus, at any time, individuals are grouped with, on average, roughly equal numbers of what we have called "constant companions" and "casual acquaintances" (Whitehead et al. 1991). The constant companions disassociate at a rate of 0.000054/day (jackknife standard error 0.00033/day), or, very approximately, once every 50 years or so, perhaps because of death or emigration from their permanent social unit by one of the members of the

Table 6.4. Numbers of Associates of Animals Identified (from High-Quality Photographs, $Q > 3$) on More Than Eight Days during Studies off the Galápagos Islands

ID	Unit[a]	Days Identified	1 d	2 d	3 d	4 d	5–9 d	10–14 d	15–18 d
#234	A	11	43	8	4	7	9	0	
#235	A	9	27	11	12	7	3		
#255	A	13	31	10	10	9	7	0	
#804	B	10	6	9	6	2	9	0	
#806	B	9	4	11	2	3	9		
#807	B	10	11	5	7	4	7	0	
#810	B	10	9	7	6	1	8	0	
#3700	T	18	0	0	0	0	0	7	1
#3701	T	13	0	0	0	0	2	6	
#3702	T	13	0	0	0	0	1	7	
#3703	T	14	0	0	0	0	0	8	
#3704	T	17	0	0	0	0	0	7	1
#3705	T	14	0	0	0	0	0	8	
#3706	T	13	0	0	0	0	2	6	
#3707	T	13	0	0	0	0	1	7	
#3708	T	15	0	0	0	0	0	8	0

[a]Members of unit A were identified in 1985, 1987, 1988, 1989, 1991, and 1995; members of unit B in 1987, 1988, 1994, 1995, and 1996; and members of unit T in 1998 and 1999.

pair, while the casual acquaintances disassociate at a rate of 0.097/day (jackknife standard error 0.069/day), or about once every 10 days.

Constant companionship is reflexive (A can be reasonably considered a constant companion of herself), symmetric (if B is a constant companion of A, then A is one of B), and usually transitive (if B is a constant companion of A, and C is one of B, then C is one of A). So, theoretically, constant companionship is an *equivalence relationship,* and the population can be partitioned into sets of mutually constant companions, which I call *social units.**

This pattern of constant companions and casual acquaintances is illustrated by the associates of some of our better-studied whales (table 6.4). For all of these, except members of unit T, there are some individuals identified with the focal individual on many days (the constant companions) and others identified with the focal individual just once or twice (the casual

* Actually, dividing animals into these sets is not straightforward (see Box 6.2). As the property of casual acquaintanceship is not reflexive or transitive, the population cannot be partitioned on this basis.

acquaintances). Unit T, identified in 1998 and 1999 when whale density off the Galápagos Islands was very low, was never seen on the same day as other females and immatures, and all associates of members of this unit are constant companions—other members of unit T. In contrast, members of unit A were identified with many different casual acquaintances.

So, within the group in which she is swimming, a female sperm whale generally has two types of associates. With some, her constant companions (members of her social unit), she shares a long-term relationship, but with the others she does not. This difference is reflected in her pattern of associations within the group. For the nineteen social units that were delineated during our eastern tropical Pacific studies (and grouped at some time with members of other units—thus excluding unit T), we compared the associations* among constant companions and casual acquaintances when grouped together (Christal and Whitehead 2001). For sixteen of the nineteen units, a proportion significantly greater than expected by chance, the mean association index among members of the unit was greater than that between unit members and other grouped animals (table 6.5), indicating an association preference for constant companions (Christal and Whitehead 2001). The three units with the reverse pattern were not consistently distinct in any way, and in two of the three exceptions (units F and M), the difference between the mean association indices was small (Christal and Whitehead 2001).

The general pattern is shown very clearly for group A2B, a group followed for a week in 1985 (see box 6.3). Individuals were associated much more frequently with members of their own social unit (A2 or B), both in terms of being found within the same cluster together and of being photographed within 10 min of each other, than with members of the other unit (Christal and Whitehead 2001; fig. 6.10). In this group, the two constituent units tended to be at opposite ends of the rank. As unit A2 was much smaller than unit B (five animals versus seventeen), its members were generally displaced to the flanks (fig. 6.11). However, neither unit had an obviously preferred side of the rank. For instance, on 27 May 1995, in the morning A2 was on the right of the foraging rank, but by midday it had switched to the left (fig. 6.12).

The final issue I will touch on in this section is whether there are preferred relationships between units when groups are formed, a pattern clearly shown by savannah elephants (*Loxodonta africana*), which have a social structure

---

* For operational reasons, we defined "association" as "identified within 10 minutes of" for this analysis. For two whales to be associated under this criterion, they had to be within a few hundred meters of each other and have reasonably synchronous dive cycles. Frequently, whales associated under this criterion will have been members of the same cluster.

Table 6.5. Characteristics of Sperm Whale Social Units Identified in the Eastern Tropical Pacific

Unit	Members	Days[a]	Mean Pairwise Association Index[b]		Permutation Test for Within-Unit Associations (*P*)[c]	Clan[d]
			Within Unit (Constant Companions)	Between Unit and Other Group Members (Casual Acquaintances)		
A	24	21	0.342	0.219	0.021	Reg.
B	22	17	0.386	0.19	0.01	Reg.
C	3	2	0.5	0.183	—	Reg.
D	15	3	0.248	0.19	0.765	+1
E	18	13	0.439	0.406	0.528	Reg.?
F	12	7	0.35	0.362	0.046	+1
G	11	4	0.191	0.137	0.448	+1
H	4	7	0.667	0.325	—	Reg.
I	7	6	0.417	0.33	0.712	Reg.
J	9	4	0.323	0.201	0.498	Reg.
K	13	9	0.292	0.22	0.594	Reg.
L	14	15	0.381	0.171	0.547	Reg.
M	3	3	0.5	0.518	—	
N	6	6	0.578	0.115	0.167	+1
O	8	9	0.13	0.231	0.946	Reg.
P	9	8	0.676	0.36	0.108	Reg.
Q	6	3	0.491	0.198	—	+1
R	11	5	0.331	0.226	—	
S	6	3	0.34	0.299	—	Reg.
T[e]	9	18			0.093	Short

[a]Number of days when two or more unit members were identified.

[b]Defined as the frequency of identification of a pair of individuals within 10 min, over only those 2 hr samples during which both individuals were identified.

[c]Results of permutation tests for differences among association indices of dyads within the unit (Christal and Whitehead 2001). For some units, there were insufficient data to perform these tests.

[d]Acoustic clan membership (see section 7.2): Reg. = regular codas favored; "+1" = +1 codas favored; Short = short codas favored; blank cell = unknown.

[e]Unit T was never observed with other females or immatures.

very similar to that of sperm whales (Weilgart et al. 1996; see section 8.5). The sparseness of our data makes this a difficult topic to address. However, if this were a major effect, it would be shown by a fall in the intermediate association rate with time lag, as dyads from preferentially associating units would be more likely to be associated a day or two after a previous associa-

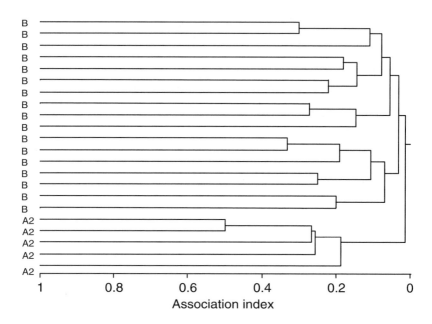

Figure 6.10 Average linkage cluster analysis of the associations within the group A2B, followed from 28 May to 3 June 1995, defining association as "in the same cluster" and using a simple ratio association index with 2 hr time units (from Christal and Whitehead 2001). The *x*-axis is an estimate of the rate at which pairs were photographed within the same cluster within a 2 hr period, given that one of them was identified.

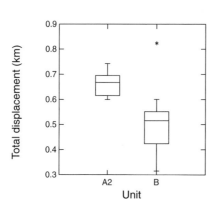

Figure 6.11 Average total displacement on either side of the mean track for members of units A2 and B during the week of 1995 in which they were tracked moving together. (From Christal and Whitehead 2001.)

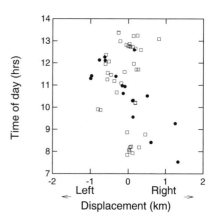

Figure 6.12 Left-right displacement about the mean track of group A2B for members of unit A2 (solid circles) and unit B (open squares) with time of day on 27 May 1995. Points are slightly "jittered" so that they are not coincident. (From Christal and Whitehead 2001.)

Table 6.6.   Associations of Delineated Social Units in the Eastern Tropical Pacific

	A	B	C	D	E	F	G	H	I	J	K	L	M	N	O	P	Q	R	S	T
A	*7*	1	1					2	2		1									
B	1	*6*									1									
C	1		*2*																	
D				*2*																
E					*6*										1					
F						*4*														
G							*2*					1								
H	2							*3*	1											
I	2							1	*5*			1								
J										*3*										
K	1	1									*5*	1				1			1	
L							1		1		1	*6*			1				1	
M													*3*				1			
N														*3*			1			
O					1							1			*3*					
P											1					*3*				
Q													1	1			*3*			
R																		*4*		
S											1	1							*3*	
T																				*3*

SOURCE: Calculated from data in Christal 1998, 169–71, with the addition of data from unit T.

NOTE: This table gives the number of identification periods (series of days separated by at least 30 d) during which pairs of social units were identified together on at least 1 day (at least two members of both units had to have been identified on 1 day) and the total number of identification periods for each unit on the diagonal in italics. See table 6.5 for more information on these units.

tion than a month or two later, when the two units would likely have split. There is no such fall in the intermediate association rate (see fig. 6.8), suggesting that preferential grouping between units was not a sufficiently important effect for us to find many groups containing the same pairs of units after lags of months or years. I have tabulated the associations among delineated units in table 6.6. There are only two pairs of units identified on the same day at least 30 days apart (A and H on 4 days between 5 and 18 January 1987 and on 9 March 1987; A and I identified on 11 March 1989 and 5 April 1991). Any association of pairs of units over three or more identification periods would have caused them to be merged by the unit delineation process (see box 6.2). While five units were never identified with any of the other delineated units, two had associations with five partner units (table 6.6). This apparent variation in unit association rates was not significantly greater than expected if the number of associating units for any unit was proportional to the number of periods in which it was identified.* However, when units are allocated to acoustic clans (see table 6.5), there are indications that units preferentially group with units of their own clan (see section 7.2).

6.5   Social Units

What Is a Social Unit?

For a female sperm whale, the social unit is probably the most significant element of her social environment. However, for human observers, units, along with clans, are the least apparent of the sperm whale social structures that have been recognized. Because units are well embedded within groups in the short term, we need long-term data to delineate and study them — data that are not easily pulled from the ocean. Social units are sets of whales who live and move together over periods of years. Our evidence (such as the fairly constant intermediate association rate of figure 6.8) suggests that animals rarely, and perhaps never, leave their unit to later return. However, as discussed below, animals can either be born into or immigrate into a unit, and they may leave it through death or emigration.

Unit Size

We have been able to delineate twenty social units using the methods described in box 6.2. The size distribution of these units is shown in figure 6.13. Units clearly vary dramatically in size, from trios to units of over twenty whales who travel together. The estimated mean unit size is 10.5 (SD 5.8) animals, and the mean typical unit size experienced by the whales themselves is 13.6 (SD 6.2). These estimates agree well with those obtained from the lagged association rate analysis (Christal et al. 1998).

---

* A chi-square statistic $[X^2 = \Sigma(\text{observed} - \text{expected})^2/\text{expected})]$ was calculated under the null hypothesis (the number of associating units is proportional to the number of periods in which a unit was identified) for the real data and compared with that from 1,000 random data sets in which the observed numbers of associating units for each unit were drawn from Poisson distributions with the expected numbers as means. The real value of X^2 was greater than 64% of the random values, indicating a one-tailed P value of .36.

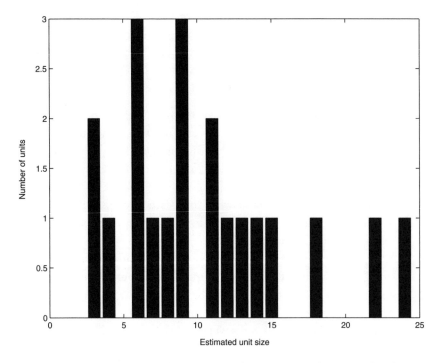

Figure 6.13 Frequency distribution of estimated sizes of sperm whale social units identified off the Galápagos Islands. (Data from Christal et al. 1998, with the addition of data for unit T observed in 1999.)

Box 6.2 **Delineating Social Units**

Social units, as I define them, are long-term structures. They are not revealed in brief observations. Two animals in a visually distinct group, or even in an even more visually distinct cluster, may be constant companions—and so members of the same social unit—or casual acquaintances—and so members of different social units. With long-term data, some relationships become clearer: a dyad seen in the same group over 3 different years is very likely to consist of constant companions. This is the basis on which we have delineated social units (see Christal et al. 1998).

We defined a social unit as a set of individuals, each of which was identified within 12 hr of at least two of the others, during at least two identification periods separated by at least 30 days. However, if, using this criterion, only one animal identified in at least three identification

Box 6.2 continued

periods (separated by at least 30 days) was a member of a unit, then all animals associated with it in at least two such identification periods were considered unit members (Christal et al. 1998). This is an imperfect way to delineate units for several reasons:

- It has several arbitrary elements. For instance, if the association criterion is decreased from 12 hr, we end up with rather more, but smaller, units. However, the 12 hr criterion is less likely to split large units than are finer measures, and so is less biased (Christal et al. 1998).
- There were some transfers between units (see below), complicating this definition and leading to five individuals being allocated to more than one unit (Christal et al. 1998).
- Any long-term preference of separate units to group together preferentially might cause them to be lumped into one unit. However, as noted above, the lack of a fall in the intermediate association rate with time lag (see fig. 6.8) suggests that such preferences are quite minor, and thus that few units were artificially lumped.

In some respects, these imperfections are relatively minor. I believe that the unit delineation method that we have used nearly always allocates true constant companions to the same unit and casual acquaintances to different units.

Social Structure of Units

We have used two methods to look for preferred or avoided companionship within units. Christal et al. (1998) found only a very small (and statistically insignificant) trend for pairs of unit members (constant companions) that were identified within 10 min of each other during one identification period to be preferentially identified within 10 min in a subsequent identification period.

A more direct and powerful analysis looked at the distribution of association indices among pairs of unit members (with "association" defined as identification within 10 min of each other). Preferred or avoided companionships are indicated by particularly high or low indices, respectively, and thus a high standard deviation of the association indices within a unit is expected if there are preferred and/or avoided companionships. For each delineated unit, Christal and Whitehead (2001) compared the calculated standard deviation of the association indices with the expected distribution if there were no preference or avoidance of particular companions. The expected distributions were produced using the Monte Carlo permutation method of Bejder et al. (1998) with a modification described by Whitehead (1999b). The results are summarized in table 6.5.

For eleven of the fourteen units for which this test was practicable, there was no statistically significant preferential association or avoidance among members. Of the exceptions, the Monte Carlo test for unit F was barely significant at $p < .05$, leaving units A and B. These were two of the best-studied units, but also the largest and among the least stable (see below). The association indices within units A and B hinted at a correlation between patterns of association indices and past or future changes in unit membership, but no clear pattern was found (Christal and Whitehead 2001). The generally homogeneous associations among unit members are exemplified by the associations of our best-studied unit, T. The association indices for all thirty-six dyads of the nine members of this unit, which are illustrated using a cluster diagram in figure 6.14, had a coefficient of variation of only 9%. Christal and Whitehead (2001) suggest, then, that there may be two general types of sperm whale units: large, relatively unstable units with some internal structure of associations (such as units A and B), and smaller units with a stable membership within which associations are quite homogeneous (such as unit T).

In some species, such as killer whales (*Orcinus orca*), animals within social units preferentially associate with their genetic relatives (Bigg et al. 1990), a pattern predicted by kin selection theory. However, among members of sperm whale unit B, the unit with the clearest evidence for internal structure (see table 6.5), Christal (1998, 108) found only small and statistically insignificant correlations between the genetic relatedness of a dyad and its association index. For the five members of unit T who have been genetically typed, such a correlation is even less evident (Whitehead 2003). These findings reinforce the emerging picture of rather homogeneous relationships among females within sperm whale units.

Stability of Units

Social units are defined and delineated on the basis of stable membership over periods of more than a month or two. But how stable are they over periods of years? This is difficult to pin down, as there are not many units for which there are good long-term identification records, and the absence of an animal from the photoidentification record of its unit could mean that it had transferred to a new unit, died, or was present but not photographed. With

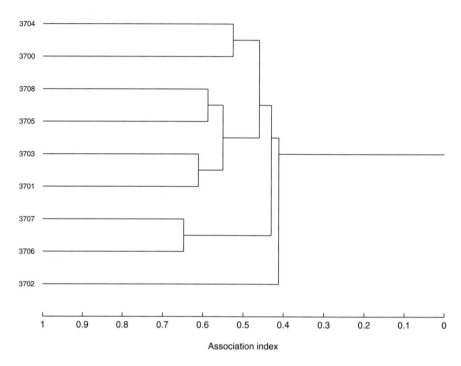

Figure 6.14 Average linkage cluster analysis showing the homogeneity of the associations among the members of unit T. Here, association is defined as identification within 10 min of each other during any day. Thus, the *x*-axis is an estimate of the proportion of days on which pairs were photographed within 10 min of each other.

these qualifications in mind, I will summarize the photoidentification record of unit F, shown in table 6.7. Unit F was identified during five periods over 9 years, often grouped with members of other units, especially in April 1987 and March 1989. There are gaps in the record, with some individuals not being identified in some periods. These gaps could indicate recruitment (by juveniles), immigration into the unit, or mortality or emigration from it (e.g., by whale #732). However, they also can be plausibly ascribed to incomplete sampling, and there is no concrete evidence here for any change in unit membership over the 9 yr period. For some other units, however, there is such evidence.

Jenny Christal looked in detail at our data and found good evidence for one unit splitting as well as for some transfers between units, and one case in which there was either a transfer between units or a merger of units (Christal et al. 1998). Her results are summarized in figure 6.15 and below.

Table 6.7.   High-Quality ($Q \geq 4$) Photoidentifications of Members of Unit F, and Associates, during Five Periods between 1987 and 1996

ID	03/01/87	16–19/04/87	05/03/89	28–29/04/96	17–18/06/96
#728	2	2	2	1	*
#731	2		3		
#732	2	2			
#733	1	7	3		
#745	1	1	2		
#748	1	3		2	
#798		3	2	2	1
#921		5		1	*
#923		4		3	
#940	1	3	2		
#960	*	3	1	1	*
#962	1	1			
Others	5	41	33	4	3

NOTE: Asterisks indicate an individual identified during this period, but from a photograph of lower quality ($Q < 4$).

Split: Five members of unit A, the largest unit that we delineated, apparently split from their companions sometime between 1991 and 1995 to become unit A2.

Transfers: Three individuals (two shown in figure 6.15) completely changed their sets of constant companions during their photoidentification history.

Merge/transfer: Whales #793 and #795 joined unit B between 1987 and 1988. This may have been either a merger of a two-member unit with the original unit B or two transfers into it.

Of the five animals that either transferred between units or were members of a two-animal unit that merged with another, three were known or presumed to be female, and one of them was estimated (using photographic measurements) to be about 5 years old when she transferred (Christal et al. 1998). There is no information on the sex or age of the other two transferring animals.

Given the difficulty of recognizing such events in our sparse identification record, it seems that, among Galápagos sperm whales, changes of unit membership are not very exceptional (Christal et al. 1998). Furthermore, it is clear that at least some of these changes involve transfers by females, and in one case, by a young female.

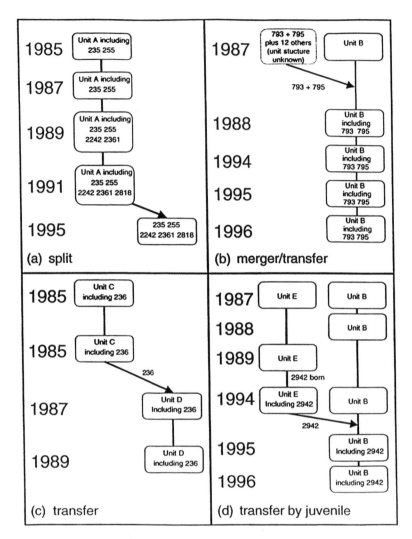

Figure 6.15 Diagrams of unit membership dynamics. Units are represented by boxes, which are linked between years to indicate the stability of unit membership. Notations such as "unit B" are used to represent all original members of that unit, except in part A, where "unit A" is used to denote members of the original unit for which individual identification numbers are not shown. (A) Unit A split between 1991 and 1995: five members of A (unit A2) were identified together in 1995 apart from the rest of the unit. (B) Two individuals (#793 and #795) transferred into or merged with unit B between 1987 and 1988. (C) Transfer of #236 from unit C to unit D. (D) Whale #2942, photographically measured and estimated to be 5–6 years old in 1995, transferred permanently from unit E to unit B. (From Christal et al. 1998, fig. 3.)

6.6 Genetic Structure of Groups and Units

The preceding description of what we know of the groups and units of female sperm whales is based on photoidentifications and observations at sea, but genetics is another important source of information, and the insights that genes can provide are growing rapidly. According to widely accepted kin selection theory (Hamilton 1964), the social relationship between a pair of animals often depends on their genetic relatedness. Thus, genetics has the potential to provide a direct window into the function and evolution of social behavior. However, genetics has another role, which has proved particularly important with cetaceans: allowing long-term patterns of affiliation, breeding, and dispersal to be deduced. This is perhaps most clearly shown in the work of Amos (1993), who was able to unravel many details of social structure in long-finned pilot whales (*Globicephala melas*) using molecular genetic methods. But genetic studies have also been very informative about the social structure of the sperm whale, and the potential for additional insight is huge.

To date, genetic studies of sperm whale social structure have concentrated on three types of markers (see appendix, section A.5): those that allow sex determination (Richard et al. 1994), the mitochondrial DNA (mtDNA) passed through the maternal line (Dillon and Wright 1993), and nuclear microsatellites, which allow relatedness among individuals to be estimated (Richard et al. 1996b). Because of the difficulty of distinguishing social units at sea, most analysis has so far been at the level of the group. Also, we have never been certain that all members of a group (or unit) were sampled. Even in the case of mass strandings in which all stranded animals were sampled, some members of the social group may not have stranded. Despite these provisos, genetic studies have provided important information on the genetic structures of groups and units, dispersal from units, and correlations between genetic relatedness and social relationships within units.

Genetic Structures of Groups

There have been a number of investigations of the genetic structures of groups of female and immature sperm whales. Many of the initial results concentrated on the maternally transmitted mitochondrial genome. Despite the remarkable homogeneity of the mtDNA of sperm whales (see section 7.4), the studies agree that there is significantly more similarity within groups than between them (Richard et al. 1996a; Lyrholm and Gyllensten 1998). This pattern is apparent in table 6.8, which gives the distribution of

Table 6.8. Distribution of Mitochondrial Haplotypes of Individuals in Sperm Whale Groups Sampled around the South Pacific

Date Group First Seen	Area	#1	#2	#3	#4	#5	#6	#7	#8	#9	#10	#11	#12	#13
04/05/89	Galápagos	4												
03/02/91	Ecuador	3	1	13								1		
22/02/91	Ecuador	9	4				1				3	4		
08/03/91	Ecuador	1	17	2										
14/04/91	Galápagos			6				1		2				
13/09/92	Western South Pacific	1	2						1					
15/09/92	Western South Pacific					6	1							
11/09/92	Western South Pacific	4												
03/01/92	Western South Pacific	2	1										1	
11/01/93	Western South Pacific	1		2			2	1						
19/03/93	Chile	1	1	2				1		1				
14/04/93	Peru	1		8										
19/04/93	Peru	1			3	2								
28/05/95	Galápagos	17												1

SOURCES: Data from Dillon 1996, except 28/05/95 from Christal 1998, using sloughed skin; haplotypes are numbered as in Dillon 1996, except "#13" represents Lyrholm and Gyllensten's (1998) haplotype "10". Only groups with at least four samples are included.

mtDNA "control region" haplotypes for fourteen groups sampled during our South Pacific studies. Many of the groups are strongly dominated by one haplotype. However, at least twelve of the fourteen groups contain multiple haplotypes, and so must consist of members of more than one matriline, if we make the standard, and almost certainly reasonable, assumption that almost all offspring have the same mtDNA haplotype as their mothers.* Similarly, of the fourteen groups analyzed by Mesnick (2001), thirteen contained two to four haplotypes, and only one (a stranding of ten animals in Tasmania) was monotypic. Another aspect of sperm whale genetics illustrated by table 6.8 is the low level of geographic structure compared with so-

* An individual could have a different mtDNA haplotype than its mother because of mutation, although this would be rare (probability of about 0.005%, using the suggestions for substitution rate in Lyrholm et al. 1996), or because of paternal transmission of the mitochondrial genome, which has been found in mammals (e.g., Gyllensten et al. 1991) but seems to be very rare.

cial structure—there is no obvious tendency for animals from particular geographic areas to have particular haplotypes (see section 4.1). Thus, the analyses of mtDNA at the level of the group seem consistent with the photoidentification data. In her group, a female has some long-term associates and some matrilineal relatives, and some casual acquaintances and matrilineal nonrelatives. At this point, reasonable hypotheses would seem to be that the long-term associates are genetic relatives—and thus that the unit is a matrilineal structure—but the casual acquaintances are generally unrelated.

At first glance, the analyses of nuclear microsatellites seem to support such conjectures. Richard et al. (1996a), using data from three groups sampled off Ecuador, found, in concordance with the mtDNA data, that animals in the same group were significantly more genetically related than those in different groups, and this finding was borne out by the wider study of Lyrholm et al. (1999). However, closer inspection of the microsatellite data revealed some very interesting discrepancies—discrepancies that are consistent across independent studies.

Richard et al. (1996a) noticed fewer first-order (parent-offspring or potentially full-sib) relationships within groups than expected given the currently accepted International Whaling Commission demographic parameters (see section 4.3) and a matrilineal model (i.e., with all females remaining with their mother until she dies). Bond (1999) found a similar shortfall of first-order relationships in groups sampled off the Azores, as did Christal (1998, 88–91) in the well-studied A2B group off the Galápagos Islands (table 6.9). In these studies, females were estimated to have an average of

Table 6.9. Mean Number of First-order Relationships per Individual Estimated from mtDNA and Microsatellite Data (Estimated SE in Parentheses) in Four Sperm Whale Groups

| Group | Estimated Mean Number of First-Order Relationships | | |
	One Female with Females	One Female with Males	One Male with Females
03/02/91	0.55 (0.92)	−0.01 (0.18)	−0.03 (0.37)
08/03/91	0.20 (0.28)	0.12 (0.20)	0.49 (0.87)
22/02/91	>0.74 (0.49)	>0.17 (0.16)	>1.36 (1.00)
28/05/95 (A2B)[a]	0.37	0	0
Expected[b]	0.94	0.23	0.83

SOURCE: Data from Richard et al. 1996a, plus analysis of data for group A2B, 28/05/95, using same methods.

[a]Only three sexed males in this group.

[b]Expected mean number of first-order relationships per individual in matrilineal groups, using International Whaling Commission population parameters.

about 0.5 first-order female relatives present with them in their groups, about half the 0.94 predicted by the International Whaling Commission's demographic parameters. Additionally, both Richard (1995, 68–72) and Christal (1998, 88–91) found a lower level of genetic relatedness within groups than expected from a strictly matrilineal model. These discrepancies have several potential explanations: undersampling of young animals, dispersal of some females from their mothers, a higher mortality rate than assumed by the International Whaling Commission's model, a lower birth rate than assumed by the model, or some combination of these factors.

In the eastern tropical Pacific waters there is a basis for expecting that the three demographic factors just mentioned were operating: females do sometimes transfer between social units (see section 6.5); heavy whaling off northern Peru, which ended in 1982 (Ramirez 1989), increased mortality directly; and birth rates seem unexpectedly low (see section 4.3). Calculations, following the methods of Richard et al. (1996a), based on a variety of demographic scenarios showed that any one of the factors—either the demographic factors or undersampling of young animals—would have to have been fairly severe to reduce expected numbers of living parent-offspring relationships to the levels found in real groups (table 6.10), but if they were operating together, then this could be easily done. Thus, given the uncer-

Table 6.10. Expected Number of Female Parents plus Offspring Alive with a Female at Any Time (i.e., First-Order Relationships) under Various Demographic Scenarios

Mortality Rate	Birth Rate	Minimum Sampling Age	Expected ♀ – ♀ Parent-Offspring
IWC	IWC	1	0.94
IWC + 0.05	IWC	1	0.61
IWC	IWC × 0.5	1	0.71
IWC	IWC	2	0.89
IWC + 0.05	IWC × 0.5	2	0.48
Killer whale	Best et al.	1	1.31
Killer whale + 0.05	Best et al.	1	0.69
Killer whale	(Best et al.) × 0.5	1	0.93
Killer whale	Best et al.	2	1.29
Killer whale + 0.05	(Best et al.) × 0.5	2	0.56

NOTE: Expected numbers were calculated using the methods of Richard et al. 1996a, based on either the International Whaling Commission's (IWC) accepted parameters (International Whaling Commission 1982) or possibly more realistic alternatives using mortality rates for killer whales (Olesiuk et al. 1990) and age-specific birth rates (from Best et al. 1984) (see section 4.3). Variations were combinations of increasing the adult mortality by 0.05/yr, halving the pregnancy rates, and increasing the minimum age for genetic sampling from 1 to 2 years.

tainties about the demography of the eastern tropical Pacific animals and the rates at which we were sampling young animals, the apparently low numbers of parent-offspring relationships and, by extension, the low levels of genetic relatedness within sampled groups do not actually tell us much about separations between mothers and daughters, except that they are not very frequent.

In an even more unpredicted development, both the studies of Richard et al. (1996a) and Christal (1998) found higher than expected levels of relatedness among group members who did not share mtDNA haplotypes. This finding indicates paternal relatedness within groups. Given the apparently fleeting nature of the relationship between breeding males and groups of females (see section 6.11), something most interesting is going on. Possibilities include consistently directed female choice, perhaps including copying in mate selection, strong dominance relationships among males on a breeding ground, or (very unlikely, but perhaps possible) significant paternal inheritance of mtDNA.

Genetic Structures of Units

Because distinguishing social units is difficult, requiring long-term data on associations, we currently have genetic data on only three, units A2, B, and T (see table 6.5), all from the Galápagos Islands. The results of genetic analyses of these three units are summarized in table 6.11. Two of the units contained at least two mtDNA haplotypes, there were few parent-offspring pairs, and there were low levels of relatedness within the units. Thus, these units were not perfect matrilines. Obviously, genetically unrelated animals are forming long-term social relationships (Christal 1998; Mesnick 2001). An important caveat here is that these units were studied off the Galápagos Islands in the mid- to late 1990s, after most other females had left the area (see box 4.2). The low density of whales during this period made it easier to

Table 6.11.   Genetics of Three Social Units

	A2	B	T
Number of members	5	17	9
Members genetically sampled	3	12–15	5
Haplotypes	2#1, 1#13	15#1	2#1, 3#11
Parent-offspring pairs	0	2	0
Mean relatedness	−0.0007	0.0005	−0.0552

SOURCES: Data from Christal 1998; Mesnick 2001.

distinguish unit members, but it may also mean that the units present were not representative in some ways.

But do these data, and our observations of splits of units and transfers between them (see fig. 6.15), also indicate the breakup of matrilines, in the sense of females living in different units from their direct living female ancestors or descendants? Crucial to this question is the distribution of matriline sizes. I define "matrilines" as sets of animals with a common oldest living female ancestor, the pattern usually found in the strongly matrilineal, and well-studied, "resident" killer whales off Vancouver Island (Ford et al. 2000, 25–26). If the mean matriline size for sperm whales is about 10.5, the mean size of a unit (see section 6.5), then the genetic data strongly suggest that matrilines commonly break up. However, simulations using the best available demographic parameters show that the mean size of a matriline is much smaller than this—about 3 animals—and that when mortality is increased or pregnancy rates decreased, the mean matriline size is further reduced (table 6.12). Even in the best of demographic circumstances, many matrilines consist of single animals (who have no direct female ancestors or descendants alive). Thus, stable units, with a mean size of about 10.5, must normally consist of more than one matriline. They could be extended ma-

Table 6.12. Mean Matriline Sizes (Numbers of Animals with a Common Living Direct Female Ancestor) under Various Demographic Scenarios

Mortality Rate	Birth Rate	Mean Matriline Size (SD)
IWC	IWC	2.7 (2.3)
IWC + 0.05	IWC	1.6 (0.9)
IWC	IWC × 0.5	2.0 (1.4)
IWC + 0.05	IWC × 0.5	1.7 (1.2)
Killer whale	Best et al.	3.7 (3.2)
Killer whale + 0.05	Best et al.	1.7 (1.1)
Killer whale	Best et al. × 0.5	2.2 (1.7)
Killer whale + 0.05	Best et al. × 0.5	1.3 (0.6)

NOTE: Calculations were based on either the International Whaling Commission's (IWC) accepted parameters (International Whaling Commission 1982) or possibly more realistic alternatives using mortality rates for killer whales (Olesiuk et al. 1990) and age-specific birth rates (from Best et al. 1984) (see section 4.3). Variations were combinations of increasing the adult mortality by 0.05/yr and halving the pregnancy rates. One hundred simulations of 250 yr were carried out using each combination of parameters, each starting with one 10-year-old female (runs in which the population went extinct were discarded). The mean matriline sizes represent living females and males less than the age of dispersal (6 yr; Richard et al. 1996a) at the end of the 250 yr. Changes to the survival and pregnancy rates (adding 0.05 and halving, respectively) were introduced during only the last 50 yr to simulate the effects of whaling.

trilines (i.e., animals with a fairly recent, but dead, common female ancestor) or maternally-unrelated matrilineal subunits. The genetic data indicate that the latter is commonly the case, but not that matrilines are breaking up. The transfers between units (see section 6.5) could well be of animals that are the only remaining members of their matriline.

6.7 The Functions of Groups and Units

Why Groups?

In sections 6.4 and 6.6, I have gone to some length in trying to describe the principal properties of a sperm whale group. There is clearly much variation with geographic area and, probably, with time in mean group size, although for many attributes we have usable data for only one area—the eastern tropical Pacific. However, it is worth reemphasizing the general nature of a group of females and immatures: about two social units, containing approximately twenty whales, usually spread out over a kilometer or so of ocean in a rank, moving in a coordinated fashion and maintaining their association over periods of days. Why? Why do units frequently coordinate their movements with other units, and why do the members of the group often form ranks?

Sperm whale groups, like groups of other animals, are fundamentally spatial structures. Spatial aggregations of animals may result purely from external factors—for instance, if animals aggregate at a food source or a predator refuge—or may be at least partially precipitated by the attraction of individuals toward one another. Making the reasonable assumption that animals do not usually behave maladaptively, Connor (2000) calls these nonmutualistic and mutualistic groups respectively (fig. 6.1), because in the first case animals do not receive benefits from one another's presence, while in the second case they do. Thus, these two types of groups can be distinguished either by examining the costs and benefits of spatial clustering or by looking for behavior that increases or maintains spatial proximity directly. If the animals actively maintain proximity, rather than being concentrated by external factors, then the group is assumed to be mutualistic.

The larger spatial structures that we have recognized for sperm whales ("concentrations" and "aggregations") seem to be principally nonmutualistic (see sections 6.2 and 6.3), but I believe that, at the group, we reach a level at which animals gain benefits from associating with one another. Animals traveling together in a structured formation could be independently reacting to external stimuli—following a prey school, for instance. However, sperm whale groups stay together for periods of days or more, traveling hundreds of kilometers. During this time, feeding success fluctuates, and from time to time, the members of the group cease feeding and gather to socialize. Given this behavioral evidence, the nature of the sperm whales' prey (see section 2.3), and other elements of their environment, the hypothesis that there is no mutual attraction within sperm whale groups can be ruled out. I conclude that sperm whales are actively maintaining their groups, which are thus mutualistic. So what are they gaining from one another?

The benefits and costs of animal grouping have been well discussed by behavioral ecologists. There is reasonable consensus that the primary benefits of grouping are resource acquisition and defense against predators, but that cooperative monopolization of resources can also be important in certain circumstances (e.g., Wrangham and Rubenstein 1986; Lee 1994; Connor 2000). In the sperm whale's three-dimensional fluid environment, it is unlikely that resources can be monopolized (see section 2.9), so sperm whale scientists have looked at both increasing feeding success and diminishing predation risk as likely benefits of the group (e.g., Best 1979; Gordon 1987b; Whitehead 1989a).

I have considered four ways in which grouping could increase feeding success (Whitehead 1989a): (1) cooperative chasing and capture of prey; (2) gaining useful information about the distribution of prey from the behavior of others ("local enhancement"); (3) the capture of prey that try to escape from other members of the group; and (4) avoiding interference among animals feeding in the same area by moving in a structured rank.

Most sperm whale prey, including the histioteuthid squids, which seem to be the major dietary item off the Galápagos Islands (see table 2.2), are small and likely to put up little fight or flight in the face of hungry sperm whales. The larger squids, such as *Architeuthis* and *Dosidicus* (see figs. 2.12 and 2.13), are a different matter (Clarke et al. 1993). However, even the largest of the giant squids (at about 500 kg) is only about 5% of the mass of a female sperm whale (even less for the large males that are their principal predators). Top carnivores, such as killer whales, lions, and wolves (*Canis lupus*), do not usually capture their prey cooperatively unless the prey approaches, or exceeds, the mass of an individual predator.* Thus, I do not think that cooperation at the prey capture level (factor 1) is likely to be an important force for grouping in sperm whales (Whitehead 1989a).

Although sperm whale cephalopod prey do seem to form schools (Clarke

---

* Killer whales may herd smaller prey cooperatively (e.g., Similä and Ugarte 1993), but in these cases each prey item is attacked by only one killer whale at a time.

1980), these do not usually seem to be large enough to allow a group of sperm whales to feed for a substantial time (Whitehead 1989a; see section 2.3), so local enhancement from other whales finding schools that can be communally utilized seems an unlikely benefit at this scale. However, the vocalizations of foraging sperm whales (such as the "creaks" that may indicate homing in on a prey item; Gordon 1991b) seem to encode information on their feeding success, and so the rank formation of the group may assist animals in directing their movements, or the movement of the rank as a whole, toward regions of greater prey density over the scales of kilometers at which they can hear each other's vocalizations. Thus local enhancement (factor 2) may be a benefit over such scales.

We know little about the flight responses of squid, but if they can hear the sperm whale's echolocation clicks before the whale detects them and take evasive action, then capture of prey that escape from other sperm whales (factor 3) may also be a benefit of foraging in a rank (Whitehead 1989a). There is evidence that at least some squids possess a sense of hearing (Hanlon and Budelmann 1987).

Perhaps the best argument for a benefit of foraging in a rank concerns the avoidance of interference by animals feeding in the same area (factor 4). Most of the sperm whale's prey swim slowly, renew their concentrations slowly, and, as may be inferred from the whales' diving behavior (see section 3.2), appear to favor a particular depth in any area at any time. For these reasons, swimming behind another whale will not be profitable, and a rank with individuals aligned side by side may be a sensible formation for animals foraging in the same vicinity.

However, this explanation for foraging in a structured rank begs another question: why are the animals in the same vicinity in the first place? The answer is not simply a matter of whale aggregations coinciding with prey aggregations, as groups form ranks and travel over long distances where there are no other sperm whales for many kilometers (such as the group A2B followed in 1995; see box 6.3). There should be another benefit to forming groups, and this benefit is probably related to the second major force favoring mutualistic groups: predation.

Box 6.3 **A Week in the Life of a Sperm Whale Group**

In May–June 1995, we followed a group of female and immature sperm whales near the Galápagos Islands continuously for a week

Box 6.3 continued

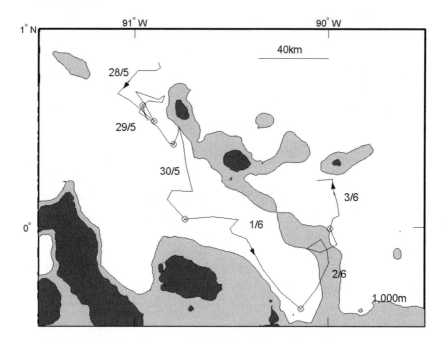

Figure 6.16 The track of group A2B near the Galápagos Islands, 28 May–3 June 1995. The land-mass of the islands is represented by dark shading; water less than 1 km deep is represented by light shading. Midnight positions are marked by a cross within a circle.

(fig. 6.16). The group contained twenty-two individuals (all photo-graphically identified) and had a consistent membership throughout the week. Eighteen of the whales had been identified during previous years, and their patterns of association in those previous sightings (as well as a subsequent sighting in 1996) enabled us to divide the group into two distinct units, A2, with five members, and B, with seventeen members. I will refer to the entire group as the "A2B" group.

In general, the characteristics of the group were representative of sperm whales studied off the Galápagos. These characteristics include its size of twenty-two animals, its composition, including two units, and its lack of first-year calves, which have been very rare in our studies. The low reproductive rate of sperm whales in the Galápagos area is ex-emplified by this group, which contained a minimum of eight females of reproductive age, yet no animals of less than about 6 years of age.

Also typical were the group's brief encounters with a mature male. The same mature male was seen in association with the group for brief

Box 6.3 continued

periods on two consecutive days. On 31 May, the male was first seen at 16:30 at one end of a rank of approximately twenty group members. The male remained at the surface for 20 min after all members of the group had fluked, then he fluked, and was not seen again that day. On the following day, 1 June, the same male was observed 400 m away from the group at 12:45, but was not seen to associate directly with any members of the group, and appeared to leave the area shortly after 13:00.

The movements of A2B were also typical. The daily distance moved correlated negatively with feeding success as indicated by defecation rates ($r = -.71$; see section 3.4). Feeding success was highest on 29 May, the day on which the group traveled the shortest straight-line distance (21 km), and lowest on 1 June, when the group made its greatest 24 hr displacement (82 km).

Some rather unusual features of the behavior this group included the lack of encounters with other groups. Groups in the Galápagos have often been observed to join with other groups to form temporary aggregations lasting for periods of hours. This group did not associate with any other groups during the period of observation. There was a very low density of groups of female and immature sperm whales in Galápagos waters at the time (see box 4.2).

Another unusual feature was A2B's incursion into waters less than 1 km deep. A2B spent most of 2 June in waters shallower than 1 km (fig. 6.16), whereas sperm whales in this area usually stay in deeper waters (see fig. 2.5). However, the depth in this area exceeded 850 m, and because of the underwater topography, the only way the group could have continued to travel in waters deeper than 1 km was to return the way they had come, through an area in which they had found poor feeding success.

(This account is adapted from that of Christal and Whitehead 2000)

Sperm whales are attacked, and sometimes killed, by predators (see section 2.8). They defend themselves from these predators communally (see section 5.9). Thus, it seems very likely that sperm whales in a group are safer than sperm whales alone, especially if they are young. Many of the benefits of grouping that reduce predation risk increase with group size: vigilance is probably increased and its cost decreased (e.g., da Silva and Terhune 1988), predators may be more easily deterred by the defensive efforts of more members, and there may also be additional advantages resulting from dilution of the predators' successes among a larger group, more confusion for the predators, and better likelihood of cover behind group members (Wrangham and Rubenstein 1986). As group size increases beyond a certain point, these benefits decelerate and eventually are outweighed by growing costs, such as feeding competition. These factors produce an "optimal group size" at which individual fitness is maximized. However, the dynamics of group membership may mean that this optimal group size is not observed;* instead, groups may converge on a stable equilibrium size that is usually larger than the optimal group size (Pulliam and Caraco 1984). Perhaps the optimal group size, or stable group size, for female and immature sperm whales is about twenty.

Under this scenario, units (which are usually smaller than this optimal or stable group size) agglomerate into groups, perhaps principally with other units of their own vocal clan (see section 7.2), to reduce predation risk, as well as perhaps to gain feeding benefits, either from information about food concentrations provided by group members or from prey items that flee from the approach of other group members. The group then spreads out into a rank to minimize interference.

It appears that most of the social behavior of the sperm whale (including close clustering at the surface, caring for calves, communal defense, and stranding) takes place at the level of the group, although this conclusion may be partially an artifact of the difficulties in recognizing the constituent units that form the group.

Why Social Units?

Social units seem to be the fundamental elements of sperm whale society. As such, it is worth reiterating what we know about their structure. They consist of about eleven long-term companions and often include more than one matriline. Although there are some transfers between units and some splitting of units, we do not know whether these events represent the breakup of matrilines, in the sense of females dispersing away from living members of their direct female line, but it is clear that such dispersal, if it does exist, is quite unusual. It may well be that transfers and merges generally occur among units of the same clan (although we know of one transfer, of whale #236, between units of different clans).

---

* For instance, if the optimal group size is twenty, and all animals are in groups of this size, newly arriving animals will do better to join existing groups than to remain alone, pushing up the mean group size beyond its optimum level.

Long-term relationships, especially among nonkin and when embedded within a higher-level, nonterritorial, fission-fusion society, are rare and interesting. Physically, dispersal to another unit is trivial for a female sperm whale, as groups usually consist of merged units. There may be social barriers to dispersal, but the ones traditionally invoked to explain cooperative breeding, such as shortage of mates and costs of reproduction following dispersal (Emlen 1991), would not seem to apply in the case of female sperm whales transferring to another unit. Without substantial costs of dispersal, it makes sense to look for benefits to individuals of maintaining long-term bonds and a unit structure.

If, as discussed above, foraging success is not well supported as the *major* factor behind group formation, it would seem even less likely to promote the long-term relationships found within units. Social units are long-term structures, and so their benefits may also be long-term. If communal memory is used to direct movements toward areas of better feeding profitability as the spatial patterns of food availability change (see section 3.7), then it makes sense for animals to form long-term relationships with those "sages" who have expertise in such matters, probably mainly older individuals. For a sage, it is beneficial in terms of inclusive fitness to allow relatives to accompany her. In the case of nonrelatives, the benefit is not so clear. Perhaps the "two heads are better than one" synergy prevails, or there may be few costs to having long-term unrelated companions. Or perhaps there are benefits, but in another area.

The most likely such benefit is reduction of predation risk. Many of the simpler benefits of group living that reduce predation risk (such as confusion and dilution) do not require long-term relationships. However, active babysitting of the young (see section 6.9) and communal defense against predators (see section 5.9) may both be more efficiently managed among animals with long-term relationships. In these cases, unrelated animals can potentially provide one another with mutual benefits by forming a long-term relationship.

There are other potential benefits to long-term relationships among females. One such benefit is allosuckling, for which there is some evidence in sperm whales (Best et al. 1984; Gordon 1987b; see section 6.9). Suckling one another's calves could benefit related females through kin selection and unrelated ones through reciprocal altruism.

Under this scenario, there are benefits to forming long-term social units (communal knowledge and care and protection of the young) and to being in a group (combating predation and minimizing interference). If the optimal size of a group is somewhere around twenty animals, why don't the females simply form units of this size? Perhaps optimal group size varies with time and place, and the fairly dynamic group memberships allow the animals to adapt to current conditions. Alternatively, it could be that groups are the result of units that are feeding in the same area "making the best of a bad job" by moving together to minimize interference.

Social systems with long-term relationships, or "bonds," are normally expected to contain considerable substructure, and even conflict (e.g., Wrangham 1980), as animals try to exploit the bond to their own advantage and the detriment of their partners, or to test it (Zahavi 1977). Consequently, the strengths of bonds between group members should vary substantially. Clarke and Paliza (1988) suggested that tooth loss in female sperms indicates "fighting for dominance," but members of the Scientific Committee of the International Whaling Commission did not find their argument convincing (International Whaling Commission 1988). Neither do I. In contrast to the expectations based on theory and terrestrial studies, there is little evidence for substructure or conflict within sperm whale social units. Why? In the case of conflict, this may be a result of our inability to see most of sperm whale behavior, but there is good evidence for a large degree of homogeneity within some sperm whale units. This may be a consequence of the structure of a habitat in which there are few opportunities for one female to exploit another. C. L. Mitchell et al. (1991) compared two species of squirrel monkeys (*Saimiri* spp.), finding little structure within the social groups of the species for which food patches are not defensible, and dominance hierarchies and long-term bonds in the other species, whose food could more easily be monopolized. It will be particularly interesting to see whether the low correlations between genetic and social relatedness within units are borne out when additional units are studied, given that units clearly contain both close matrilineally related dyads and pairs who have no close connection through the maternal line.

6.8 The Cluster

The cluster, which may consist of anywhere from two to more than forty animals swimming along side by side in the same direction at the same speed (fig. 6.17), is perhaps the most visually obvious expression of sperm whale sociality. In our studies from around the South Pacific, 62% of females and immatures (4,077/6,542) were clustered with one or more other whales as they showed their flukes at the start of foraging dives (see also fig. 6.18). During social or resting periods, almost all females and immatures were clus-

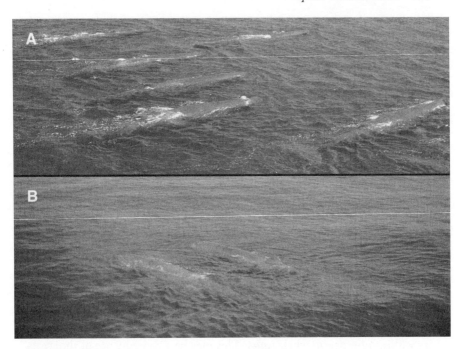

Figure 6.17 Clusters of sperm whales, (A) large and (B) small.

tered, and the cluster sizes were larger, frequently approaching the group size (see fig. 5.9 and section 5.5). Overall, the estimated mean cluster size observed during scan samples off the Galápagos Islands in 1985 and 1987 (including singletons, but only clusters < 1,000 m from the boat and excluding large males and calves, $n = 17,965$ clusters) was 2.4 (SD 3.1) animals, and the mean typical cluster size experienced by the whales themselves was 6.3 (SD 7.3). The large discrepancy is a result of the long tail in the distribution of cluster sizes (fig. 6.18). The maximum cluster size recorded during our 1985 and 1987 studies was 52 animals. Similar distributions of sizes of clusters, or entities with other names but which seem to correspond to clusters, are reported from studies of females and immatures off Sri Lanka (mean 3.3, SD 2.3 from data in figure 2 of Gordon 1987b), in the northern Gulf of Mexico (mean 2.2, SD 1.6; Weller et al. 2000), and off Dominica (Gordon et al. 1998).

Cluster formations vary, from ranks with the animals side by side and equally spaced to more irregular formations. Some examples from Gordon's (1987b) work in Sri Lanka are shown in figure 6.19.

Nearest-neighbor distances within clusters are available for 18 pairs of animals (sometimes within larger clusters) from photographic measurements

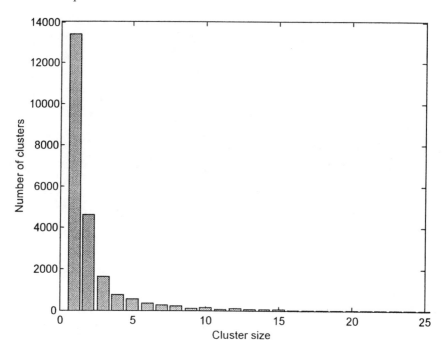

Figure 6.18 Frequency distribution of sizes of sperm whale clusters as recorded during 5 min scan samples off the Galápagos Islands in 1985 and 1987, excluding large males and calves as well as clusters more than 1,000 m from the boat and the fifty recorded clusters with more than twenty-four members.

taken off the Galápagos Islands in 1985 (see Waters and Whitehead 1990b). The median estimated distance between neighbors was estimated to be 6.5 m, and the interquartile range 3.9–11.4 m. These are very much less than inter-cluster separations (ca. 200 m; see section 6.4), and thus the cluster is a clear spatial structure within the group. The inter-individual, within-cluster separations are usually less than the limit of underwater vision off the Galápagos Islands (ca. 10–20 m; H. Whitehead, unpublished Secchi disk measurements). This finding suggests that cluster formations may be maintained visually, even at night, as sperm whales are often visually apparent among phosphorescent plankton (H. Whitehead, personal observation). The white mouth of the sperm whale (see fig. 5.15) may act to enhance visual signaling among cluster members in addition to, or instead of, its possible role in feeding (see section 5.4).

Sperm whales can frequently be observed changing their direction of movement while swimming at the surface in order to cluster with others

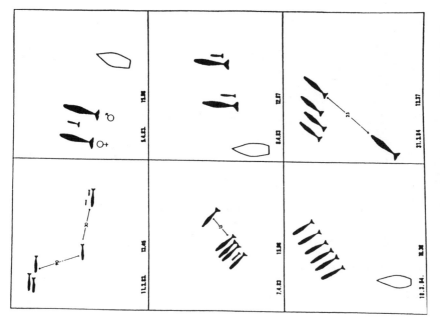

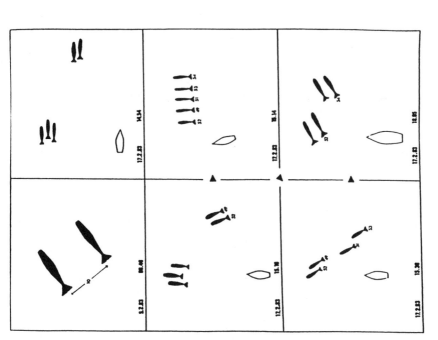

Figure 6.19 Sketches of the spatial distribution and orientation of individuals in a selection of typical clusters (called "pods") encountered off Sri Lanka. Some ranges (in m) are indicated. Arrows indicate sequential order of sketches of the same cluster. (Adapted from Gordon 1987b, fig. 1.)

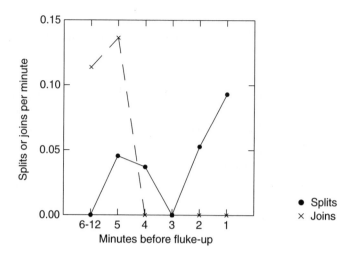

Figure 6.20 Rates of joining and splitting from clusters including a focal female/immature, plotted against minutes before she fluked up. (Data collected off the Galápagos Islands by J. Christal in 1995.)

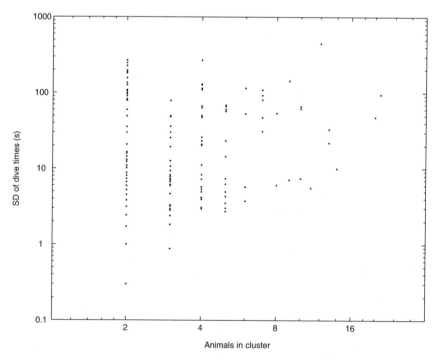

Figure 6.21 Synchrony in diving among members of clusters in 1995, expressed as the standard deviation of fluke-up times plotted against cluster size.

(e.g., fig. 6.19). Cluster membership is quite dynamic, especially during so-cial or resting periods. During surface periods not terminated by fluke-ups, a cluster of one to five female or immature whales joins, or is joined by, other whales about once every 4–8 min, and, if it contains more than one whale, splits at about the same rate. During the roughly 8 min surface periods be-tween deep dives (terminated by fluke-ups), clusters are more stable, gener-ally not changing membership, especially if they contain three or fewer ani-mals (fig. 6.20). Joins, if they do occur, usually happen during the first minutes of the surface period, and splits when other cluster members dive just before a focal animal (fig. 6.20). The joining data indicate that animals preferentially join those others at the surface whose dive cycles are aligned with theirs within about 2 min.

The temporal synchrony of dives among cluster members is indicated numerically in figures 6.21 and 6.22. The degree of synchrony, as expressed by the standard deviation of the fluke-up times, seems to depend little on cluster size (fig. 6.21). However, it is radically different from the expected distribution if cluster members dived at random times throughout one

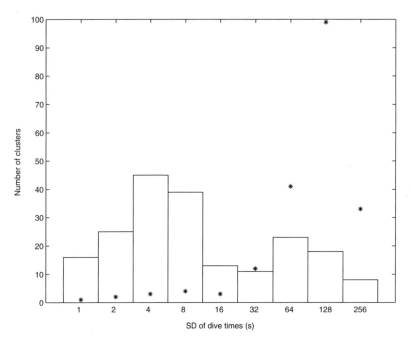

Figure 6.22 Distribution of synchrony in diving among members of clusters in 1995, expressed as the standard deviation of the fluke-up times. A distribution obtained by randomly placing each dive time within an 8 min period (the approximate surface time of the whales; see table 5.6) is shown by the asterisks.

Figure 6.23 Synchronous dives of sperm whales.

another's roughly 8 min surface period (fig. 6.22). Cluster members usually fluke up within 10 s of one another; the median synchrony is 6.4 s ($n = 198$) versus 123 s for randomly simulated data, and fluke-ups quite frequently took place within 2 s of one another (figs. 6.22 and 6.23). I interpret the bimodal distribution of the synchrony of fluke-ups within clusters (fig. 6.22) as the result of those cluster members who are naturally due to begin their dives within about 60 s synchronizing them to within about 10 s.

Thus, female and immature sperm whales clearly change their behavior to join clusters and harmonize their behavior with that of other cluster members. An ensuing question is whether particular animals have special roles in this process. Our data are insufficient to look at this question in much detail, but it is possible to examine one aspect: leadership in diving. I first looked to see whether particular animals tended to follow or lead in dive sequences, and then whether, within particular dyads, one animal or the other tended to lead.

There are twenty-two individuals for whom we have reasonably good data on preferred diving order (at least eight occasions on which there were more dives by other whales in the 1 min preceding the focal animal's dive than in

Table 6.13. Preferred Diving Order of Whales Studied off the Galápagos Islands

ID	Unit	More Following Dives	More Preceding Dives
#235	A	7	3
#255	A	4	5
#487	A	4	5
#754	B	8	2
#795	B	3	6
#800	A	3	5
#804	B	4	10
#807	B	11	4
#809	B	3	5
#810	B	4	9
#814	B	8	3
#833	L	3	5
#836	B	5	3
#2041	E	7	2
#2080	E	4	5
#3700	T	3	5
#3701	T	3	5
#3702	T	3	5
#3703	T	5	5
#3706	T	6	4
#3707	T	8	0
#3708	T	3	6

NOTE: For each photographed fluke-up of each individual (separated by at least 2 hr), the number of animals photographed fluking up in the previous 1 min was compared with the number in the following 1 min. Listed are those animals for which there were at least eight occasions on which the total of preceding dives was different from the number of following dives. The null hypothesis that the probability of having more following dives was equal to that of having more preceding dives was tested against the alternative that individual whales might have tendencies to precede or follow (likelihood ratio test, $P = 0.0537$, from 10,000 random allocations of occasions to leading or following).

the following 1 min, or vice versa). Although many of the animals seemed equally likely to lead or follow, there were some apparent followers (whales #804 and #810 in unit B) and leaders (#235 in A, #807 and #814 in B, #2041 in E, and #3707 in T), and the overall pattern was nearly significantly different ($P = .054$) from that expected if all individuals had the same probability of being leaders or followers (table 6.13). Particularly notable was the behavior of #3707 in unit T, who was always a leader and never a follower.

Turning our attention to dyads of identified individuals, there are fewer

Table 6.14.   Precedence in Diving among Dyads Studied off the Galápagos Islands

| ID "X" | ID "Y" | Unit | Dives within 5 Min | |
			X Precedes Y	Y Precedes X
#754	#810	B	5	0
#795	#806	B	4	2
#804	#809	B	3	4
#804	#3287	B	3	3
#806	#807	B	1	4
#807	#809	B	7	0
#807	#811	B	2	4
#809	#810	B	4	1
#810	#814	B	3	6
#814	#3290	B	3	2
#838	#841	L	3	2
#1936	#2041	E	2	3
#3701	#3702	T	4	1
#3703	#3708	T	3	3
#3706	#3707	T	1	4

NOTE: Data are shown for dyads that were photographed fluking up within 5 min of each other at least five times (only dives at least 2 hr apart are considered). The null hypothesis that the order of diving within each dyad was random on each occasion was tested against the alternative that within each dyad a preferred diving order might exist (likelihood ratio test, $P = 0.0525$, from 10,000 random allocations of precedence).

data. Shown in table 6.14 are all pairs of animals who dived together at least five times within 5 min, excluding dives within 2 hr of one another. Overall, there was a marginally significant trend for precedence ($P = .053$). This trend was largely due to the fact that, within unit B, #754 always preceded #810 (five occasions) and #807 always preceded #809 (seven occasions). #754 was an immature male, but the other three animals were all adult females, and levels of genetic relatedness within these two dyads were low (Christal 1998, 88–91).

An important question remains: Are clusters just a surface and near-surface phenomenon? Do the whales remain clustered after their dives? The evidence is rather sparse and equivocal. Studies using an array of hydrophones (Watkins and Schevill 1977a; Watkins et al. 1985) and a depth sounder (Papastavrou et al. 1989) suggested that the whales usually spread out under water. However, in some depth sounder records, traces of several sperm whales can be seen together at depth (Papastavrou et al. 1989; Gordon 1987a, 186), and a small amount of footage from "Crittercam," a video system attached to a sperm whale's back, clearly shows whales remaining in close proximity as they descend (Marshall 1998).

Whatever takes place at depth, it is clear that at the surface sperm whales alter their behavior to form clusters and synchronize their dives. Why? There seem to be two likely principal functions for the cluster. First, clusters may reduce predation risk in case of sudden attacks by sharks, killer whales, or other predators—for instance, by increasing vigilance (see above and Connor 2000 for a consideration of this idea). Second, active clustering may be a way in which animals maintain social bonds, the bonds being demonstrated by the change of behavior to increase spatial proximity and temporal synchrony.

6.9 Care of the Calf

A number of sperm whale scientists, including Best (1979), Gaskin (1982, 134), Gordon (1987b), and myself (Whitehead 1996a), have suggested that care of young calves may be a key driver of sperm whale sociality. Therefore, it is frustrating that our understanding of relationships between females and calves has progressed more slowly than that of most other elements of sperm whale social structure. The best data are principally from the earliest studies of living whales, particularly the first, off Sri Lanka (Gordon 1987b). Sperm whale calves have been very rare in the eastern tropical Pacific since we first started work there in 1985 (see section 4.3), and as whale densities dropped during the 1990s (see box 4.2), opportunities to study calves have nearly disappeared.

Young sperm whale calves—animals of less than about a year of age—can swim horizontally fairly well, even when just hours old (Weilgart and Whitehead 1986), and generally seem to have little trouble keeping up with the roughly 4 km/hr horizontal movements of their group (Gordon 1987b; Gordon et al. 1998). However, vertical movement is harder for the calves, and they do not dive to the depths at which their mothers feed for the lengths of time adults do (Best et al. 1984; Gordon 1987a, 193; see section 3.2). It is not clear whether these limitations are a result of physiological constraints, or of the need for a suckling animal to conserve energy, or both. However, the mothers continue their long and deep dives, leaving a considerable vertical gap between themselves and their calves during much of the nursing period.

Calves can be seen at the surface alone, but significantly less often than their elders (calves are alone, rather than in clusters of two to three animals,

Table 6.15. Summary of Associations of Calves with Females and Immatures off Sri Lanka from 27 March 1983 to 8 April 1984

Calf ID	Scored as "Escort"[a]	Seen Alone with	"Suckling" from[b]
#153	#166	#166	
#160	#166 (2), #179 (3)	#176, #180, #179 (2)	#179 (2), #180
#170	#179 (2), #166 (7), #169	#169, #166, #168	
#178	#179	#179	
#181	#171	#179 (2)	
#1034	#1029	#1033, #1029	

SOURCE: Gordon 1987b.

NOTE: The identification numbers here use a sequence distinct from those in the eastern tropical Pacific studies given elsewhere in this book. Numbers in parentheses indicate multiple observations.

[a]"Escorting" is not defined, but presumably means close accompaniment (<2.5 m apart by the definition of Arnbom and Whitehead 1989).

[b]"Suckling" was assumed from above-water observations and does not necessarily imply transfer of milk.

31% of the time, compared with 43% for other females and immatures; Whitehead 1996a). It became apparent fairly soon in the Sri Lankan studies that calves clustered with a number of different females and immatures while at the surface, and that a particular female or immature might be found alone with more than one calf at different times, suggesting a form of babysitting (Gordon 1987b).* Gordon's (1987b) Sri Lankan data (table 6.15) show this diversity of associations clearly. Off the Galápagos Islands in 1985, we identified two females/immatures that were each seen alone as well as escorting (less than 2.5 m from) two different calves on different occasions, as well as one calf that was recorded as being escorted by three different females/immatures at different times (Arnbom and Whitehead 1989). And off Dominica, West Indies, Gordon et al. (1998) observed an identifiable calf being escorted for extended periods by three different adults over one afternoon.

This serial accompaniment probably benefits calves by providing some protection from the serious threat of predation by sharks or killer whales. The dynamic relationships between calves and older members of their group (or perhaps their unit—we do not yet know), even when they are very young (Weilgart and Whitehead 1986), contrast with most of what is known for other cetaceans, in which infants usually accompany their mothers (Mann and Smuts 1998; Whitehead and Mann 2000). However, there are indica-

* Kleiman and Malcolm (1981) define "babysitting" as "remaining with the young in the absence of the mother."

tions that there may be a similar pattern of babysitting for some, but probably not all, calves of the northern bottlenose whale (*Hyperoodon ampullatus*), another deep diver (Gowans et al. 2001).

Calves clearly adapt their behavior to that of the other whales at the surface, but the older animals also change their behavior to accompany calves. I have shown that, after controlling for group size, groups containing calves had less synchronous dives, and shorter intervals with no adults at the surface (a 2.5 min shorter maximum interval in each hour, on average), than groups without calves (Whitehead 1996a). This decrease in synchrony does not seem to have been the result purely of changes in dive schedule by the calves' mothers (Whitehead 1996a). Thus, the group as a whole seems to change its behavior to benefit the calf—a form of alloparental care, in the sense of behavior that benefits the calf and would not be carried out if it were not there (cf. Woodroffe and Vincent 1994).

The potential significance of dive synchrony in protecting the young makes it particularly interesting that there seem to be substantial differences in dive synchrony between the two principal clans that use Galápagos waters (see section 7.3). Maybe the less synchronous "+1" clan is more effective at protecting its calves, or perhaps the "Regular" clan has other compensatory ways of reducing predation.

Babysitting is not the only form of alloparental care attributed to sperm whales (fig. 6.24). Off Sri Lanka, Gordon (1987b) observed one calf that seemed to suckle from two females at different times, and two calves of similar size (and so very unlikely to be siblings) appearing to suckle from the same female at the same time. Best et al. (1984) summarize evidence from whaling operations and strandings in which large proportions of a group were captured or stranded groups examined. In both situations, lactating females appear to considerably outnumber suckling calves (forty-one lactating females to fourteen calves for the Japanese whaling data). Additionally, in Best et al.'s (1984) study, of the twenty-two females examined that were estimated to be over 41 years old, none were pregnant or ovulating, but six (27%) were lactating. Thus, there is considerable evidence for communal suckling among sperm whales, including by nonreproductive females, but because of the difficulties of observing sperm whales at sea and the uncertainties about animals missed by whaling operations or which did not strand, that evidence is not absolutely conclusive. However, putting these results together with those on babysitting leaves the impression that females in sperm whale groups work together to raise their offspring. Mesnick et al. (2003) aptly echoes the African proverb in titling their short paper "Sperm whale social structure: Why it takes a village to raise a child."

Figure 6.24. Female with calf, turning to suckle? The female is not necessarily the calf's mother. (Photograph by Linda Weilgart.)

6.10 The Social Male?

The male sperm whale, before he disperses from his mother's social unit, seems to play a role within it similar to that of the females, including babysitting calves (Gordon 1987b). But after leaving, his social life takes a very different shape. In table 6.16 I summarize the spatial and social structures of female and immature sperm whales, as described in the previous sections of this chapter, and use them as benchmarks for comparison with the spatial and social structures of nonbreeding males, described in this section. Breeding males will be considered in section 6.11.

Concentrations?

To my knowledge, there has been no formal analysis of spatial clustering of aggregations of nonbreeding males, but concentrations about 50–200 km across seem apparent in the results of large-scale surveys off Norway (Christensen et al. 1992) and in the Antarctic (Gillespie 1997). However, Kirpichnikov (1950, cited in Gosho et al. 1984) considers that the sperm whales of the

Table 6.16. Spatial and Social Structures of Nonbreeding but Sexually Mature or Maturing Males Compared with Those of Females and Immatures

Structure	Females and Immatures	Nonbreeding Males
Clans	~10,000's of animals sharing coda dialect spread over large ocean area	No evidence
Concentrations	~100's of animals concentrated over an area ~100's of km across for months or more	Probably exist, but less pronounced than for females
Aggregations	~10's of animals aggregated over an area ~10 km across for a few hours	~20 males aggregated over an area ~20 km across for days or more
Groups	~20 animals foraging in structured formations spanning ~1 km for days	Little evidence for groups, except coordinated heading by aggregated males
Social units	~10 animals with long-term relationships over years to decades; may contain several matrilines	No evidence for long-term relationships
Clusters	~2 animals clustered at the surface over ~10 min	Occasional clustering of 2 or more males

SOURCE: Adapted from Letteval et al. 2002.

Antarctic are more evenly distributed than animals at lower latitudes. Thus, I conclude that concentrations of nonbreeding males probably exist, but seem to be less well circumscribed than those of females and immatures. We can only guess as to how many animals such male concentrations might contain.

Aggregations?

At the scale of aggregations, things become clearer. In visual and acoustic surveys at high latitudes, aggregations of males spanning ~10–30 km are commonly found (e.g., Gillespie 1997; Leaper and Scheidat 1998), and such aggregations were described by whalers (Caldwell et al. 1966). Spatial and social structure in nonbreeding males has been studied in some detail at four study sites: off the Galápagos Islands during the late 1990s, when female groups had largely deserted the area and were replaced by apparently nonbreeding males (see Christal and Whitehead 1997; Christal 1998, 112–45; see box 4.2); the Gully off Nova Scotia, where we have been studying sperm whales as a secondary goal of a project oriented toward northern bottlenose whales (see Whitehead et al. 1992); Andenes, Norway, where a study has been carried out in conjunction with the local whale-watching industry (see Ciano and Huele 2000); and, most comprehensively, at Kaikoura, New Zealand (e.g., Childerhouse et al. 1995; Jaquet et al. 2000). Some information about these studies, which together cover a wide range of latitudes, is

Table 6.17. Summary of Studies on Nonbreeding Male Sperm Whales from Four Study Areas

	Andenes, Norway	Galápagos	Gully, Nova Scotia	Kaikoura, New Zealand
Latitude	69° N	0°	44° N	42° S
Years	1987–2000	1995–1998	1988–1998	1990–2001
Mean length of animals	15.2 m	14.3 m	14.1 m	12.8 m
Mean stay in study area	13 d (SE 4 d)	27 d[a]	3 d (SE 4 d)	42 d (SE 10 d)
Population in study area at any time	16 (SE 2)	24[a]	10 (SE 4)	14 (SE 1)
Mean cluster size	1.10	1.03	1.13	1.06

SOURCE: Letteval et al. 2002.

[a]Because of few data, no standard error is available in these cases, so the estimates should be viewed very cautiously.

given in table 6.17. Off Kaikoura and the Galápagos Islands, and in the Gully, the males studied were about 13–14 m long, or roughly in their early twenties (Ohsumi 1977), whereas at the higher latitude of Andenes, the animals were larger, averaging 15 m—roughly the size at physical maturity (Ohsumi 1977).

In all four of these study areas, male sperm whales form aggregations. Off Andenes and Kaikoura, and in the Gully, their distributions are quite well circumscribed by the deeper waters of a prominent submarine canyon, and at any time the aggregation is about 10–30 km across (McCall Howard 1999; Ciano and Huele 2000; Jaquet et al. 2000). The males are found in different parts of the three canyons (which span 40–60 km) at different times (e.g., Whitehead et al. 1992; Jaquet et al. 2000), but they are quite sedentary. The Galápagos aggregations are different. We encountered, and then tracked, discrete aggregations of males spanning a few kilometers within a much larger study area. Over time scales of hours and days, some Galápagos male aggregations remained in the same general area. In contrast, the aggregation shown in figure 2 of Christal and Whitehead (1997) moved about 50 km in a northwesterly direction over 17 hr.

Using mark-recapture methods (Letteval et al. 2002), we have been able to estimate the number of males in each aggregation and the mean residence period of an individual male within it (see table 6.17). The aggregations seem to contain about ten to twenty-five males at any time, and individual males enter and leave the aggregations roughly every 3–50 days, although some males seem more resident in each study area than others, staying for months or longer, or repeatedly returning to the same area over years, while others pass quickly through (e.g., Christal 1998, 121; Ciano and Huele 2000; Jaquet et al. 2000).

Groups or Units?

Although Gaskin (1970) and others (e.g., Best 1979) describe "bachelor schools" of nonbreeding males, there is little evidence from modern studies of living animals for males, in the absence of females, forming structured groups of the type that are apparent for females and immatures. Although members of the male aggregations followed by Christal (1998, 126–29) off the Galápagos Islands usually showed consistent headings, mostly within about 20° of the modal heading of the aggregation on that day (fig. 6.25),

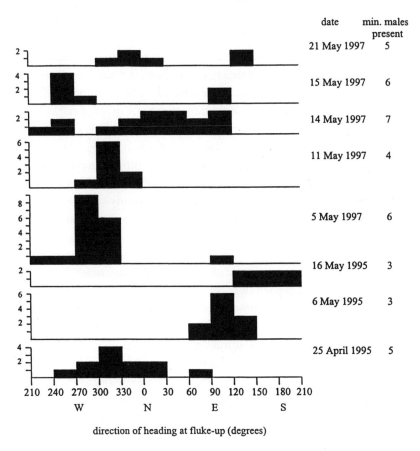

	date	min. males present
	21 May 1997	5
	15 May 1997	6
	14 May 1997	7
	11 May 1997	4
	5 May 1997	6
	16 May 1995	3
	6 May 1995	3
	25 April 1995	5

Figure 6.25 Consistency of headings of large males at their fluke-ups within aggregations of males off the Galápagos Islands. (From Christal 1998, 129.)

there was no indication of a structured rank. The heading consistency could result either from the males changing their behavior to coordinate with one another or from their all being influenced by a common external factor, such as a moving prey distribution or a desire to swim in a certain direction relative to a current or the nearest depth contours.

The mean sizes of the "bachelor schools" described and illustrated by Gaskin (1970) decrease with mean male size (mean bachelor school size = $117.26 - 8.16 \times$ mean length of male in m; from Best 1979). These "bachelor schools" may be aggregations (in the sense used above) with some clustering (see below). Rice (1989) notes that males in "bachelor schools are usually more loosely grouped" than females. However, the larger-sized (about five to fifteen animals) bachelor schools reported by Gaskin (1970) contained generally smaller animals ($\sim$12–14 m) than the aggregations of nonbreeding males studied off Andenes, the Galápagos, Kaikoura, or in the Gully, and were observed in deeper, more open waters. Thus it is possible that more structured groups of smaller males exist, perhaps especially away from the continental shelves where nonbreeding males have been principally studied. An interesting mass stranding on the coast of Oregon in 1979 contained twenty-eight females and thirteen males, with all the males 14–21 years old (Rice et al. 1986). This stranding may have included one or more groups of females plus one or more groups of subadult males.

In all of the data collected from the four areas of male aggregation (Andenes, the Gully, Kaikoura, the Galápagos), there is only one instance of a dyad of identified individuals found clustered on more than one day (off Kaikoura over an interval of 3 days), and there are no indications of preferred companionship using looser temporal measures (seen on the same day or within 2 hr) given the membership of the aggregation (Letteval et al. 2002). Additionally, three males were marked by Japanese scientists in the same "bachelor school" of about forty whales and killed 2 days later, all in different "bachelor schools" of ten, five, and two animals (Ohsumi and Masaki 1975). Six juvenile males (ca. 12–14 m long) entered the confined waters of Scapa Flow, Scotland, for about a month in 1993, and were frequently seen clustered together, or in pairs, during this time (Goold 1999). However, they were the only whales in Scapa Flow, so once again, while there was aggregation, there is no evidence for preferred companionships.

Thus, among males, there is no evidence of the long-term relationships, or social units, that are characteristic of female sperm whales and their dependent young. Male sociality is thus much less evident than that of the females. But these are difficult animals to study in any detail, and it is quite probable that undetected forms of preferred association exist. For instance, scientists studying savannah elephants, which have social structures remarkably similar to those of sperm whales (see section 8.5), long thought that associations among mature males, when they were not breeding, were essentially random. However, records over longer time scales have shown that some dyads interact repeatedly and intensely for short periods separated by long intervals (Douglas-Hamilton et al. 2001). Such important, but cryptic, relationships may well exist among mature male sperm whales.

Clusters?

Most nonbreeding males are seen alone at the surface (Letteval et al. 2002). However, these males do sometimes actively cluster (Caldwell et al. 1966). The mean cluster size ranges from 1.03 to 1.13 in the different study areas (see table 6.17). Sometimes pairs of males will cluster on successive surface intervals between dives (Christal 1998, 134; Letteval et al. 2002). Although they cluster less than females, the nonbreeding males behave much as the females and immatures do when clustered. For instance, the median standard deviation among fluke-up times of clustered males off the Galápagos was 4.2 s (Christal 1998, 135–36), which is comparable to the 6.4 s for females (see section 6.8). Under the unusual conditions of competition for the location in which a longline was being hauled aboard a fishing vessel, a well-coordinated cluster of six male sperm whales was able to displace a group of five killer whales (Nolan et al. 2000; see section 5.9), showing that clustering by males can, at least sometimes, be useful.

Strandings

Strandings consisting only of males are not uncommon. For instance, Rice (1989) knew of a total of thirteen all-male strandings in which more than one animal beached itself (with 3–37 animals each, mean 12.5 animals). As shown by S. Iverson's description of a mass male stranding on Sable Island, Nova Scotia, (fig. 6.26), which is quoted in the prologue, this is at least sometimes, and maybe usually, the result of animals changing their behavior to beach alongside other males. Bond (1999) found that among two all-male strandings on the coast of Scotland, most animals were genetically unrelated to one another, although some potential half-sibling relationships were identified.

Figure 6.26 Two male sperm whales that stranded on Sable Island, Nova Scotia, in January 1997. A third male, seen offshore, swam in to join them. (Photograph by Shelley Lang, courtesy of Shelley Lang.)

Male Sociality

Reviewing the comparisons in table 6.16, it is apparent that while nonbreeding male sperm whales show spatial clustering over a range of scales from a few meters to hundreds of kilometers, there is rather little evidence that they actively join with one another to form mutualistic groups. The major exceptions are cluster formation, which is clearly the result of active changes in behavior, and strandings. Cluster formation in males may have similar benefits to the very comparable, but much more common, clustering of females, perhaps giving some protection against potential predators. The reduced incidence of clustering in males may correlate with the lower vulnerability to predators of these larger animals. However, clustering in males may also be related to bonding—bonding that we have so far been unable to pick up in other ways, perhaps because our data are too few or collected over the wrong spatial scales (males can certainly hear each other over at least 10 km). The active mass stranding of males together certainly suggests an important role for social behavior among these animals.

6.11 Mating Systems

Who Are the Breeding Males?

In cavalier attendance upon the school of females, you invariably see a male of full grown magnitude, but not old; who, upon any alarm, evinces his gallantry by falling in the rear and covering the flight of his ladies. In truth, this gentleman is a luxurious Ottoman, swimming about over the watery world, surroundingly accompanied by all the solaces and endearments of the harem.
—Melville 1851, 501

Confronted with an animal as sexually dimorphic as the sperm whale (fig. 6.27), it is natural that both "macho" whalers and evolutionary biologists have assumed that, in male sperm whales, size is correlated with breeding success. However, some groups of females do not contain large males, but do include smaller immature males; large males are often seen alone on the breeding grounds; and there has been some evidence for reduced testis size and seminal activity in some large males. These observations led some whalers and scientists to believe that the largest males are to some extent "outcasts" and that smaller males do much of the breeding (Caldwell et al. 1966; Berzin 1972, 250–51). However, Best (1979) and Best et al. (1984) show that there is a clear overall increase in the density of spermatozoa with body length. The lack of large males with some groups of females, and observations of large males alone on the breeding grounds, are easily explained by the roving behavior of breeding males, described below. Additionally, as the number of large males was reduced by whaling, there seems to have been a lowering of pregnancy rates in some areas (Clarke et al. 1980; Whitehead et al. 1997; see section 4.3). Thus, I follow Best (1979) and R. Clarke et al. (1994) in concluding that while smaller males (10–12.5 m, ~12–27 years old) may be fertile, most successful breeding is likely to have been the work of larger, older males, especially before modern whaling reduced their relative abundance. We really need the results of genetic paternity testing to fully clarify the identity of breeding males, although given the spatial scales involved, such testing will be difficult. Current genetic results, however, suggest that females do not usually breed with their own close relatives, and that a male's natal group may at least sometimes be well separated geographically from the areas in which he breeds (Lyrholm et al. 1999).

Knowledge—or more accurately, the lack of it—about the movements

Figure 6.27 Females and immatures react to the presence of a male. (Photograph by Flip Nicklin, courtesy of Minden Pictures.)

of large males between their low-latitude breeding and high-latitude feeding grounds is discussed in section 3.5. However, it is known that some males may spend a month or more within the Galápagos Islands concentration of females (individual males were identified over spans of 27 days, 34 days, 36 days, and 70 days within a year), and the overall distribution of observed time spans of mature males within the Galápagos concentration was not significantly different from that of females, indicating similarities in residence patterns (Whitehead 1993). Off Crete, in the eastern Mediterranean, males have longer residence periods in the 20 km × 70 km study area than female social units (A. Frantzis et al., unpublished data). In contrast, off Dominica, West Indies, males spent significantly shorter time spans in the study area than females and immatures (means of 6.4 days and 24.8 days, respectively; Gordon et al. 1998). In this study, three of nine identified large males were found in consecutive study years (Gordon et al. 1998). Off the Galápagos Islands, male return rates were lower. Although six of the fifty-four identified large males were found in more than one year, only one was seen interacting with groups of females in two or more years: #507, who was identified on one day in 1985 and on one day in 1991 (Whitehead 1993). The other Galápagos Islands multi-year males were identified exclusively within all-male aggregations (Christal 1998, 121).

Behavior of Breeding Males

Individual breeding males showed no tendency to remain within particular sub-areas of the Galápagos concentration in any of our study years (Whitehead and Waters 1990; Whitehead 1993). Thus, breeding success among males does not seem to be based on territorial defense.

While on the breeding grounds, males may be observed alone or with groups of females (Kahn 1991; Whitehead 1993; Gordon et al. 1998). When alone, they move much as they do at high latitudes (or off the Galápagos Islands in the late 1990s, after the females had left), swimming slowly but steadily and diving. Both when alone and when with females, males on the breeding grounds make their loud and distinctive "slow clicks" about 74% of the time, but they may also make the standard "usual clicks" (Whitehead 1993; see section 5.2).

We have few data on the movements of males alone, but figure 6.28 shows the movements of a male tracked in the central Pacific on 2 October 1992 from 07:00 until 14:00. He had been followed acoustically during the previous night, and in the early morning was moving principally eastward, but

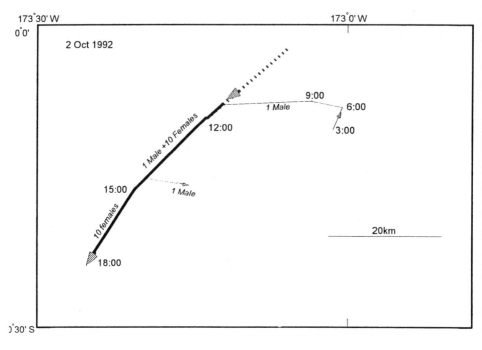

Figure 6.28 The track of one mature male sperm whale (thin line) as he approached, accompanied, and then left a group of ten females (thick line) in the western tropical Pacific on 2 October 1992. The tracked animals are represented by solid lines, and inferred movements by dashed lines.

at about 09:00 he turned sharply to the west, heading relatively fast (~7.5 km/hr) on an intercept course with a group of ten females and immatures, whom he met at about 11:00. He accompanied them as they moved southwest until he left them at 14:00. As the females moved on a fairly straight course while accompanied by the male and afterward (fig. 6.28), I infer that he was approximately 20 km from them when he changed course to make the intercept.

Although whalers and earlier scientists believed that males (the "harem-masters") accompanied groups of females for a breeding season (e.g., Ohsumi 1971), Best (1979) began to doubt this when he noticed that, whereas females killed within the same group usually had similar patterns of cyamid (external parasite) infestation, accompanying large males frequently hosted no cyamids, or different species. His conclusion that relationships between groups of females and mature males are fairly transitory was borne out and clarified by our studies of living animals (e.g., see fig. 6.28). We found that mature males usually spent only a matter of hours with each group of females before moving on, but that they might reencounter the same group several times within periods of a few days (Whitehead 1993). We never saw reassociations between a male and a group of females over intervals of more than one month, but within shorter periods there were indications of preferential associations between particular males and particular females (Whitehead 1993). For instance, female #854 was identified eight times on 29 April and 5 May 1989, always with male #530.

The behavior of males when accompanying groups of females is not very obviously different from that of the females (Whitehead 1993): they move at similar speeds, dive to similar depths for similar periods of time, defecate (and so eat), are found in similar-sized clusters, and do not obviously dive before or after the females in their cluster. However, they are much less likely to breach or lobtail than are the accompanying females (Waters and White-head 1990a).

When the Sexes Meet

When large males joined mixed groups, they were often observed to be the focus of intense attention. Calves, immature whales and mature females were all seen to push up against the mature males and roll along their bodies. On several occasions, males were seen holding young calves gently in their mouth. Extended penises were often seen, although copulation was not witnessed.
—Gordon et al. 1998, 556

Interactions between large males and females are interesting in many ways. From an evolutionary perspective, a particularly important question is, who controls the choice of mating partners? Are copulations forced by males, chosen by females, or determined by other processes? We rarely see copulations themselves, so we must make inferences from the whales' more general associative behavior.

As Gordon et al.'s (1998) description of observations made in the clear Caribbean waters states so explicitly, females and immatures react strongly to the presence of a mature male (see fig. 6.27). For the murkier waters off the Galápagos Islands I cannot give such a rich description, but will substitute statistics. The rates at which females and immatures sidefluked and spyhopped—activities indicative of maneuvering at the surface—roughly doubled with the addition of one mature male, as did the rate of producing communicative codas, and these rates showed even greater increases when there were two large males present (Whitehead 1993).

Within groups, large males often alter course to join females (or maybe immatures), females alter course to join large males, and, occasionally, both a female and male may alter course to approach each other (Whitehead 1993). Additionally, females may swim away from males, and males from females (H. Whitehead, personal observation off Chile, 2000). Thus, clustering between mature males and females seems not to be the result of the actions of one sex alone, but rather can be instigated by either or both. However, among male-female pairs who were seen associated on two or more occasions, there was a strong tendency for the female to dive first (female before male on eleven occasions; male before female on three), suggesting that in these cases the male is attempting to maintain contact with the female, rather than vice versa (Whitehead 1993). There is no indication at all in the modern data that males are the "harem's lord" (Melville 1851, 501) or "harem leader" (Berzin 1972, 250). Instead, it seems that associations between large males and breeding females are regulated by mutual consent, perhaps mediated by the presence or activities of other whales, especially other large males (see below).

There are a number of descriptions in the literature of sperm whales mating (Caldwell et al. 1966; Gambell 1968; Best et al. 1984; Di Natale and Mangano 1985; Ramirez 1988, and references therein). Two animals, usually one substantially larger than the other and assumed to be the male, are seen belly to belly for periods of a few seconds to half a minute, either vertically or horizontally. In the descriptions of horizontal mating, the female is usually described, or illustrated, as being beneath (e.g., Best et al. 1984; Ramirez 1988). In none of these cases was intromission observed. During our work at sea, we have several times observed animals belly to belly for periods of a few seconds or more, usually horizontally. In some of these cases one animal was a large male, which may have been above or beneath the female. I do not think it justifiable to consider these instances copulations, and have the same reservations about the literature reports just summarized. We need clear, careful, and close observations of sperm whale mating—observations like those made by Gordon et al. (1998; see the quote at the beginning of this section) off Dominica, who saw much physical contact between males and females and extended penises, but no intromission.

Sperm competition is likely to be important in mating systems in which a female frequently mates with several males at one estrus. Sperm competition has been implicated in the breeding systems of some cetaceans, especially the right whale (*Eubalaena* spp.) with its 1 t testes (Brownell and Ralls 1986; Connor et al. 2000a). Connor et al. (2000a) suggest that as male sperm whales seem to have large testes, females may mate with more than one male. However, R. Clarke et al. (1994) examined the sperm whale anatomical data more thoroughly, finding no attributes of the vagina consistent with sperm competition, relatively small testes, and a relatively small penis. Their conclusion that sperm competition is unlikely to be an important element of the sperm whale's mating system is borne out by the observations at sea — or, rather, by the lack of them (intromission is often observed in right whales; Brownell and Ralls 1986), as well as by the indications of pronounced female choice and/or male dominance from genetic studies and pregnancy rates (see below).

The generally gentle and affiliative nature of the association between a mature male and a group of females was obvious in our observations as well as in those of Gordon et al. (1998). Infanticide is considered to be an important driver of some mammalian social structures (e.g., Packer and Pusey 1983; Van Schaik and Kappeler 1997), and there is some evidence for it among cetaceans (Patterson et al. 1998). Thus, the observation by Gordon et al. (1998) of large males "holding young calves gently in their mouth" is especially interesting, suggesting to me that infanticide is not an important factor in sperm whale sociality.

Relationships between Breeding Males

He was not so very much smaller than the big one, and before I realized that there was anything on the programme, here he was, coming for the big bull, fire in his eye, I could imagine, and jaw dropped. When he was a hundred feet away, he turned over, nearly on his back, apparently, for I saw his jaw projecting above the surface of the water.

The big bull was aware of the other just in time to slip out of the way, but not in time to escape entirely. The jaw closed on his small, and I saw the wounds made by the teeth, which tore out great pieces of blubber and flesh. By what seemed agreement, the two big whales turned about as soon as they could and went at each other full tilt. Their jaws locked, and they wrestled there for a minute, each seeming to try to break the jaw of the other, and tearing and thrashing the water into boiling fountains of spray. As we found out later, great gobs of flesh were torn from the sides of their heads. After a while they broke their hold, I could not see how, and they backed off and went at it again.

This time the fight was fiercer than before, and it was impossible to see what was happening, or to see anything but white water. This round was a little longer than the first. The performance was repeated two or three times, and then I saw the boiling white water gradually become quiet. The two great bodies lay there for a few seconds, head to head; then the smaller of the whales moved off slowly away from the school. He seemed to have lost all interest in the cow, and the bigger one, satisfied that the other had definitely given up the fight, let him go in peace.
—Hopkins 1922, 330–31*

More than one large male can sometimes be seen with a group of females, or clustered together with females. Even more frequently, two or more slow clicks can be heard together. As densities of large males have been low off the Galápagos Islands, such instances were not common in our data. However, two or more sets of slow clicks were heard together half as often as expected (and statistically significantly less often than expected) given the overall rate of hearing slow clicks, suggesting that the large males within a breeding concentration to some extent avoid one another (Whitehead 1993).

However, this avoidance is not absolute. Off the Galápagos Islands, I estimated that, if one male was within acoustic range of our hydrophone (perhaps 10 km for slow clicks), there was a probability of 0.22 that another was (Whitehead 1993). Large males on the breeding grounds may approach each other much more closely than this. Two or more large males have been recorded in the same cluster with females and immatures (Caldwell et al. 1966; Whitehead 1993), and this was seen fairly frequently during 2000 off northern Chile.[†]

Relationships among breeding males who are clustered with, or in close proximity to, one another and associating with females seem generally neutral. During the Galápagos studies we observed only one incident of possibly aggressive interaction between males. Two males positioned side by side within a cluster that included nine females and immatures showed "much

---

* It is not clear whether this account is fact or fiction, or (perhaps most likely) something between the two.

[†] During 765 hours when such information was recorded off Chile in 2000, in 76% no males were observed, in 21% one male was observed, and in 3% two or more males were seen. These numbers underestimate the presence of males with groups of females, as they were sometimes present, and heard, but not observed.

Table 6.18. Scars on the Heads of Male Sperm Whales in the Antarctic with Length and Sexual Status (from Kato 1984)

Males		Tooth Scar "Stage"			
Length	Sexual Status	None	Indistinct	Distinct	Heavy
9.5–11.9 m	Immature	8	25		
	Mature				
12.0–13.7 m	Immature		2		
	Mature		1	1	
14.0–15.2 m	Immature			3	
	Mature			2	6

thrashing of flukes at or beneath the surface" (Whitehead 1993). They both dived together with the other members of the cluster, and whether any aggression was present is uncertain.

However, relationships between large males are not always amicable. There are a number of accounts in the older whaling literature of male sperm whales battling with one another (reviewed by Caldwell et al. 1966; Clarke and Paliza 1988; see the quote at the beginning of this section). Kato (1984) and Best (1979) showed that, in the Southern Hemisphere, as males grow and become sexually mature, they accumulate on their heads more, and heavier, scars of the kind that can be presumed to have been made by other large males (table 6.18; fig. 6.29; see also fig. 8.1). These wounds seem to have been incurred on the low-latitude breeding grounds (Kato 1984). In the older reports, the males are described as battling with their jaws (Caldwell et al. 1966), which agrees with the scarring patterns. Clarke and Paliza (1988) interpret tooth loss in sperm whales as a result of interspecific fighting, but this interpretation is not widely accepted (International Whaling Commission 1988). However, despite the evidence from scars and the early whalers' reports, there are almost no modern accounts of male sperm whales fighting, causing some puzzlement in the scientific literature (Best 1979; Clarke and Paliza 1988). The exceptions are observations by Zenkovich (1960), who simply reported that "fights between big sperm whale males were observed twice" in 1957, and a brief observation of our own off Chile in 2000 (see box 6.4). The latter had the flavor of a sumo contest: two huge, and apparently lumbering, contestants suddenly showed unexpected quickness and agility. The outcome was settled very swiftly as one seemed to overpower the other, who quickly conceded.

Figure 6.29 Fresh wounds on the head of a large male who was observed interacting with groups of females off northern Chile.

Box 6.4 **An Observation of Two Male Sperm Whales Fighting**

During a study of the behavior of sperm whales off the northern coast of Chile in 2000, large males were frequently observed interacting with groups of females. There were often one or two such large males with a group. In very calm conditions, at 14:10 on 21 July 2000, in position 21°04' N; 71°39' W, we were following a large (15–16 m) male and a smallish female (ca. 9.5 m). The female repeatedly "shallow dived" beside the male, sometimes coming up on one side of him, sometimes on the other. The male swam steadily at the surface at perhaps 3 km/hr. This pattern was observed for about 8 min, after which we saw another large male about 300 m ahead of the first. He swam fast toward the pair, lobtailing (see section 5.6) vigorously on three interspersed occasions. The males came together with much splashing. The tail of one was observed above the surface, with the other's jaws around it (fig. 6.30). Quickly they separated. The contact lasted very roughly 15 s. One male (we were not sure which) left the area very rap-

Box 6.4 continued

Figure 6.30 Two large males fighting, as observed off northern Chile in 2000. (Copyright Emese Kazár.)

idly. The other remained with the female and a juvenile who suddenly appeared. No pictures were taken. At least one of the males in the area was later seen with deep, fresh scars on his head (see fig. 6.29), but we do not know if he was one of the participants in the fight.

The scarcity of observations of males fighting can perhaps be explained by two factors: fights are rare, and fights are quick. It may be that our brief observation off Chile is more typical of fights between males than the protracted affair described by Hopkins in the quote at the start of this section. It is also likely that males, who can clearly do each other much damage, have other ways of settling disputes or deciding precedence. Their slow clicks (see section 5.2) may have characteristics allowing them to assess one another's size and potential fighting ability over long ranges (Weilgart and Whitehead 1988; Gordon 1991a), and so might result in dominance hierarchies being set up within a breeding concentration, as suggested by Watkins et al. (1993).

I wish to stress one lacuna in the relationships between breeding male sperm whales: There is no evidence for male coalitions, which are such a prominent part of the social structure of a number of terrestrial mammals (including primate and cat species) as well as some populations of bottlenose dolphins (Connor et al. 2000b) and some other cetaceans, such as the northern bottlenose whale (Gowans et al. 2001).

The Puzzles of Female Pregnancy Rates and Shared Paternity

Two pieces of evidence from other areas of sperm whale biology have implications for the whales' mating system. First, as noted in section 4.3, the pregnancy rate in the eastern tropical Pacific region has dropped since the proportion of mature males was reduced by whaling. The natural implication is that these events were linked, with females failing to conceive because of low rates of insemination (Clarke et al. 1980; Whitehead et al. 1997). It is possible that the low pregnancy rates in this area have nothing to do with the mating system, and are rather the result of disease or poor female body condition because of insufficient food intake. However, the Galápagos females seemed to have reasonable feeding success (see section 2.6), at least in non–El Niño years, and were not obviously thinner than those that I observed in other parts of the world where calving rates are higher. Thus, a failure in the mating system is implicated.

Off the Galápagos Islands between 1985 and 1995, the proportion of large males in the population was about one-fourth of that predicted by International Whaling Commission models and found in the catches of the Yankee whalers working in the area 150 years earlier, and the calving rate had fallen to about one-third of that expected from International Whaling Commission models and found in other parts of the world (Whitehead et al. 1997). However, the groups of Galápagos females are attended by large males on about 75% of days during the primary breeding season (Whitehead 1993). Why is this not enough to ensure reasonable pregnancy rates? Resolutions to this paradox include the possibility that the presence of several large males is necessary to ensure successful breeding behavior in either the males or the females, or that females will mate only with certain males (Whitehead and Weilgart 2000).

Pronounced female choice could also explain the puzzle of paternal relatedness within groups and units (see section 6.6). Off the Galápagos Islands, we identified three to eight mature males during the principal field seasons when females were present in reasonable numbers (1985, 1987, and 1989), and as some males were identified on only one day, there were likely to have been other males present that we did not encounter (Whitehead 1993). Additionally, it seems as though only a few males return to the same concentration of females in successive years (see above). Thus, the probability that two calves in the same group would have the same father would seem to be small, unless there is a mechanism biasing paternity toward just one or two males within a concentration and these are the animals that are likely to return in subsequent years. Female choice—and, in particular, copying of the choices of other females—or dominance hierarchies would produce such a bias, and if the successful males are those most likely to return to an area, the puzzle of paternal relatedness within groups and units may be resolved.

Thus there is some, quite indirect, evidence for male dominance and/or female choice within sperm whale mating systems, in addition to the behavioral observations of males fighting and females selecting particular male companions. To satisfy our interest in sexual selection in a highly dimorphic social species, as well as to look at its effect on population biology, we need to work out ways of studying mating tactics in this species.

6.12 Summary

1. Sperm whales typically share their ranges with, and may have relationships with, several thousand other sperm whales.

2. Female and immature sperm whales often form concentrations a few hundred kilometers across, containing very approximately 750 whales. These concentrations probably roughly coincide with areas of increased food resources.

3. Sperm whales are aggregated over scales of about 2–20 km. The sizes of these aggregations vary considerably spatially and temporally, and they have dynamic membership over periods of hours, but the modal aggregation size is about thirty animals. Large aggregations probably usually indicate a concentration of food resources.

4. Female and immature sperm whales form groups that move together in a coordinated fashion over periods of days. Mean group size is about twenty to thirty animals, although there is much variation.

5. A socializing group of sperm whales may be tightly clustered at the surface. However, when foraging, the whales usually spread out to form a rank, which can be a kilometer or more long and is usually aligned perpendicular to the direction of travel. There is no indication that individuals have preferred positions along the rank.

6. Groups often consist of two or more social units. Members of units are long-term companions, over at least years. The units forming a group disassociate very roughly once every 5 days. Within a group, animals preferentially associate with members of their own unit.

7. Unit size is very variable, but averages about ten animals. Except in the largest units, there is no evidence for preferred companionships among unit

members, and there seems to be little correlation between the association index of pairs of unit members and their genetic relatedness.

8. Over periods of years, there are some mergers of units, splits of units, and transfers between units. Genetic evidence shows that while there are close matrilineal relationships within units, units frequently contain more than one matriline. However, this should be expected, as the expected matriline size given the life history parameters of sperm whales is about three animals. Observed splits of units, and transfers between them, are not necessarily divisions of matrilines.

9. Females may benefit from unit membership through communal care of their offspring and the sharing of information. I suspect that units agglomerate into groups to reduce the risk of predation on their calves and themselves, as well as to gain feeding benefits, either from information about food concentrations provided by group members or in the form of prey items that flee from other members of the group. In this scenario, the group often spreads out into a rank to forage, principally to minimize interference.

10. During their surface periods, sperm whales actively cluster together, swimming along about 6 m apart and synchronizing their fluke-ups. This practice probably enhances safety from predators and may reaffirm social bonds.

11. Female and immature sperm whales seem to actively change their dive schedules to improve the protection of calves, an apparent form of babysitting, and there is good (but not conclusive) evidence for allosuckling.

12. Males at high latitudes may form concentrations about 50–200 km across, but these concentrations seem less pronounced than those of the females. Aggregations of males in areas 10–30 km across have been studied at various latitudes. These aggregations contain a mean of about ten to twenty-five males at any time. In these male aggregations, there is no sign of the long-term relationships and social units that are characteristic of female sperm whales and their dependent young.

13. Although they are most often seen alone at the surface, nonbreeding males sometimes cluster with other males, and may mass strand together, suggesting some form of social structure.

14. Large breeding males stay on the low-latitude breeding grounds for at least periods of months, roving between groups of females. They usually spend just a few hours with any one group. Their behavior appears not unlike that of the females whom they are accompanying, although the males show less aerial behavior. In contrast, females are generally more active in the presence of large males.

15. Although mating has never been convincingly described, consortships between large males and females seem largely a matter of mutual consent. It seems unlikely that sperm competition, infanticide, or male coalitions are important elements of male reproductive strategies. Female choice and/or male dominance hierarchies, mediated by the "slow click" vocalization and the occasional fight, may be significant determinants of breeding success.

7 Sperm Whale Cultures

Click-Click-Click-Click-Click (Unit A)
Click-Click-Click-Click-pause-Click (Unit G)
Click-Click-pause-Click-Click (Unit T)

7.1 The Quest for Sperm Whale Culture

Why look for culture in sperm whales? The idea that some aspects of cetacean behavior are culturally determined has a fairly long history (e.g., Norris and Dohl 1980; Osborne 1986; Shane et al. 1986), but two factors have sparked a more rigorous, quantitative search. In our studies off the Galápagos Islands in 1987, we sometimes had the impression that the different groups we followed had consistently different ways of behaving. Most memorably, there were the "sadistic upwinders" (actually unit A), who seemed to move fast upwind whenever it came on to blow, making our work both uncomfortable and frustrating. Traits specific to certain groups among a well-mixed population are likely to be culturally determined (see section 1.8). However, such impressions in the field often fall apart under quantitative statistical analysis, and it was many years before I carried out such an analysis on our visual records of the Galápagos sperm whales (Whitehead 1999c).

In the meantime, Linda Weilgart and I analyzed the coda repertoires heard off the Galápagos Islands and elsewhere in the South Pacific (Weilgart and Whitehead 1997). We found that groups of sperm whales showed significant differences in their usage of different types of coda vocalizations (see section 5.2 for an introduction to codas and section 7.2 for more details on dialects). Comparing two groups, one might make more short codas containing fewer than five clicks, and the other might have a predominance of codas with a relatively longer interval before the final click ("+1" codas, see table 5.2). Group repertoires seemed stable over days, and probably over longer periods (Weilgart and Whitehead 1997). How could this have happened? The group-specific repertoires were not the result of particular individuals dominating the repertoire with their distinctive coda types (see section 5.2)—the repertoire is a group property. Parallel individual learning from the environment among individuals within groups seemed unlikely to explain this convergence, as groups with different repertoires were found in the same habitat. Genetic determination was also unlikely, as some members of groups are genetically unrelated (see section 6.6).

So, following the elimination procedure outlined in section 1.8, we con-

286

cluded that social learning—animals learning their coda repertoire from other members of their group, perhaps especially their mothers—was the most plausible explanation for the group-specific coda repertoires, and thus that the repertoires were likely a form of "culture" (Weilgart and Whitehead 1997). This conclusion was, rather ironically, buttressed by our finding that, among a small number of sperm whale groups (six), similarity of coda repertoire was significantly correlated with similarity in the mitochondrial matrilineal genes (Whitehead et al. 1998). To achieve such a pattern would require that coda repertoire be determined either by an unusual, largely matrilineal, form of genetic inheritance or, more plausibly, by quite stable matrilineal cultural transmission (Whitehead et al. 1998).

A problem with this coda analysis was that it was done at the level of the temporary group, not the permanent social unit, and the unit seems to be the most likely forum for cultural homogeneity (see chapter 6). The strong indications of stable group-specific coda repertoires were most easily explained as unit-specific cultural repertoires with many groups containing just one unit, or having their repertoire dominated by that of just one unit, but there were clearly uncertainties (see Christal 1998, 152; Mesnick et al. 1999).

Thus, in the next phase of research on sperm whale culture, we looked for consistent differences in vocal and nonvocal behavior between social units studied off the Galápagos Islands. Because sperm whale units are highly mobile and share their ranges with many other units with which they intermingle to form groups, environmental causation plus individual learning can more easily be ruled out as a cause of such patterns than when considering more sedentary animals such as apes. As social units at least sometimes contain sets of unrelated animals (Mesnick 2001; see section 6.6), genetic causation is also very unlikely. Demographic differences between social units could also produce differences in behavior—for instance, if groups with many young whales have characteristic ways of life—but this seems an unlikely explanation for most of the patterns discussed below. Thus, the standard difficulties of identifying cultures among nonhumans using the ethnographic method are relatively minor for sperm whales.

However, the study of sperm whale culture possesses its own substantial challenges. These challenges include the problems of quantifying, or even seeing, the behavior of an animal that spends so much of its time beneath the surface. In Whiten et al.'s (1999) powerful presentation of chimpanzee (*Pan troglodytes*) cultures, thirty-nine very specific types of behavior were characterized as customary/habitual/absent at each site. But with sperm whales, we have only rough quantitative measures of a few general types of behavior. Another challenge lies in identifying unit-specific behavior in temporary groups. Fortunately, the discovery of a supra-unit structure, the acoustic clan (see below), obviates some of this difficulty.

7.2   Coda Dialects of Units and Clans

Coda Repertoires

A coda repertoire is simply the set of codas emitted by—or, in practical terms, recorded from—a set of whales in a particular circumstance. Our most basic level of analysis is the repertoire in a continuous hydrophone recording, which usually lasts between a few minutes and an hour and may contain anywhere between two and several hundred individual codas. We amalgamate these recordings to produce repertoires for days, which are useful because we almost always follow the same group during a day (see box 6.1). Photographic identifications can sometimes be used to show that the same group was present on different days, and recordings for those days can then be lumped to produce a more representative coda repertoire for the group. We can also look at coda repertoires for groups containing a particular unit, but obtaining a definitive unit-specific coda repertoire requires special circumstances in which we know that only one unit was present (as with unit T). Further amalgamation gives coda repertoires for geographic areas. Ideally, we would like to be able to describe the coda repertoires of individuals, but that is not yet possible.

A coda repertoire is fully described by the acoustic record of all the codas in it. However, this is not usually an appropriate or efficient manner in which to convey the contents of a repertoire to other scientists, the interested public, or a statistical analysis package. Instead, codas are allocated to types, and each coda type is given a name based on the number of clicks and the relative inter-click intervals it contains, such as "6R," "4+1," "8L" (see table 5.2). In this way, a repertoire can be described by the number of codas of each type that it contains: 12 × "3R," 5 × "5R," 1× "4+1," 3 × "8L." Codas can be assigned to types using "rules of thumb" (e.g., Moore et al. 1993; Weilgart and Whitehead 1993) or objective statistical methods (Weilgart and Whitehead 1997; Rendell and Whitehead, in press; box 7.1).

Alternatively, each coda can be displayed as a point in multidimensional space, so that a four-click coda is represented on a plane with the first dimension giving the proportion of the length of the coda between click one and click two and the second the proportion between click two and click three (the third dimension is redundant in this representation, as the third relative inter-click interval is simply one minus the sum of the first two, although for the quantitative methods described in box 7.1 all three dimensions are used). Then all the four-click codas in a repertoire can be plotted on one such diagram, and a complete repertoire by a series of such plots, with increasing dimensionality, containing one plot for all the codas of a particular length. The plots with high dimensionality can then be simplified by statistical techniques such as principal components analysis. This kind of representation is shown in figure 5.2, and a whole repertoire in figure 7.1.

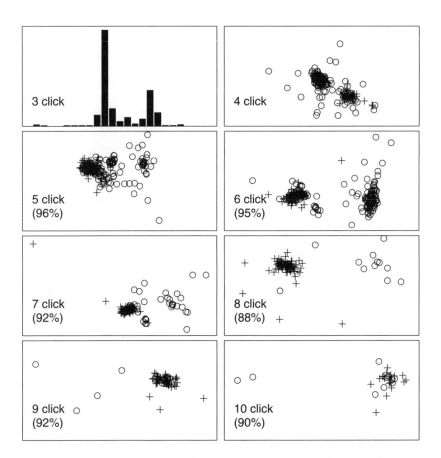

Figure 7.1 Two coda repertoires represented in multivariate space. For three-click codas, a histogram of the proportion of the coda between the first and second clicks is shown; for four-click codas, the proportion between the second and third clicks is plotted against the proportion between the first and second; and for codas of five to ten clicks, the first two principal components of the relative inter-click intervals are plotted against each other (as in fig. 5.2, with the percentage of the total variance accounted for by the first two principal components noted at the bottom left). Each symbol represents one coda: plus signs represent codas from group A2B (see box 6.3); open circles, codas from unit T.

Once a coda repertoire has been represented as either a numerical distribution of coda types or points in multidimensional space, two repertoires can be compared numerically. We have investigated a number of methods for doing this (Rendell and Whitehead, in press). The one we have found most useful is summarized in box 7.1. Thus, we end up with a numerical measure of the similarity between two coda repertoires, high if the repertoires contain many similar codas and low if they do not. These similarity measures are fundamental to the analysis of how coda repertoires vary geographically, temporally, and socially.

Box 7.1 Describing and Comparing Coda Repertoires

Categorizing Codas

It seems that much of the information contained in a coda is represented by the relative positioning of the clicks, as codas can be more easily categorized into types using relative inter-click intervals than absolute ones (Moore et al. 1993; Weilgart and Whitehead 1993). In this way, we obtain multidimensional representations of coda repertoires of the kind shown in figures 5.2 and 7.1. Such plots, showing fairly discrete clusters of points (codas), confirm the impression of aural analysis that codas often have stereotyped patterns that should be amenable to classification.

After a lot of experimentation, we have settled upon a hierarchical divisive method of categorizing into types a set of codas with a particular number of clicks (Rendell and Whitehead, in press). For, say, a set of five-click codas, a hyperplane is drawn through the set of coda points in the four-dimensional space of inter-click intervals, dividing the codas into two subsets, one on each side of the hyperplane. The hyperplane is positioned so as to maximize the difference between the sets. Hyperplanes are then drawn through each of the subsets, diving them each into two, and so on. If the division of a subset into two makes no significant difference to the residual variance of the coda points, division stops. This process was carried out for the whole catalogue of 14,065 codas recorded from around the South Pacific and the Caribbean, resulting in seventy types of codas. Each type was then given a name, following the naming criteria of earlier papers (Weilgart and Whitehead 1993, 1997).

Box 7.1 continued

Comparing Coda Repertoires

Once their codas have been categorized, two repertoires can be compared by examining the proportions in each repertoire of each coda type, and, for instance, calculating a Spearman correlation coefficient between these proportions (Weilgart and Whitehead 1997). However, a more direct method is to compare the multivariate representations of the repertoires without intermediate categorization. In this method, the similarity of two repertoires is the mean similarity between randomly chosen codas from each repertoire (for example, how coincident are the plus signs and open circles in figure 7.1?). But what is the similarity between two codas?

Once again, we have tried many possibilities (Rendell and Whitehead, in press). Luckily, all the reasonable measures that we have tried give very similar results when we use them to compare coda repertoires. Therefore, we have settled on the following measure: If codas have different numbers of clicks, their similarity is zero. If they have the same number of clicks, then their similarity is

$$s = \frac{0.01}{(0.01 + \text{maximum difference between corresponding relative inter-click intervals})}$$

So, if two five-click codas have relative inter-click intervals of {0.11 0.23 0.27 0.39} and {0.19 0.26 0.25 0.30}, the maximum difference between corresponding relative inter-click intervals is 0.09 (0.39 − 0.30), and the similarity between the codas is

$$s = \frac{0.01}{(0.01 + 0.09)} = 0.10.$$

This similarity varies from $s = 0$ if codas have different numbers of clicks to $s = 1$ if they have the same number of clicks and identical patterning. The constant that sets precision, 0.01, was chosen because it roughly corresponds to the discrimination of the analysis system, but, in practice, reasonable variation in this constant has little bearing on results (Rendell and Whitehead, in press).

Using this method, two repertoires are compared by calculating the mean of s between all pairs of codas with one chosen from the first repertoire and the other from the second.

Coda Repertoires of Units: The Discovery of Clans

Luke Rendell has looked at the coda recordings from the Galápagos Islands that were made in the presence of known social units (those listed in table 6.5) (Rendell and Whitehead 2003). In most of these recordings there is a possibility that other social units were present, so the recordings may include codas made by non–unit members. The coda repertoires from these known-unit recordings, and the similarities between them, are shown in figure 7.2.

There is a very pronounced pattern here. Off the Galápagos Islands, the units that we recorded had three types of repertoires:

> "Regular" units, which possessed repertoires mainly consisting of co-das with five to eight regularly spaced clicks (5R, 6R, 7R and 8R)
>
> "+1" units, whose repertoires were dominated by codas with a pause before the final click, especially 4+1 and 5+1 codas
>
> "Short," just unit T, which favored codas with fewer than five clicks and those with a pause after the first two clicks, particularly 2+1, 2+2, and 2+4

Ford (1991) uses the term "clan" to describe a set of killer whale (*Orcinus orca*) pods that have similar vocal repertoires—their "dialects"—and there seems to be an equivalent situation here, with stable sperm whale units possessing stable vocal repertoires, but some pairs of units sharing very similar repertoires, whereas others have distinctly different ones. Thus we also use the term "clans" for the sets of sperm whale social units with similar coda repertoires, and call a clan's repertoire its "dialect."

Our division of the coda repertoires of units into clans is strongly supported by the statistical analysis, with 100% bootstrap replicate support for the two principal nodes of the dendrogram in figure 7.2, and thus for the three clans. The repertoires of units recorded over periods of years, such as units A and B, stayed similar, and within the same clan, when entered separately into the analysis.

The pattern in figure 7.2 seems too good to be true for a situation in which units are frequently merged to form groups and we record only at the level of the group (see section 6.4). Why don't we get intermediate repertoires when units of different clans group together? Well, we may occasionally, but it seems as though units from different clans rarely mix. Table 6.6 shows eighteen situations in which two units were identified together on at least one day during an identification period (identification periods are separated by at least 30 days). These observations break down as follows:

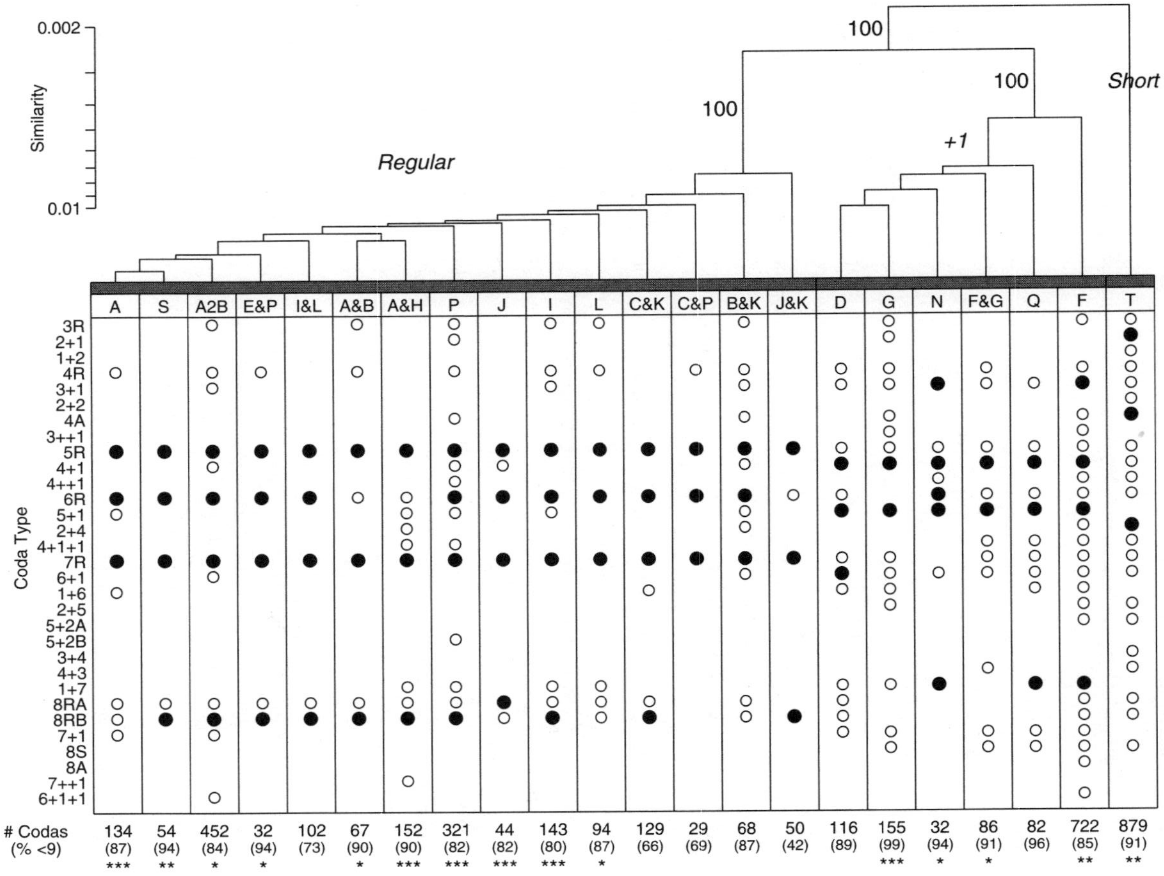

Figure 7.2 Coda repertoires of sperm whale groups containing known social units recorded around the Galápagos Islands, compared using multivariate similarity (dendrogram) and k-means classification methods (table). Unit codes are from table 6.5; more than one code is given when members of more than one unit were identified within 2 hr of the recording. Numbers next to the dendrogram branches are the number of bootstrap resamples in which that branch was recreated (/100). Circles in the classification table indicate coda types present in a repertoire; shaded circles indicate coda types that made up 10% or more of a repertoire. The numbers below each column are the number of codas in each repertoire and (in parentheses) the percentage of those codas with fewer than nine clicks and hence included in the table (all codas were included in the hierarchical cluster analysis shown at the top). An asterisk indicates groups recorded on more than one day; a double asterisk indicates 30 or more days between first and last recording; and a triple asterisk indicates more than 1 year between first and last recording. (From Rendell and Whitehead 2003.)

"Regular" clan unit with "Regular" clan unit: 14 unit pairs

"Regular" clan unit with probable "Regular" clan unit (unit E): 1
  unit pair

"+1" clan unit with "+1" clan unit: 1 unit pair

"Regular" clan unit with unknown clan unit (unit M): 1 unit pair

"Regular" clan unit with "+1" clan unit: 1 unit pair

Thus, we have only one observation of units of different clans on the same day. This was 27 April 1993, when units G and L, from the "+1" and "Regular" clans, respectively, were seen together off the coast of mainland Ecuador. Two members of unit G were identified between 06:32 and 06:35, eight members of unit L between 06:48 and 07:46, and twelve members of unit G between 08:10 and 12:50, including the two first sighted. Members of other units were also identified throughout this time. Thus these two units were certainly in the same area, but they may not have been associating or coordinating their movements closely. This brief association contrasts with the prolonged grouping of units from within the "Regular" clan, such as the A2B group in 1995 (Christal and Whitehead 2000; see box 6.3).

A question that arises from the discovery of these strong clans is whether there are differences in repertoire between units of the same clan, or, alternatively, whether the strong unit-specific dialects that we have described (Weilgart and Whitehead 1997) are entirely explained by clan membership. An examination of this question is particularly complicated by the fusion of units into groups, which were what we recorded. However, a conservative test compares whether coda repertoires from pairs of days on which the same unit was recorded were more similar than those from pairs of days on which the same clan was recorded, but there was no unit in common. For the "Regular" clan the result was marginally statistically significant ($P = .061$, Mantel test), while for the "+1" clan there was no significant sign of unit-specific dialects within the clan ($P = .734$, Mantel test; L. Rendell, unpublished data), although in the latter case the sample size was very small.

The Geography and Genealogy of Acoustic Clans

We have extended the acoustic clan analysis to include recordings from across the South Pacific and Caribbean. The results are displayed in figures 7.3 and 7.4. Although there are a couple of ambiguities, the general picture reveals about four or five clans across the whole South Pacific Ocean. The clans generally have wide geographic ranges and overlap with one another, although each appears to be based in one part of the South Pacific. As the South Pacific contains on the order of 90,000 sperm whales (perhaps

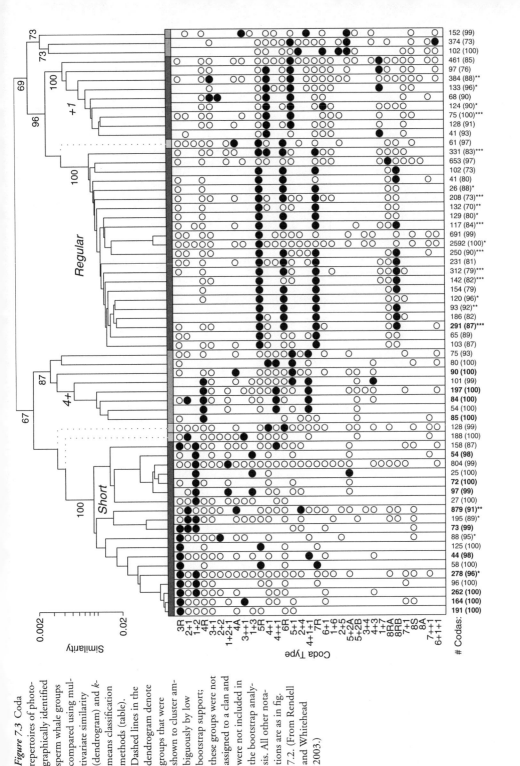

Figure 7.3 Coda repertoires of photographically identified sperm whale groups compared using multivariate similarity (dendrogram) and *k*-means classification methods (table). Dashed lines in the dendrogram denote groups that were shown to cluster ambiguously by low bootstrap support; these groups were not assigned to a clan and were not included in the bootstrap analysis. All other notations are as in fig. 7.2. (From Rendell and Whitehead 2003.)

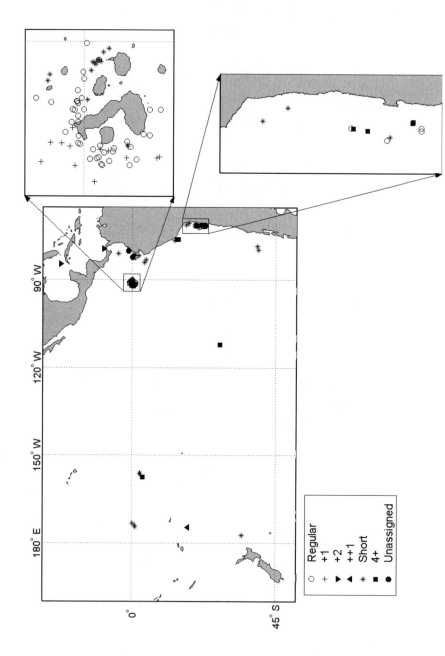

Figure 7.4 Map showing locations where different acoustic clans were recorded in the South Pacific and the Caribbean. (From Rendell and White-head 2003.)

one-fourth of a world population of 360,000; see section 4.2), each clan is likely to have many thousands of members. In box 7.2, I summarize the repertoire and distribution of each of the clans.

Box 7.2 **Sperm Whale Clans of the South Pacific**

"Regular" Clan

Units in the "Regular" clan preferentially make codas formed from four to eight regularly spaced clicks. The "Regular" clan was the dominant clan off the Galápagos Islands, containing units A, B, C, H, I, J, K, L, O, P, and S (unit E also seemed to have this repertoire, but only nine codas have been recorded from this unit). Members of this clan are also found off northern Chile and mainland Ecuador. This clan's principal habitat seems to be the eastern tropical Pacific. Off the Galápagos Islands, groups containing units of this clan were often found close to the islands, usually traveled tortuous paths, and dived relatively synchronously.

"+1" Clan

The "+1" clan, which characteristically adds a pause before the last click in a coda, has been recorded only in the waters off the Galápagos Islands and mainland Ecuador (units D, F, G, N, and Q). In neither area were its units numerically dominant. It may be a small clan, or it may have an important part of its range in an unsampled part of the Pacific—perhaps the central equatorial Pacific, which traditionally held a large number of sperm whales (Townsend 1935) but has not yet yielded any coda recordings. A mid-Pacific origin for this clan would fit with the "+1" units seeming to process through the waters off the Galápagos Islands between 1987 and 1996 as opportunities opened up farther east. Off the Galápagos Islands, groups containing units of this clan were usually found away from land, tended to have direct movements, and dived fairly asynchronously.

"Short" Clan

The units of the "Short" clan preferentially make codas containing only three or four clicks. There may be some substructure within this clan, with some units preferring 3R and others 1+2 among the three-click codas (see fig. 7.3). The "Short" clan is the most widespread, with all but two of the recorded western Pacific groups being from this clan,

Box 7.2 continued

but it is also found in the eastern Pacific off the west coast of South America (see fig. 7.4). Unit T, the only unit identified off the Galápagos Islands in the late 1990s after all other females and immatures had left, was of the "Short" clan.

"4+" Clan

Codas from the "4+" clan often start with four regularly spaced clicks. Recorded principally off Chile and southern Peru, this clan has been found off Easter Island in the central Pacific and as far west as Christmas Island. This clan seems to be based in the southeastern Pacific.

Added to the geographic analysis (figs. 7.3 and 7.4) are two repertoires recorded from the Caribbean. These repertoires, dominated by long codas, cluster with each other and separately from the Pacific clans, indicating that other oceans may possess separate clans.

Off the Galápagos Islands we have recorded three clans, at very different rates: predominantly the "Regular" clan, but a fair number of "+1" units, and just one known "Short" clan unit (T). Particularly interesting is the sequence in which these clans have appeared off the Galápagos Islands during our studies (fig. 7.5). In 1985, our first year, only "Regular" units were identified, although codas from the "+1" clan were recorded on two consecutive days on which unit membership could not be determined. The "+1"s started to appear in numbers in 1987 and formed a substantial part on the population through the mid-1990s. But female and immature sperms were leaving the Galápagos area throughout this time (see box 4.2). In the late 1990s, only one unit was identified: unit T from the "Short" clan. But unit T is not an anomaly: similar "Short" clan repertoires were recorded off the Galápagos Islands in 1985 and 1995, as well as right across the South Pacific (see figs. 7.3 and 7.4 and box 7.1). The impression from the Galápagos Islands, then, is of a disorderly sequence of clans moving through the islands, first the "Regular" clan, then the "+1" clan, and finally the "Short" clan. This pattern is consistent with the picture of sperm whales moving generally eastward across the southeastern Pacific in the 1980s and 1990s to fill the void in the waters of the Humboldt Current created by intense whaling based in Peru (see box 4.2). The apparently greater diversity of clans in the eastern South Pacific than in the west (fig. 7.4), also fits with this one-way movement hypothesis.

So, sperm whale populations, or at least the populations of females and

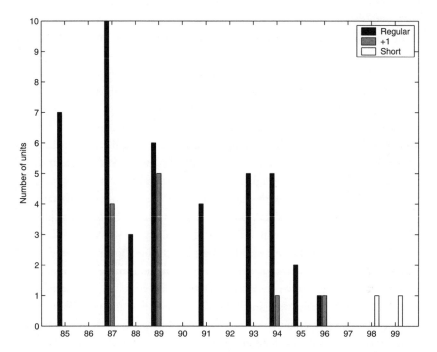

Figure 7.5 Number of units from the three acoustic clans identified off the Galápagos Islands each year.

immatures, seem well structured by these clans—in fact, more clearly structured than they are by genes (see section 4.1). But there are many reasons for asking how genetically based the clans are. The first is the basic issue of what determines an individual's coda repertoire: Is it genes or social learning (culture)? It would require a very strange matrilineal genetic system to produce the vocal dialect patterns indicated by the coda results, but perhaps it is possible (see Janik 2001 and section 8.11). However, such a scenario would be almost completely ruled out if the clans had overlapping matrilineal mitochondrial haplotypes. And they do. We have genetic data from fifteen animals from four groups of the "4+" clan, sixty-seven animals from ten groups of the "Short" clan, and nineteen animals from one group of the "Regular" clan. Here are their haplotypes:*

4+: 1 #1, 7 #3, 5 #5, 1 #6, 1 #7
Short: 18 #1, 23 #2, 17 #3, 2 #6, 1 #7, 1 #8, 1 #11, 1 #12, 3 #13
Regular: 18 #1, 1 #13

* The haplotype delineations are as in Table 6.8. "7 #3" means seven animals with haplotype #3, etc.

Several haplotypes (#1, #3, #6, #7, and #13) are found in more than one clan, and different haplotypes exist in the same clan. So much for genetic causation of coda dialect.

Genetic data can also tell us about the formation and evolution of the clans. Unfortunately, because of genetic similarities within groups (see section 6.6), the sample size is basically the number of groups (4, 10, and 1 in the different clans) for which genetic data are available. Permutation tests (in which the clan memberships of the groups were scrambled) indicated no significant difference between clans in the distribution of mitochondrial haplotypes ($P = .59$), although this test has little power. If this nonsignificant pattern is confirmed with a larger sample size, it may indicate that individuals quite frequently switch acoustic clans. One of the animals that was observed to transfer between known units, #236, moved from "Regular" unit C to "+1" unit D between 1985 and 1987 (see fig. 6.15), so perhaps inter-clan transfers are not rare. Alternatively, larger sample sizes may show genetic differences between clans. Such differences would explain the mtDNA-dialect correlation indicated in our earlier analysis (Whitehead et al. 1998). Future analyses of mtDNA may be able to give us an estimate of the rate at which females change clans.

The distribution of nuclear DNA markers, which are inherited from both parents, will be determined both by how often females move between clans and by whether males mate within or outside their natal clan. My analysis of the distribution of five nuclear microsatellites collected from 111 individuals in twelve groups belonging to three clans (K. Richard, unpublished data) showed no sign of differences between clans (permutation test, $P = .931$). Although a more powerful analysis using a larger sample size might produce a different result, this finding indicates either that females transfer between clans reasonably frequently or that males sometimes mate with females who are not of their natal clan.

Clans: The Broader Context

The scenario that emerges for sperm whales, of clans with quite different vocal repertoires using the same area, and sometimes mixing, is quite similar to the current view of the resident, fish-eating killer whales near Vancouver Island (Ford 1991). But there are differences. The most obvious are in the types of vocalizations (patterns of clicks versus pulsed calls) and the scale: whereas clans of killer whales contain about a hundred animals (Bigg et al. 1990) and seem to have ranges of the order of 1,000 km, sperm whale clans probably number in at least the thousands and may stretch 10,000 km, right across the

South Pacific. Also, the "pods" within killer whale clans have clearly distinctive dialects, but this is not yet well established for the sperm whale social units, which are probably the most equivalent social structure.

The sperm whale coda dialect is almost certainly a form of culture, and it seems to have a great deal of stability. This apparent stability is especially interesting because stable cultures are very rare in nonhuman animals (Boyd and Richerson 1996). However, theoretical studies show that stability is a prerequisite if culture is to have much effect on genetic evolution (Laland et al. 1996), and stable communication repertoires specific to particular social groups can promote other group-specific cultural traits.

7.3 Nonvocal Cultural Traits of Sperm Whales

Candidate Nonvocal Cultural Traits

What might these other group-specific cultural traits be? We can see so little of sperm whale behavior that this is difficult to imagine. But, given what we do know of sperm whale behavior, and extrapolating from information on killer whales, humans, and other terrestrial animals with important cultures, especially chimpanzees (Whiten et al. 1999), some good candidates might be foraging techniques, food processing methods, movement patterns, use of habitat, defensive tactics against predators, and social conventions. We have systematic data on sperm whales in only a few of these areas, and these data are usually few and imprecise. However, there do seem to be some interesting differences between units, and especially between clans.

Unit-Specific Nonvocal Cultural Traits

Using our Galápagos data, I have looked for differences between units* in eighteen visually observable measures of group behavior. These measures concerned speeds, aerial behavior, cluster size when foraging, heading consistency, diving synchrony, foraging formations, temporal aspects of socializing, and movement measures, which were calculated for each day spent watching a group (Whitehead 1999c). Group/unit identity accounted for over 50% of the variance in two measures: the total distance traveled during daylight and the straight-line displacement between 06:00 and 18:00 posi-

---

* Actually, this study was carried out at the level of the group, but only groups that seemed to be numerically dominated by one unit were considered.

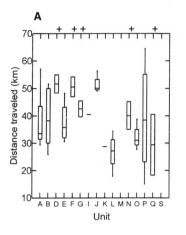
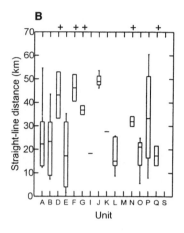

Figure 7.6 Variation in movement measures within and between social units studied off the Galápagos Islands between 1985 and 1995: (A) distance traveled through the water in 12 hr and (B) straight-line distance between positions 12 hr apart. Units marked "+" are from the "+1" clan; all others with data are from the "Regular" clan.

tions (after correcting for defecation rate, which is closely related to this measure; see section 3.4). The first of these effects was statistically significant, even given the low power of the tests, whose significance levels were corrected for the multiple comparisons carried out (Whitehead 1999c). For the units listed in table 6.5, these two measures are shown in figure 7.6. For instance, unit J consistently moved much farther, both through the water and in terms of displacement, than unit L, even given differences in feeding success, as illustrated in figure 7.7. Such differences are likely to be cultural, but this conclusion must be tentative, as they could be caused by unaccounted environmental variation or by demographic differences (e.g., maybe J mainly contained youngsters and L elderly whales). Cultural differences between units in the other measures of visually observable behavior, which showed less strong effects in the original analysis (Whitehead 1999c), must remain even less substantiated. Additionally, the variation among units in movement patterns could actually be the result of differences between higher-level entities—the clans, as discussed below.

Clan-Specific Nonvocal Cultural Traits

In both 1987 and 1989, units from the "Regular" and "+1" clans were frequently identified off the Galápagos Islands (see fig. 7.5). I have looked to see whether they used the study area any differently, or behaved any differently.

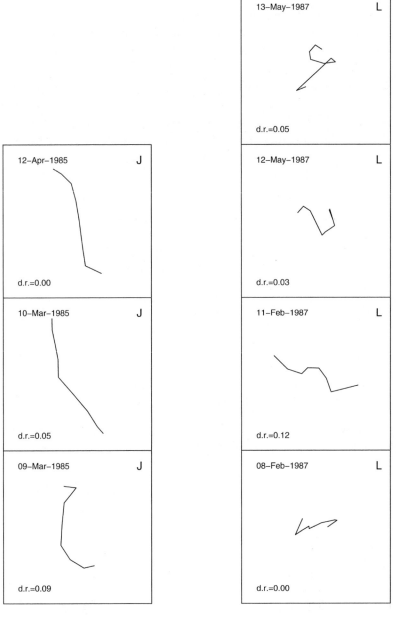

Figure 7.7 Twenty-four-hour tracks of units J (left) and L (right) off the Galápagos Islands ("d.r." is defecation rate for that day, in feces observed per fluke-up; north is up the page; resolution = 3 hr; boxes are 1° × 1° or 111 km square).

Shown in figure 7.8 are the dates on which units in the different clans were identified in the two study years. Units from the different clans seemed to enter the study area in clumps, some "Regular" units and then some "+1" units, but there was no obvious tendency for the clans to consistently use the Galápagos study area in different seasons. Although units from the two clans used overlapping parts of the study area, the "Regular" units were generally more inshore and the "+1" units farther from land, especially to the north and west of the islands; this pattern was consistent in the two years (fig. 7.9).

I also looked for behavioral differences between units of the "Regular" and "+1" clans observed off the Galápagos Islands in 1987 and 1989, using

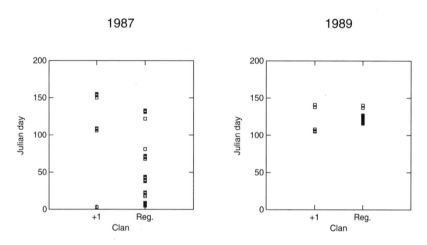

Figure 7.8 Identifications of units of the "+1" and "Regular" clans in the Galápagos Islands study area by Julian day (1 January = 1) in 1987 and 1989.

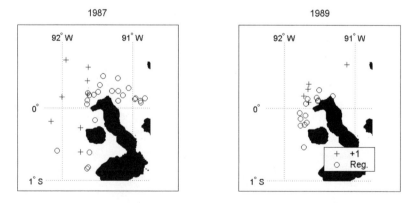

Figure 7.9 Midday positions of units of the "+1" and "Regular" clans off the Galápagos Islands for days in 1987 and 1989 on which units of either clan were identified.

the same eighteen measures as in the search for unit-specific nonvocal cultural traits. Two measures showed clear and consistent (between years) differences between the two clans (fig. 7.10): consistency of movement among 3 hr periods over 12 hr and straight-line distance moved over 12 hr. In both 1987 and 1989, the "+1" groups moved consistently farther (a mean of 35.1 km versus 19.3 km displaced over 12 hr) and straighter than the "Regular" groups. So, in Galápagos waters, the "+1" groups seemed much more "on the move" in both 1987 and 1989. Contrasting 24 hr tracks from units of the two clans are shown in figure 7.11. The difference is also apparent in figure 7.6, with the "+1" units moving generally farther than the "Regular" units, although there are a couple of exceptions: "Regular" unit J moved much more like a directed "+1" unit than a wiggly "Regular" unit, and unit Q showed the reverse pattern.

The division of a population into phenotypes with different movement patterns is not unusual, some examples being caribou (*Rangifer tarandus*) and Galápagos giant tortoises (*Geochelone elephantopus*) (Swingland et al. 1989; Bergman et al. 2000). However, an important difference in the case of sperm whales is that the division of the population was initially made using the cultural trait of coda vocalizations, rather than the movement differences being used a priori to define phenotypes.

One other behavioral measure, diving synchrony, showed a significant difference between clans (fig. 7.10), with the "Regular" clan groups generally diving much more synchronously, even when group size and the presence of a calf were accounted for (see Whitehead 1996a, 1999c). However, this result must be treated cautiously, as only one year's data (for 1987) were available, and there is a possibility of false positives arising from the multiple tests for behavioral differences between clans.

Movement patterns and diving synchrony are collective behaviors, and so are unlikely to be the result of individual learning. Given the improbability of clan-specific behavior being genetically caused (see section 7.2) or the result of demography (as clans contain huge segments of the sperm whale population), they are almost certainly forms of culture. However, the measured attributes could be subsidiary consequences of some other fundamental cultural difference between the clans, rather than cultural differences in their own right. For instance, if the clans specialized on different types of prey,* this could lead to differential habitat use, movement patterns, and synchrony of dives.

---

* Unfortunately, we have only four squid beaks from defecations of members of the "+1" clan off the Galápagos Islands in 1987 and four in 1989, so we can say little about differences in diet among clans.

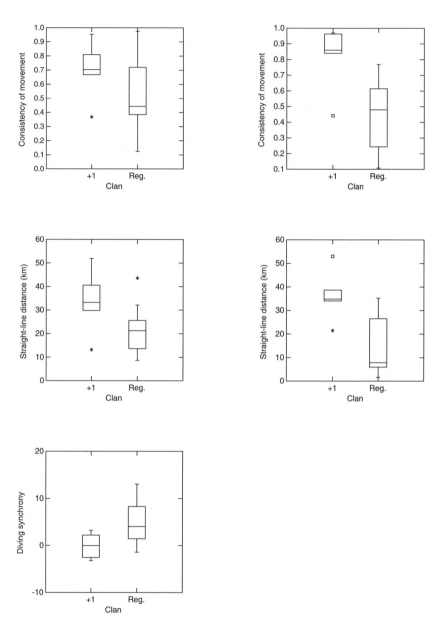

Figure 7.10 Behavioral differences between "+1" and "Regular" clans off the Galápagos Islands during 1987 and 1989: consistency of movement direction among 3 hr intervals during 12 hr of daylight (*t* test, *P* = .005); straight-line distance between positions of whales at 06:00 and 18:00 on any day (*t* test, *P* = .005); and diving synchrony, measured as the difference between the maximum time that no whales were visible at the surface during any hour and that predicted from a model including the number of animals in the group and the presence/absence of calves, averaged over days with at least 5 such hours (*t* test; *P* = .015). (See Whitehead 1999c for more details of methods.)

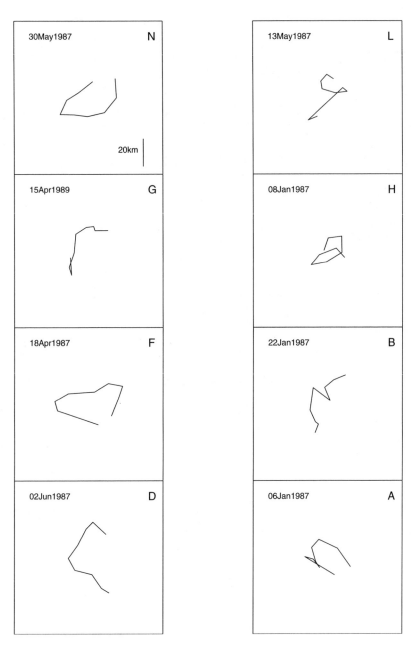

Figure 7.11 Twenty-four-hour tracks of units from the "+1" clan (left) and the "Regular" clan (right) off the Galápagos Islands (north is up the page; resolution = 3 hr; boxes are 1° × 1° or 111 km square). All tracks are from 1987 except G, which was not identified in 1987.

Movement patterns and synchrony of dives are particularly interesting measures of clan-specific differences, as movement behavior affects use of habitat and, potentially, feeding success (see section 3.7), while diving synchrony seems to be important in protecting the young (see section 6.9). Thus, these cultural differences, probably unlike the specifics of coda dialects, have considerable potential to directly affect fitness. Have we any evidence of such effects?

Fitness Consequences of Cultural Differences between Clans

The two most likely channels through which culture might affect the fitness of sperm whales are feeding success and predation. Although we cannot look at either of these measures directly, there are proxies: defecation rate (see section 2.6) and marks and scars (see section 2.8).

In figure 7.12, I show the defecation rates measured for the "Regular" and "+1" clans off the Galápagos Islands in 1987 and 1989. 1987 was a warm-water El Niño year of poor feeding success (see fig. 2.18), whereas in 1989 the waters were cool and feeding was good. The contrast between the defecation rates of the clans is remarkable. Under the poor conditions of 1987, the straight-moving "+1" units did rather better, but with the plenty of 1989, the wiggly "Regular" units were defecating, and presumably feeding, substantially more frequently.

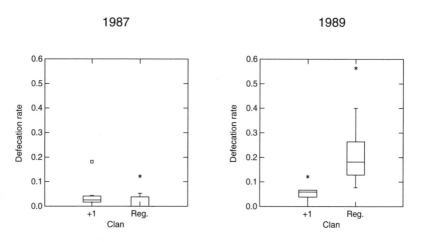

Figure 7.12 Distribution of defecation rates, per observed fluke-up, during days on which units of the "Regular" and "+1" clan were identified off the Galápagos Islands in 1987 and 1989.

In contrast, there were no significant differences in the numbers of marks and scars on the flukes of the members of the two clans. They possessed similar mean numbers of small nicks (2.8 for "Regular" clan; 3.3 for "+1" clan), distinct nicks (1.3; 1.1), "scallops" (1.3; 1.0), "waves" (3.4; 3.9), missing portions (0.2; 0.2), and toothmark scars (0.2; 0.2),* and in no case were the means significantly different (t tests, $P > .05$).

Fitness is ultimately about surviving offspring. Therefore, if the culturally determined strategies of clans are important for fitness, their effects might be manifested in differential reproductive rates. Although units of "+1" clans contained generally more calves than those of "Regular" clans (0.059 and 0.019 calves/adult, respectively, in 1987 and 1989), this difference was not statistically significant (permutation test, $P = .12$).

Clans: A Question of Identity?

What do the clans mean to sperm whales? I suspect that they may largely be about identity (see McGrew 2003), so that for a sperm whale, membership in a clan has a connotation comparable to that of nationality in humans. So, being a "+1" sperm whale, like being a Slovenian, means not only being a member of a group with distinctive ways of communicating and behaving, but also knowing that one is a member of that group, which is different from other groups. Like humans in multicultural settings, sperm whales seem to show affinity for their own clan. Clan identity may be an important part of how sperm whales see themselves, if they see themselves. Group identity has benefits for an animal: a well-proven way of behaving and a pool of companions who behave similarly who can be used as models and colleagues in cooperative endeavors. Clan identity may also have disadvantages, however—perhaps especially a reluctance to adopt behavior patterns outside the clan's repertoire. Thus, in the El Niño year of 1987, the "Regular" clan units continued their wiggly ways (see fig. 7.10), while intuitively, and apparently in practice, straight movement would seem to have been more profitable (see fig. 7.12). In contrast, in the cool, productive year of 1989, it was the "+1" clan that suffered from being "culturally narrowminded."[†] If global warming systematically changes the environment of the eastern tropical Pacific, perhaps the clans will be differentially affected.

---

* Scar types are defined and illustrated in Whitehead (1990a) and Dufault and Whitehead (1995a).
[†] Thanks to Wolfgang Wickler for this term.

7.4   Low Mitochondrial DNA Diversity and Culture: Cultural Hitchhiking

During the 1990s, molecular geneticists entered into the world of sperm whale science as they did into the studies of so many other species. Their research was principally aimed at two areas: the geographic and the social structures of sperm whale populations (see sections 4.1 and 6.6). Initially, their interest focused on just a few of the most tractable areas of the genome, especially the maternally inherited "D-loop" or "control region" of the mitochondrial DNA. The D-loop was chosen for several technical reasons, one of which was that it is believed to be selectively neutral. Neutral sites evolve relatively rapidly, and so tend to be diverse and informative in studies of relationships among conspecifics—the basic information needed to determine social and population structures.

However, the first results of studies of the sperm whale D-loop startled the geneticists. Dillon (1996, 96) found only twelve different D-loop sequences, or haplotypes, among 182 samples collected from around the world, while Lyrholm et al. (1996) identified just thirteen haplotypes in a similar, but independent, worldwide study using 37 samples—and most of these haplotypes were common to the two studies. Part of this lack of diversity could be ascribed to an apparently low mitochondrial mutation rate for large cetaceans (Martin and Palumbi 1993). But the genetic (nucleotide) diversity of the sperm whale D-loop, 0.0028 in Dillon's (1996, 101) study and 0.0040 in Lyrholm and Gyllensten's (1998), was less than 20% of that of the similar-sized humpback whale (*Megaptera novaeangliae*): 0.0260 (Baker et al. 1993). Why?

Lyrholm and Gyllensten (1998) considered several possible explanations for the curiously low mtDNA diversity of the sperm whale, but only from the sperm whale's standpoint. However, when searching for such explanations, it often helps to take a broader comparative perspective. In figure 7.13, I show estimates of the D-loop nucleotide diversity of several cetacean species and subspecies. More numerous and widespread populations are expected to have generally greater diversity. Thus, in figure 7.13, D-loop nucleotide diversity is plotted against the range of latitudes a species occupies, which should capture much of the variation in genetic diversity due to population size and geographic spread.* Overall, D-loop diversity does increase

---

* In an earlier presentation of some of these data I used population size as a covariate (Whitehead 1998), but population size estimates are very approximate for most cetacean species and have been heavily affected by human activities, so I think latitudinal range is probably a better measure of the expected variation in genetic diversity due to demographic factors.

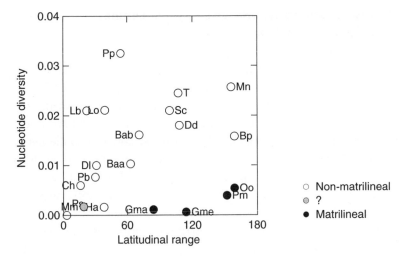

Figure 7.13 Nucleotide diversity in the D-loop of the mtDNA for species and subspecies of cetaceans: minke whales (*Balaenoptera acutorostrata acutorostrata,* Baa; *B. acutorostrata bonaerensis,* Bab), humpback whale (*Megaptera novaeangliae,* Mn), beluga whale (*Delphinapterus leucas,* Dl), bottlenose dolphin (*Tursiops* spp., T), northern right whale dolphin (*Lissodelphis borealis,* Lb), common dolphins (*Delphinus delphis,* Dd), harbor porpoise (*Phocoena phocoena,* Pp), Hector's dolphin (*Cephalorhynchus hectori,* Ch), dusky dolphin (*Lagenorhynchus obliquidens,* Lo), striped dolphin (*Stenella coeruleoalba,* Sc), narwhal (*Monodon monoceros,* Mm), sperm whale (*Physeter macrocephalus,* Pm), killer whale (*Orcinus orca,* Oo), pilot whales (*Globicephala macrorhynchus,* Gma; *G. melas,* Gme), vaquita (*Phocoena sinus,* Ps), fin whale (*Balaenoptera physalus,* Bp), franciscana (*Pontoporia blainvillei,* Pb), and northern bottlenose whale (*Hyperoodon ampullatus,* Ha). The four species known to have at least partially matrilineal social systems are marked, as is the narwhal for which matrilineality has been suggested but not shown. (From Whitehead 2003.)

with latitudinal range among cetaceans, from the homogeneous D-loops of the vaquitas (*Phocoena sinus*), which are restricted to the northern part of the Gulf of California, to the widespread humpback whale, whose D-loop is one of the most diverse among cetaceans. From this perspective, the sperm whale has dramatically low diversity given its wide habitat. But there are three other species with substantial geographic ranges and abnormally low D-loop diversity: the killer whale and the two pilot whale species (*Globicephala melas, G. macrorhynchus*). These are the only species, of those whose D-loop diversity is plotted in figure 7.13, that have a known matrilineal social system, in the sense of most females remaining grouped with their mothers throughout their lives.*

Correlation does not show causation, but the strong interspecific relationship between matrilineal social systems and low diversity of matrilineally

---

* The narwhal (*Monodon monoceros*) may or may not have a matrilineal social system.

transmitted genes certainly suggests that the two are related. Explanations for the low D-loop diversity in these four species that do not involve matrilineal social systems, such as "genetic bottlenecks" due to low population sizes or selection within the mitochondrial genome, do not account for this correlation and so seem unsatisfactory. While all four matrilineal species have been exploited, they have not been brought to the levels of just hundreds of animals for several generations that would be necessary to reduce mtDNA diversity (Whitehead 1998). Such bottlenecks could have occurred in the time before whaling (e.g., Lyrholm et al. 1996), but why just in the four species of matrilineal whales, which are currently particularly widespread and reasonably numerous (Whitehead 1999a)? The other standard explanation for low genetic diversity is selection. Janik (2001) has invoked this explanation for the matrilineal whales, suggesting that the energy-producing mitochondria are under increased selection in deep-diving animals because of their physiological challenges, and that this selection results in lower diversity. This sounds reasonable for the sperm whale, and possibly for the pilot whales, which reasonably regularly dive a few hundred meters beneath the surface (e.g., Mate 1989), but not for the killer whale, which is basically a creature of the upper water layers (Baird 2000).

More convincing are demographic explanations that incorporate the animals' social structure, so that the low diversity of the matrilineally transmitted mtDNA is seen as a consequence of their matrilineal way of life. There have been several attempts at such explanations (Siemann 1994; Amos 1999; Tiedemann and Milinkovitch 1999). These models vary in content, but have the general characteristic of assuming that demographic processes are occurring, to at least some extent, at the level of the matrilineal group. Thus the effective population size is essentially reduced to the number of groups— a process that, on the surface, appears at least feasible. However, in each case, the models either make assumptions or result in predictions that are not consistent with what is known of the species' biology (Whitehead 1998, 1999a). So they are not satisfactory explanations for the low mtDNA diversity of the matrilineal whales.

My suggested explanation for the low genetic diversity of the matrilineal whales is cultural hitchhiking (Whitehead 1998). This explanation depends on the matrilineal social systems of the species, and it is supported by the discovery of strong matrilineally based cultures for killer whales (Boran and Heimlich 1999; Rendell and Whitehead 2001b) and indications of them for sperm whales (see above). I suggest that selection on matrilineally transmitted cultural traits has reduced the diversity of the similarly transmitted mtDNA. The idea is simply that if a set of females possess a cultural trait that gives them a selective advantage, and pass this trait on to their female descendants, then their matrilines, and the trait, will spread. But, as mtDNA is being transmitted in parallel with the cultural trait, the mtDNA haplotypes of the animals who originated the trait will also spread, eventually dominating the genetic structure of the population and lowering diversity. I call this process "cultural hitchhiking," as it is analogous to genetic hitchhiking, in which the diversity of a neutral gene is reduced by selection on a linked gene (Maynard Smith and Haigh 1974). In computer models, it works: cultural hitchhiking reduces diversity, both in the case of a single advantageous trait (Whitehead 1998) and in the probably more realistic scenario in which there are multiple small cultural innovations (some possibly with negative effects on fitness) that build on one another (Whitehead 2003). Thus, cultural hitchhiking is a feasible explanation for the low mtDNA diversity of the sperm whale and the three other matrilineal species.

However, there have been objections to this explanation. In addition to the alternative explanations for low mtDNA diversity, discussed above, these criticisms focus on three arguments: that the whales' social systems may not be sufficiently stable; that the cultural traits may not be sufficiently stable; and that the whales' cultural traits may not be important enough, in fitness terms, to drive such a process. Here I answer each of these classes of objections.

> *Unstable social systems:* In the original formulation of cultural hitchhiking, in which a single advantageous innovation is transmitted matrilineally, if more than a few females do not copy their mother's cultural phenotype, then the process breaks down (Whitehead 1998). If animals move between matrilineal social groups, either mtDNA genes or cultural traits may be transferred, and so the tracks of mtDNA and cultural transmission become twisted. Consequently, results showing imperfect matrilineality in sperm whale social units (see section 6.6) and transfers between them (see section 6.5) have been interpreted as counterindicating cultural hitchhiking (Mesnick et al. 1999, 2003). However, this interpretation assumed that units were the level of social structure at which cultural selection was taking place—a reasonable assumption while these units were the only fairly stable social structures known in sperm whales. With the discovery of clans (see section 7.2), which may be stable enough to maintain different mtDNA haplotype distributions, cultural hitchhiking becomes more feasible: it seems quite reasonable, although it is far from proven, that sufficient female sperm whales remain within their mother's clan to allow reduction in mtDNA diversity

through cultural hitchhiking in the multiple innovations case (Whitehead 2003).

Unstable cultural traits: Deecke et al. (2000) studied the stability of the vocal dialects of the "resident" killer whales off Vancouver Island, finding that some aspects of their dialects evolved over time scales of decades. They then suggested that if the same were true for other aspects of killer whale cultures, those cultures would not be able to drive cultural hitchhiking. However, whereas my first model of cultural hitchhiking assumed stable cultures (Whitehead 1998), the more recent one (Whitehead 2003, unpublished data) allows cultures to evolve and still reduce genetic diversity.

Which cultural traits? What cultural traits might have a sufficient effect on fitness to drive cultural hitchhiking? The cultural trait of sperm whales for which we have the best information is the coda dialect (see section 7.2), which is quite likely to be selectively neutral as long as the individual adopts and understands the dialect of its clan or, perhaps, unit. Selectively important cultural traits are harder to study, but we now have good indications that the clans have significantly different movement strategies, which should affect feeding success, and possibly diving synchronies, which should affect babysitting (see section 7.3). Thus, we have suitable candidate cultural traits that could drive cultural hitchhiking.

Cultural hitchhiking as a driver of low mtDNA diversity in sperm whales remains a hypothesis. However, since I originally proposed the hypothesis in late 1998, three developments have strengthened it. My new model, which allows evolving cultures and permits higher levels of non-matrilineal cultural transmission, considerably widens the conditions over which cultural hitchhiking might occur (Whitehead 2003, unpublished data). The discovery of clans with distinctive and functionally important behavioral patterns suggests another, and probably more plausible, level of stable social organization at which such processes might operate. The lack of statistically significant differences in mitochondrial haplotype distributions between clans argues against this, but the sample sizes are small, and it is quite likely that real differences exist at the level needed for cultural hitchhiking to operate. Furthermore, one of the predictions of the cultural hitchhiking hypothesis seems to be met. In matrilineal social systems, if males mate across cultural boundaries, then, while maternally inherited mtDNA diversity may be reduced, biparentally inherited nuclear DNA diversity will not be affected, as these genes will not remain correlated with cultural variants. In contrast, a population bottleneck should reduce variation in both mtDNA and nuclear genes, although to different degrees (e.g., Amos 1996). Thus, Lyrholm et al.'s (1999) finding of substantial variation in the biparentally inherited nuclear microsatellites of sperm whales supports the cultural hitchhiking hypothesis.

I am not convinced that cultural hitchhiking has been the process that caused the low mtDNA diversity of sperm whales, but these recent developments have strengthened my feeling that it is the most probable of the competing explanations. If this is indeed what happened, it represents a case of gene-culture coevolution, a process so far known only from humans (see de Waal 2000, 270–71).

7.5 Summary

1. Certain sets of sperm whale social units possess very similar coda repertoires, allowing the units to be classified into "clans." These clans almost certainly represent cultural variants. About four or five clans have been identified across the South Pacific, with each clan generally spanning many thousands of kilometers and probably containing tens of thousands of animals.

2. The sperm whales that use the waters off the Galápagos Islands are principally from two acoustic clans, the "Regular" clan (which favors regularly spaced clicks) and the "+1" clan (which favors codas with a pause before the final click).

3. Only one brief association between units of different clans has been identified, although members of different clans may use the same area on different days.

4. Acoustic clans are not perfectly matrilineal, and there is one record of an individual who switched clans.

5. Off the Galápagos Islands, the two principal clans showed consistently and significantly different habitat use and movement patterns. In 1987, an El Niño year, the "+1" clan had higher feeding success, but with cooler weather in 1989, it was significantly outperformed by the "Regular" clan. The data also suggest that units of the two clans have substantially different diving synchrony.

6. Selection on matrilineally transmitted cultural traits could have been the cause of the low diversity in sperm whale mtDNA, which is maternally inherited, through a process I call cultural hitchhiking.

8   Social and Cultural Evolution in the Ocean

8.1   Evolution to Extremes

Evolutionarily distant from any other species, the sperm whale is a beguiling amalgam of widely shared marine mammal attributes with a suite of unusual traits, which range from the spermaceti organ to the acoustic clan. In this chapter I try to put these strange, and sometimes extreme, features in a functional and evolutionary context. How did the sperm whale make its way into such a remote area of biological trait space? How does it stay out there? What are the selective pressures that maintain its large size, sexual dimorphism, unique nose, socially homogeneous groups, and communication based only on clicks?

I will start by considering the two most remarkable organs of the sperm whale, the nose and the brain. The sperm whale possesses the largest representatives of either of these organs on Earth, and both are closely tied to its ecology and behavior. The spermaceti organ and the brain point to issues that have been important in sperm whale evolution, but they need to be considered in a broader comparative context.

In making comparisons with other species, I take two approaches. First I discuss those species that appear to share the sperm whale's habitat most closely: the deep-diving northern bottlenose whale (*Hyperoodon ampullatus*) and the elephant seals (*Mirounga* spp.). If habitat were the principal driver of behavior and social structure, we might expect parallels between the social structures of these species (especially the northern bottlenose whale; elephant seals, breeding on land, face a different set of constraints) and those of the sperms, but there are remarkably few such parallels. Then I try another approach, looking at some animals that share much of the sperm whale's behavior and social structure: the elephants (Proboscidae). The convergences between these animals and sperm whales are particularly illuminating, and lead me to propose a scenario for this "colossal convergence." *

---

* The parallels between sperm whales and elephants were discussed by Weilgart et al. (1996) in *American Scientist.* It was an editor of this magazine who coined the term "colossal convergence."

This discussion is followed by a more detailed examination of the function and evolution of the contrasting social behavior of the two sperm whale sexes, and some thoughts on the broader issues of social and cultural evolution in the deep and open ocean.

8.2   The Function and Evolution of a Big Nose

> *These proportional comparisons reveal that the most extreme hypertrophy in the sperm whale nose is in the lipid spermaceti organ. The index of hypertrophy is so far beyond any "normal" scaling component that it cannot be explained with simple allometric rules: it must indicate functional significance.*
> —Cranford 1999

Much of a sperm whale's body is its nose (fig. 8.1, see also fig. 1.6 and section 1.4). This nose, which contains the spermaceti organ and related struc-

Figure 8.1 The head of a large (17 m) male sperm whale, showing the protuberance of the spermaceti organ. (Photograph courtesy James G. Mead, Division of Mammals, National Museum of Natural History, Smithsonian Institution.)

tures, is large—proportionally sixteen times larger than the homologous structure in the common dolphin (*Delphinus delphis*) for females and twenty-two times larger for males (Cranford 1999). This enormous evolutionary development must have costs to the animal carrying it. Well muscled, it must be an energy drain, and, despite some superficial streamlining, it must increase drag as the sperm whale swims. The spermaceti organ has also led to changes in basal structures, such as the skull (Cranford 1999), that may have led to trade-offs in the efficiency of other systems. Considering these costs, the selective pressures favoring the evolution of the spermaceti organ must have been intense. What were they?

Despite some contrary opinions (see boxes 5.1 and 8.1), it seems that the sperm whale's nose functions principally as a sonar (see section 1.4)—an extremely powerful one, as shown by important recent studies (Møhl et al. 2000; Madsen et al., in press). In an environment where sperm whales compete with other sperm whales in a demanding scramble for mesopelagic squid (see section 2.7), she who finds the most squid will prosper and multiply. Any attribute that allows its bearer to detect squid more effectively will be favored.

Box 8.1 The Spermaceti Organ: A Buoyancy Control Device or a Battering Ram?

A Buoyancy Control Device?

M. R. Clarke (1970, 1978) noted that variation in the temperature of the spermaceti oil would change its density, and thus the buoyancy of the animal. Such changes could allow a sperm whale to ascend and descend, or remain motionless at depth, more easily. Hence he proposed that the spermaceti organ has evolved to assist in deep diving. However, a number of authors have found problems with this theory. These problems, as summarized by Cranford (1999), include the modest change in buoyancy achieved; the absence of anatomical structures expected in a heat exchange organ; the need for physiological exertion at depth, which deep-diving animals generally avoid; the apparently active swimming of sperm whales at depth (see section 5.4); and the difficulty of evolving an organ with this principal function, as the spermaceti organ would not be an effective buoyancy regulator until it became huge.

Box 8.1 continued

A Battering Ram?

Carrier et al. (2002) have revived the theory that the spermaceti organ evolved as a battering ram. They hypothesize that "the greatly enlarged and derived melon of sperm whales, the spermaceti organ, evolved as a battering ram to injure an opponent" in male-male competition.

There are a number of problems with this hypothesis as it stands. Specifically, consider the female sperm whale. She has an enormous spermaceti organ, constituting about 20% of her body and containing many highly derived structures. If the spermaceti organ evolved as a battering ram for use in male-male competition, natural selection was very remiss in giving females one, too, as there is no good evidence that they batter anybody or anything (see section 6.7). Any female without this monstrous structure would do better, and natural selection would arrange that it not be present, or present only vestigially, in females.

However, while the female spermaceti organ is huge, the male's is relatively huger. Could this excess size be a result of selection for its function as a battering ram? Perhaps, but the standard acoustic function of the spermaceti organ gives a perfectly reasonable, and well supported, explanation for the increased relative size of the male spermaceti organ: sexual selection on the sounds of male sperm whales, if the sounds are either used as weapons themselves or include honest signals—most likely attributes of the slow click—that determine male reproductive success (see Weilgart and Whitehead 1988; Cranford 1999; see section 6.11).

This theory of sexual selection on vocal traits is not disproved by Carrier et al.'s (2002) work, and it seems to be more plausible than their theory that selection as a battering ram led to the sexual dimorphism. Carrier et al. (2002) show that ramming could injure other males, but this does not mean that the spermaceti organ evolved to do this. Something like the tusk of the narwhal (*Monodon monoceros*) or the northern bottlenose whale's hardened head (Mead 1989b; Gowans and Rendell 1999) would seem more efficient in jousting. Carrier et al.'s (2002) other finding is that melon size relative to body size (the melon is the general odontocete structure that evolved into the spermaceti organ in sperm whales; Cranford 1999) increases with sexual size dimorphism among the cetaceans. While supporting the battering ram hypothesis, this result fits the acoustic sexual selection hypothesis just as well, if the melon sizes calculated are those of males, which they

Box 8.1 continued

probably are. In more male-biased sexually dimorphic species, in which male-male fighting, or female choice of male-specific characters, is likely to be important, honest signals of size will be increasingly important.

There are a number of factors that determine the effectiveness of a sonar. Some, such as the strength of the target (how well it reflects sound) and the properties of the intervening water, are outside the control of the sperm whale, but other important factors are available for molding by natural selection: the signal's strength, the signal's characteristics, and the detection and processing of the returning information. Detection and processing are the realm of the ear and brain, and while there is room for development here, there is also an important constraint that the animal can do nothing about; namely, background noise. So improvements in a sonar system are generally more easily achieved by making a louder or better-designed signal than by developing the reception and processing equipment. The strength and nature of the sonar signal will be determined by the anatomy of the producing system, and so an evolutionary biologist believing in the adaptationist agenda will conclude that the spermaceti organ and its associated structures are largely designed to produce powerful, highly directional signals at the right frequencies for detecting mesopelagic squid (Madsen 2002). For now, this is just a hypothesis, but it seems a very reasonable one.

It is worth noting, and considering, alternative plausible scenarios that have been proposed for the evolution of the sperm whale's remarkable acoustic system, such as prey stunning (Norris and Møhl 1983). However, we rarely hear sounds that might be appropriate as stunning weapons (see section 5.2), so this hypotheses would not seem to be sufficient to explain the acoustic output of the sperm whale at depth and its ability to capture a wide range of prey.

The click of the sperm whale certainly has multiple functions, being used frequently for communication in codas, and probably in slow clicks and perhaps some creaks (see section 5.2). The sperm whale's dependence on clicks for communication is somewhat unusual. Other highly social odontocetes, such as pilot whales (*Globicephala* spp.) and killer whales (*Orcinus orca*), seem to use pure tones (such as whistles) or pulsed calls for most communication. While sperm whales do make some non-click vocalizations, these sounds seem rare and quiet, and may not even be intended as communication (see section 5.2). It seems that the sperm whale, having developed the spermaceti organ to make powerful clicks, lost either the ability or the necessity to make other types of sounds to communicate, instead adapting the clicks for that function. The use of clicks for communication is also found in some small, and apparently less socially complex, cetaceans, such as the harbor porpoise (*Phocoena phocoena*) and *Cephalorhynchus* dolphins (Tyack 2000).

But why, given that the spermaceti organ is so huge, is it proportionally huger in the adult male? In the largest males, the spermaceti organ appears visibly swollen, protruding forward well beyond the lower jaw and the junk (see fig. 8.1). When there is sexual dimorphism in any feature that is not directly part of the reproductive system, it is usual, and reasonable, to implicate sexual selection (Stearns and Hoekstra 2000, 182). Thus, the spermaceti organ would appear to have a role in either competition among males or mate selection by females. How could this be? Cranford (1999) discusses the possibility that males may seek to injure each other during conflicts over females by making particularly intense sounds. In this scenario, the spermaceti organ is a weapon, and he who has the largest and most powerful weapon wins in the competition to get genes into the next generation. Males may also assess one another by listening to one another's vocalizations, especially the slow clicks (Weilgart and Whitehead 1988; see section 6.11). If there are features of the slow click that correlate with the size of the spermaceti organ, and thus the size of the animal, such as the inter-pulse interval (Norris and Harvey 1972; Gordon 1991a), then males can usefully use them to regulate their interactions, avoiding apparently larger males and approaching smaller ones. Alternatively, or perhaps additionally, females may use attributes of the slow click to choose mates (Weilgart and Whitehead 1988; section 6.11). In either case, there will be selective pressure for proportionally larger spermaceti organs in males.

8.3   The Function and Evolution of a Large Brain

"When you come to think of it, it's a wonder we ever get a whale. Why, they ought to kill us all, and they would if they had any brains in that monstrous head of theirs."
—Hopkins 1922, 312

"Could there be other intelligences?" he asked Moontail one day as they fed in a canyon together. "We've always believed that cetaceans are the only fully sentient beings in the world. But we know so little of the land—could there possibly be beings like ourselves who think and feel?"

"I should think it highly unlikely," she replied as she gulped down a squid.

"But why?"

"Gravity—the pull of the earth—doesn't draw quite so hard on us sea-creatures, owing to the buoyancy of the sea-water in which we live. That is why we were able to grow such big bodies, and the big brains that go with them. A creature living on land simply would not have those advantages. It certainly couldn't grow large enough to have a brain the size of ours."
—Baird 1999, 90

These sperm whales (in a retelling of *Moby-Dick* from the whale's perspective) are rather less species-centric than most humans—they do discuss the possibility of intelligence in other animals.* But Moontail, just like the Yankee whaler in Hopkins's account, finds it hard to assess the intelligence of animals that live in a totally different environment. However, as she suggests, intelligence is a function of the capacity of an animal's brain, and larger brains are generally thought to produce more sophisticated intellects.

There has been considerable discussion as to whether, when comparing the cognitive capabilities of animals, relative or absolute brain size is the better measure, or whether some other relationship between brain size and body size is more appropriate (e.g., Byrne 1999; Gibson 1999). For sperm whales, this is an issue of great significance. In absolute terms, they have the largest brains on Earth, with a mean of 7.8 kg and a maximum of 9.2 kg in adult males (Kojima 1951), compared with about 1.5 kg for a human and 5 kg for an elephant (see table 8.1 †). However, their brains are a much smaller fraction of their body size than in many smaller odontocetes, or in humans (table 8.1).

Recent research is beginning to sort out the relative/absolute brain mass controversy. One might expect that the benefits of a brain—computational power, and thus cognitive abilities—would be more closely related to its absolute size and its structure (e.g., density of neurons, number of synapses), than to its relative size (Corballis 1999). This is probably the case within the primates, in which absolute brain size, but not relative brain size (as measured by Jerison's [1973] "Encephalisation Quotient"), is well correlated interspecifically with performance on learning tasks that measure mental flex-

---

* There is no evidence that sperm whales possess language, as it is usually defined (by the presence of symbolism, grammar, and syntax), although they do communicate acoustically.

† R. C. Connor (personal communication) has noted that all measurements of the mass of sperm whale brains are from males. Because of the extreme sexual dimorphism in size and lifestyle, females may possess rather different brain measures than those given in table 8.1. For instance, in elephants, females have brains that, although smaller in absolute size, are relatively much larger than those of males (Douglas-Hamilton et al. 2001).

Table 8.1. Some Brain Masses and Body Masses of Large Mammals

Species	Brain Mass (kg)	Body Mass (kg)	Brain Mass/ Body Mass
Sperm whale (male)	7.8	33,596	0.00023
Fin whale	6.9	81,720	0.00008
Killer whale	5.6	5,448	0.00103
Bottlenose dolphin	1.6	154	0.01038
Elephant (male)	4.8	6,000	0.00080
Human	1.5	64	0.02344

SOURCES: Data from Berta and Sumich 1999, 153, except Douglas-Hamilton et al. 2001 for elephants.

ibility (Rumbaugh et al. 1996; Beran et al. 1999). In contrast, the costs of a brain, especially the fraction of metabolism devoted to brain maintenance, as well as the potential difficulties of giving birth to large-brained offspring (important in humans, but probably not in cetaceans), will depend more on relative than absolute brain size (Gibson 1999). Worthy and Hickie (1986) have noted that the relief from gravity offered by an aquatic habitat changes constraints on body size, and so standardizing brain size by body size may not be appropriate for aquatic animals, especially for the largest marine animals, such as sperm whales.

So what can we say of an animal whose brain constitutes only a small fraction of its body, but is the largest on Earth? I suggest that this brain gives the sperm whale substantial cognitive abilities, but lesser incidental penalties than incurred by smaller animals such as humans and bottlenose dolphins, whose large brains constitute a larger proportion of their bodies.

But the large absolute and small relative size of the sperm whale brain are not its only prominent attributes. The anatomy of a brain may indicate some of its strengths and weaknesses. Using the carcasses produced by the whaling industry, it has been possible to look closely at the anatomy and development of the sperm whale's brain (e.g., Kojima 1951; Oelschläger and Kemp 1998). Oelschläger and Kemp (1998) note the following unusual features:

1. A large telencephalon (the area considered to produce conscious mental processes, intelligence, personality, and sensory processing)

2. A greatly reduced olfactory system

3. An enlarged auditory system

4. Slow growth or regression of some parts of the limbic system (which controls emotions, motivation, pain, and pleasure)

5. A relatively small cerebellum (which controls fine motor skills) and

pons (which controls arousal, autonomic motor functions, and sleep) compared with other odontocetes

So, what have we got? A highly intelligent animal, with varied personalities and great hearing, but unemotional, with little athletic ability or sense of smell? Such conclusions have to be taken very cautiously (Pabst et al. 1999). However, some of these characteristics fit with what we do know of the animals. They are not usually aggressive (see sections 6.7 and 6.11), they do seem to have varied personalities (see table 6.13; H. Whitehead, personal observation), and they are certainly sensitive acoustically (e.g., Watkins et al. 1985; Bowles et al. 1994; Mate et al. 1994a), but they do not show the spectacular feats of motor coordination evident in many smaller odontocetes. It is also very likely that their sense of smell is poor. Can we then conclude that, because the other characteristics of sperm whales' brains match their abilities, they are as intelligent as suggested by the huge telencephalon?

This, I think, would be premature. Intelligence is difficult to define and hard to test rigorously in laboratory settings. There are no sperm whales in laboratories, and except for very young, and probably sick or stressed, animals, there are unlikely to be any in the foreseeable future. There are no obvious signs of great intelligence in what we can see of the animals at sea, although we can see very little. They "frequently allow . . . themselves to be slaughtered impassively" by humans (Caldwell et al. 1966), and sometimes by killer whales (Pitman and Chivers 1999), in situations in which a well-planned deep dive might seem to be the key to survival. However, supposedly intelligent humans often behave no more rationally in such circumstances, at least partially because of cultural imperatives (Boyd and Richerson 1985, 204–5; e.g., religious martyrs or kamikaze pilots). If sperm whale behavior is also substantially culturally determined (see section 7.3), we should not expect it to always maximize individual fitness, however intelligent the individuals may be. Thus, I think the case for great intelligence in the sperm whale must be left open for now, forming an intriguing and challenging target hypothesis for whale science of the future.

Staying in the speculative vein, I will move on to the evolution of this huge and rather unusually configured brain. It is easy to see why the olfactory system has been reduced and the auditory system enhanced in an environment where smell is not very useful, but sound is the most effective way both to sense the environment and to communicate (Tyack 2000).

What about the evolution of large brain size? If relative size is the measure to be considered, then the sperm whale's brain has nothing to answer for—it is not relatively large in mammalian terms (e.g., Tyack 1999; see table 8.1), and its large absolute size is simply explained by allometry, the consequences of a large body and proportional growth. However, as I have noted above, both evidence and theory are shifting attention toward absolute brain size as a correlate of intelligent behavior (Gibson 1999), and brains have substantial metabolic and other costs. From this perspective, the very large brain of the sperm whale is likely to be adaptive.

Because humans have large brains, there has been a great deal of thought and debate about their evolution (more than there has about the evolution of large noses!). Proposed explanations for the size of the human brain include tool use, the need to remember the locations of dispersed food sources or caches, and social (or Machiavellian") intelligence. Of these explanations, social intelligence has generally come to the fore: some social animals may have evolved large brains and substantial intelligence so that they can make more "rational" decisions—for instance, about alliances and assistance—in complex social circumstances (Byrne 1999). We can rule out tool use as an evolutionary driver of the sperm whale's large brain. Memory of resource distributions in space and time, and as related to other variables, is likely to be important to sperm whales (see section 3.7), and, like human fishers, they may benefit from an ability to infer resource distributions from sketchy environmental information. However, in the fluid ocean, they will not need the same precise memory as, say, a nut-caching squirrel. Sperm whales do have complex social arrangements (see chapter 6), so a social intelligence function for Earth's largest brain seems indicated. The sperm whale's neocortex is particularly large (Oelschläger and Kemp 1998), and this is the area of the brain most closely associated with social intelligence in primates (Barton 1996).

8.4 Sperm Whales and Other Deep Divers

The comparative method has proved a powerful technique for illuminating past evolutionary processes (Stearns and Hoekstra 2000, 316–17). Correlations across species between variables can indicate evolutionary cause and effect, or suggest coevolution. Hence, in an attempt to unravel the social evolution of the sperm whale, I will look at the social structures of species that share its habitat and possess similar anatomical and physiological constraints: other mammalian deep divers. If these species have social systems that are similar to that of the sperm whale, and that are not common in anatomically similar species that inhabit other environments, then some feature or features of their shared environment are indicated as important determinants of social evolution. In contrast, if anatomically similar species in

Figure 8.2 Elephant seals, with a male in the foreground. (Photograph by Michael Goebel, courtesy of Michael Goebel, US-AMLR Program, NOAA.)

similar habitats possess radically different social systems, then we must look more broadly for the factors that have driven or constrained the evolution of the sperm whale's social system.

There are not many mammals that dive deeply and make their living principally on mesopelagic and bathypelagic squid and fish. Apart from the sperm whales, the two most prominent groups of mammalian deep divers are the beaked whales (Ziphiidae) and elephant seals (see section 2.4). While the two elephant seal species (fig. 8.2) have been studied intensively, at least by marine mammal standards, and we know about many aspects of their biology, the ziphiids are much more of a mystery. New beaked whale species are discovered every few years (e.g., Reyes et al. 1991), and some of the species have never been identified alive. Often found far from shore, undemonstrative, and boat-shy, most ziphiids are hard to study.

However, the northern bottlenose whale (fig. 8.3) is something of an exception. It was for many years the object of a whaling industry, during the course of which a number of scientific studies were carried out (e.g., Benjaminsen 1972; Benjaminsen and Christensen 1979; Mead 1989a). Reasons

Figure 8.3 Northern bottlenose whales at the surface.

for the whalers' attention included the northern bottlenose's size (7–9 m long), ease of sighting, and curiosity toward boats (Mead 1989b). These attributes have allowed us to make a 12-year study of a population of northern bottlenose that is based in a submarine canyon, the Gully, off Nova Scotia (e.g., Gowans et al. 2000; Hooker et al. 2002). This research has included an analysis of social organization (Gowans et al. 2001), the only such study of any beaked whale. Thus, if we wish to compare the sperm whale's social structure with that of a beaked whale, the northern bottlenose is the only current candidate. As one of the largest ziphiids, the most sexually dimorphic, and probably the deepest diver (Hooker and Baird 1999), the northern bottlenose is perhaps the most obvious candidate for a pseudo-sperm beaked whale (Mitchell 1977). The northern bottlenose whale's vocalizations at depth, principally broadband clicks with peak frequencies of about 24 kHz and an inter-click interval of about 0.4 s (Hooker and Whitehead 2002), are also fairly similar to those of the sperm whale (see section 5.2).

   Some basic information about these deep-diving mammal species is summarized in table 8.2. They are all large, sexually dimorphic, and feed principally upon cephalopods. Southern (*Mirounga leonina*) and northern (*M. an-*

Table 8.2. Some Basic Characteristics of Deep-Diving Mammals

	Sperm Whale	S. Elephant Seal	N. Elephant Seal	N. Bottlenose Whale
Length (adult male)	17 m	4.7 m	4.4 m	8.8 m
Length (adult female)	11 m	2.6 m	2.6 m	7.5 m
Dive depths	~500 m	~450 m	~400 m	~1,000 m
Proportion of cephalopod prey	70%	75%	60%	70%

SOURCES: Data from Benjaminsen and Christensen 1979; Le Boeuf and Laws 1994; Pauly et al. 1998; Hooker and Baird 1999.

gustirostris) elephant seals often travel several thousand kilometers from their breeding sites (Stewart and DeLong 1995; Hindell and McMahon 2000), and so at least equal sperm whales in their mobility. The movements of northern bottlenose whales are less well known, but information from the Gully population strongly suggests that their movements are much more limited than those of the sperms, spanning perhaps just a few hundred kilometers (Hooker et al. 2002; T. Wimmer, unpublished data).

In contrast to the ecological and movement parallels between sperm whales and elephant seals, their social structures are radically different. Breeding elephant seals cluster on land, where there is pronounced female defense polygyny (Le Boeuf and Laws 1994). However, once the animals return to the water, there is little sign of social structure, although there has been little search for it. Mothers have left their pups before the pups enter the water for the first time (Le Boeuf and Laws 1994), and there is no evidence for preferred companionships, or even clustering in the absence of prey, while at sea. The elephants seals dive almost continuously, with none of the prolonged pauses at the surface that are characteristic of sperm whales (Le Boeuf et al. 1988). Perhaps there is some, so far undetected, social structure among elephant seals when they are at sea, but there is clearly nothing like the multi-level, permanent social unit structure found in sperm whales. Another great contrast between the elephant seals and sperm whales is in life history: elephant seals rarely reach the age of 20 (e.g., Bester and Wilkinson 1994), at which age a female sperm whale may have had only one or two calves (see section 4.3), and a male no breeding attempts at all (Best 1979).

Northern bottlenose whales initially seemed a better prospect to have a social system comparable to that of the sperm whale. They are clearly social (see fig. 8.3), forming clusters containing an average of about three animals, but up to twelve on occasion (Gowans et al. 2001). They may spend periods of up to several hours socializing at the surface (Gowans 1999, 90), and, like sperm whales, were known to try to defend one another when harpooned by whalers (Benjaminsen and Christensen 1979). Their life histories also seem more similar to those of the sperms, with animals frequently living into at least their twenties and thirties (Benjaminsen and Christensen 1979). Thus, we began our study of the Gully bottlenose whales in 1988 with a working hypothesis that their social system, driven by the importance of protecting calves during deep dives (see section 6.7), would resemble that of the sperms.

We found something very different. In northern bottlenose whales, it is the males, not the females, that form fairly permanent companionships. For instance, mature male bottlenose whale #1 was identified in the Gully during 1989, 1990, 1994, 1996, 1997, and 1998 (Gowans 1999, 117). Between 1989 and 1994, he was usually observed in the same cluster as mature male #3. In 1996 and 1997 he was observed without #3, but in 1998 they were seen together once more, butting heads in a remarkable, seemingly ritualized, display that we do not fully understand (Gowans and Rendell 1999). Overall, males #1 and #3 were observed clustered together on 44% of the days on which either was identified (Gowans 1999, 116). In contrast to such strong patterns, permutation tests could find no statistically significant indication of preferred clustering between identified females or immature animals (Gowans et al. 2001). The females seemed to have a loose network of associates among all, or at least many, of the other females in the population. While radically different from the social structure of sperm whales, this same general scenario—strong bonds between mature males, but weaker and more diverse associations among females—has also emerged from some of the detailed studies of coastal bottlenose dolphins (*Tursiops* spp.) (Connor et al. 2000b).

So, neither the elephant seals nor the northern bottlenose whale have a social system similar to that of the sperm whale. No other mammalian deep diver has sexual dimorphism as pronounced as these species, and there are no reports among pinnipeds or ziphiids of anything like the groupings of female and immature sperm whales. Thus, this investigation of the species that closely share the sperm whale's habitat and possess similar anatomical and physiological constraints does not support the hypothesis that deep diving has driven sperm whale social evolution.

But it does not prove that it is wrong. The hypothesis basically states that female sperm whales form permanent groups so that they can mutually protect one another's offspring during deep dives (see section 6.7). The elephant seal infant is safe from predators (except humans) on land during its first months, and when it does enter the ocean, it seems to have no relationship with its mother.

The northern bottlenose, like the sperm whale, is completely aquatic, and, as I have noted, the two species have strong parallels in other areas of their biology. They also face the same natural predators, particularly the killer whale (Jonsgård 1968a,b). However, the geographic structures and sizes of the populations are very different. Sperm whale populations are not strongly geographically structured over less than an ocean basin (see section 4.1), ocean basins contain many tens of thousands of animals (see section 4.2), animals range widely, covering about 70 km in a day and 1,500 km over months or years (see section 3.4), and the density of females in any area can vary widely (with CVs between years of 0.5–0.75 off the Galápagos Islands; see section 2.2). In contrast, northern bottlenose whales show significant differences in genetic structure over scales of 2,000 km (Dalebout et al. 2001), and the population using the Gully numbers only 130 animals (Gowans et al. 2000). The whales cover only about 4 km in a day (Hooker et al. 2002), and preliminary evidence tentatively suggests that home ranges may span a few hundred kilometers or less (T. Wimmer, unpublished data). The population density in the Gully is also quite stable, with a CV between years of 0.3 (using sighting rates from Hooker 1999, 62, excluding 1988, the first study year, when search methods were still being developed).

Like sperm whales, northern bottlenose mothers seem to leave their infants at the surface during deep dives, and like sperm whale infants, northern bottlenose infants move between adults and immatures, including subadult males, at the surface (Gowans et al. 2001). This is possible because females and their calves live in small, well-defined areas where other northern bottlenose whales are predictably present. So, the female northern bottlenose whale does not need to bring her permanent social unit with her on her travels to help care for her infant—she travels little, and the babysitters are already there. The whole small population is, in some senses, her social unit.

Thus, although sperm and northern bottlenose whales use similar resources, the more specialized (see section 2.4), localized lifestyle of the northern bottlenose whale, and probably other beaked whales (see Waring et al. 2001), has led to a radically different social organization. Instead of diving being the key to northern bottlenose social structure, it seems that horizontal movement is a much better predictor. Northern bottlenose whales have horizontal movement patterns and ranges similar to those of coastal bottlenose dolphins, and the two species share the same basic social system (Gowans et al. 2001), even though the northern bottlenose whale regularly dives to 1,000 m (Hooker and Baird 1999) and the habitat of the bottlenose dolphins in the best-studied populations (Sarasota, Florida, and Shark Bay, Australia) is rarely more than 20 m deep.

Among the deep-diving marine mammals, horizontal movement patterns seem to reflect niche breath: the species with the largest day ranges, sperm whales and elephant seals, encounter the greatest variety of squid species and have the widest niche breadths, while the northern bottlenose whale, with its more localized travels and distribution, has much less diversity in its diet (see section 2.4 and table 3.5). Prey specialization may be the cause of localized movement, or vice versa, or there may be some other contrast in the biologies of the deep-diving mammals that leads to a correlation between niche breadth and movement pattern (Whitehead et al. 2003).

Among the cetaceans whose social systems have been studied, the one with the social system most similar to that of the sperm whale is probably the long-finned pilot whale (*Globicephala melas*) (fig. 8.4). These animals are observed in groups of about twenty, and they live in permanent social units containing about eight animals, which are probably matrilineal and may be enmeshed within larger social structures (Amos 1993; Amos et al. 1993; Ottensmeyer 2001). Pilot whales also seem to have wide ranges spanning 1,000 km or more; dive fairly deeply, often going below 100 m (Mate 1989); and, like sperm whales, live long and reproduce slowly (Bernard and Reilly 1999). However, there are differences between the social structures of pilot and

Figure 8.4 A group of long-finned pilot whales. (Photograph by Luke Rendell, courtesy of Luke Rendell, Dalhousie University.)

sperm whales, most notably in the behavior of males. In the long-finned pilot whale, males seem to stay with their mother's social unit throughout their lives (Amos et al. 1993), whereas the dispersal of male sperm whales from their maternal units is clear and profound in its consequences for social structure and mating systems.

To find a closer parallel to the sperm whale's social system, we have to move to a radically different habitat: the terrestrial tropics.

8.5   The Colossal Convergence: Sperm Whales and Elephants

One animal stands out as possessing most of the elements of the sperm whale's social structure (as described in chapter 6): the elephant (Best 1979; Weilgart et al. 1996). There are three elephant species: the savannah (*Loxodonta africana*), forest (*L. cyclotis*), and Asian (*Elephas indicus*) elephants. Although there are some differences among them, the social structures and natural histories of the three elephant species are broadly similar (Lee 1991b), so I will concentrate here on the savannah elephant, which has been the best studied. Although the elephant is a terrestrial herbivore and the sperm whale a pelagic carnivore, these species share a remarkable number of congruent features in several areas of their biology.

Anatomy and Morphology

The elephant is the largest land animal and the sperm whale the largest toothed whale, surpassed in size only by a few baleen whales. Both species are highly sexually dimorphic (table 8.3). While the sperm whale has the largest brain on Earth, the elephant's is the largest on land (see table 8.1). Perhaps most importantly, both species possess highly modified, and extremely useful, noses: the spermaceti organ and the trunk generally give them increased foraging success compared with their potential competitors of other species (e.g., Sukumar 1991).

Life History

Elephants and sperm whales have similar life spans, ages at sexual maturity, and calving intervals, although elephants seem to nurse for longer (see table 8.3). Although males disperse from their maternal unit at a younger age (~6 years) in sperm whales than in savannah elephants (~13 years), Asian ele-

Table 8.3. Sexual Dimorphism, Life History Parameters, and Social Organizations of Savannah Elephants and Sperm Whales

	Elephant	Sperm Whale
Adult mass, male:female	4.7:2.7 t	45:15 t
Life span	~60 yr	≥60–70 yr
Age at maturity		
Females (sexual)	10–12 yr	~10 yr
Males (sexual)	~17 yr	~20 yr
Males ("sociological")[a]	29 yr	25 yr
Calving interval	4–5 yr	~5 yr
Age at weaning (mean)	5 yr	2 yr
Age at weaning (maximum)	8 yr	13 yr
Age of male dispersal from maternal unit	9–18 yr	~6 yr
Social unit size of females and offspring	~10 animals	~10.5 animals
Mean group size	~20 animals	~25 animals

SOURCES: Data from earlier sections of this book and Whitehead and Weilgart 2000, except age of male dispersal from Lee and Moss 1986, Best 1979, and Richard et al. 1996a, and elephant herd size from Lee 1991b.

[a]"Sociological maturity" is the age at which it is thought that males begin to take a substantial role in breeding.

phants have closer parallels, starting to disperse at about 6 or 7 years of age (Douglas-Hamilton et al. 2001). In both sperm whales and elephants, while males become sexually mature in their late teens, they do not take much of a role in breeding for the next decade. In both situations, there are mechanisms that keep younger males away from most breeding females. In sperm whales, this mechanism is the dramatic dispersal of males to high latitudes. Elephant males also tend to stay separate from the groups of females, but to a much lesser extent. However, the principal mechanism excluding younger male elephants from breeding is "musth," a physiological state giving a male dominance over non-musth males. Only the larger males come into musth during the primary breeding season (Poole 1989).

Ecology

Both elephants and sperm whales have wide geographic ranges and broad niches (see sections 3.4 and 2.4; Douglas-Hamilton et al. 2001), dominating their guilds in biomass and potentially structuring ecosystems (Laws 1970). The sperm whale is the most phylogenetically distinct of the extant cetaceans, sharing its range with no remotely similar animal. Although there are three elephant species and several subspecies (Douglas-Hamilton et al. 2001), they are almost entirely allopatric: no elephant shares its range with any other species with a trunk. Both sperm whales and elephants can be typed as highly successful, dominant ecological generalists.

Social Structure

Moving to social structure, the parallels continue. Like sperm whales, female elephants are seen in spatially clustered groups, called "herds," whose size varies considerably, but which in the savannah species averages about twenty animals (Lee 1991b). However, these groups are not necessarily permanent structures. They are made up of one or more largely matrilineal family units, each with about ten members, and within units there is communal care for the calves—exact parallels with sperm whale social units (see section 6.5), although there is temporary disassociation within elephant units (Lee 1991b). At higher social levels, there seem to be more substantial differences. Units of savannah elephants may form "bond groups" with other units with whom they are particularly friendly and to whom they may be related (Lee 1991b), a structure that we have not identified in sperm whales. Elephant scientists also describe "clans" of savannah elephants that use the same principal resources, but these seem somewhat different from the acoustic clans we have found in sperm whales.

When we consider males, the parallels re-emerge. On leaving their maternal unit, young male elephants may be found in all-male groups, or with other groups of females, but become more solitary during their nonbreeding periods as they age (Lee 1991b). Breeding males, like their sperm whale counterparts, move between groups of females searching for receptive animals (Lee 1991a). As elephants can be seen more easily than sperm whales, the courtship and mating process has been well described (e.g., Lee 1991a). If a male encounters an estrous female, he will try to stay with her, and there are some recognized courtship signals on the part of both partners. However, males will interfere with each other, and can be outrun by reluctant females, showing that both male-male competition and female choice are important in determining mating success.

There are other parallels between sperm whales and elephants, such as in their use of sound to communicate and their conservation problems (Weilgart et al. 1996). The similarities that I have listed above suggest a profound evolutionary convergence, an intriguing puzzle for an evolutionary biologist. What may have led these animals along similar evolutionary paths, despite such superficially disparate habitats and only a very ancient (and probably socially primitive) common ancestor?

8.6   *K*-Selection, Sociality, and Ecological Success

Let me now suggest a scenario for this colossal convergence, an evolutionary route whereby two groups of animals living in totally different habitats have come to resemble each other in a wide range of attributes more than they do either other members of their own ecological guild or their phylogenetic relatives. This scenario is probably wrong in detail, and possibly in some of its major elements, but I hope it will stimulate more thought as to why the sperm whale and the elephant should each be so remarkable, so often in the same ways.

Step 1: A Useful Nose

Both the sperm whale and the elephant have uniquely useful noses (see section 8.2; Sukumar 1991), and there is some evidence that these noses may have evolved before many of the other characteristic and parallel elements of the two species' biology. For instance, there are fossil elephants and sperm whales that are themselves much smaller than the living species, but which possessed seemingly fully formed spermaceti organs and trunks (Shoshani 1991; Cranford 1999). Their specialized noses gave these animals a feeding advantage over other members of their guilds, promoting their ecological dominance (fig. 8.5) and leading to their phylogenetic isolation as similar species became extinct. Now the sperm whales or elephants were competing principally with members of their own species for resources, and their populations were regulated by the carrying capacity of the environment—classic "*K*-selection."* One of the most characteristic attributes of *K*-selected populations is a slowing of life history processes, with low birth rates, prolonged parental care, and long life spans (Horn and Rubenstein 1984). Additionally, the highly variable biotic environment of the sperm whale (see section 2.1) would tend to lead to the evolution of life history strategies that reduce variation in individual fitness, so that individuals would be expected to "hedge their bets" by slowing life history processes (Seger and Philippi 1989; Pásztor et al. 2000).

* Although the terms "*r*-selection" and "*K*-selection" have been largely abandoned because of both theoretical and practical concerns (Stearns 1992, 206–7), the principal issues were with "*r*-selection." The term "*K*-selection" describes the evolutionary circumstances that I envisage here particularly well.

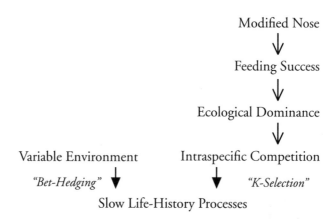

Figure 8.5 Step 1 in the evolutionary convergence between sperm whales and elephants: the development of anatomical structures that increase feeding efficiency (the spermaceti organ and the trunk, respectively) leads to a slowing of life history processes. Arrows with solid heads indicate primarily evolutionary processes; other arrows indicate primarily ecological processes.

Step 2: Sociality and *K*-Selection

With slow life history processes and low reproductive rates, it becomes particularly important for animals to exploit resources wisely and to avoid mortality. Social relationships can help with both these challenges by permitting cooperative foraging, vigilance, and defense against predators. Long lives and low reproductive rates allow such relationships to develop, as animals have the opportunity to interact repeatedly with a rather small number of conspecifics who are often relatives. In this way, a feedback loop is set up in which sociality and *K*-selection reinforce each other (cf. Horn and Rubenstein 1984; figure 8.6). In the cases of the sperm whale and the elephant, particularly important benefits of group living are likely to be communal vigilance and defense against predators, especially predators of calves (see sections 5.9 and 6.9), and traditional knowledge about a large home range, which permits efficient group movement and use of resources. This second benefit reduces the environmental variation experienced by the animals and enhances their ability to overcome environmental downturns (Sukumar 1991; Weilgart et al. 1996). A by-product of this process, in which efficient use of widespread and temporally variable resources determines the reproductive success of individuals, is ecological success of the population in competition with other species.

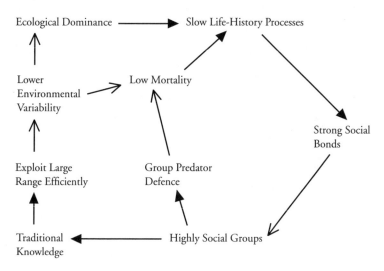

Figure 8.6 Step 2 in the evolutionary convergence between sperm whales and elephants: the feedback loop between sociality and slow life history processes. Arrows with solid heads indicate primarily evolutionary processes; other arrows indicate primarily ecological processes.

Large Bodies and Large Brains

The *K*-selection/sociality feedback loop is promoted by two other factors. First, large size reduces travel costs as a proportion of metabolic rate, allowing more efficient exploitation of a large range (Williams 1999). A large animal can also store food more effectively, and so better overcome environmental downturns, than a small one (see section 2.1). It is also likely to have a shorter list of dangerous predators, and thus lower mortality. In the ocean, large body size is helpful in another way: it increases the relative oxygen storage capacity* and thus allows longer dives. This is especially important for a deep-diving mammal that spends a significant fraction of its time commuting between the surface and foraging areas at depth. Thus, *K*-selection and large size are linked (Horn and Rubenstein 1984), and a species in the *K*-selection/sociality loop is likely either to be large in the first place or to face selective pressure to grow.

Brains are costly to their owners metabolically, and sometimes in other ways, but for a large animal, the costs are relatively less (see section 8.3).

---

* A larger animal will generally have a higher ratio of oxygen storage capacity (which will be approximately proportional to body mass) to metabolic rate (approximately proportional to body mass$^{0.75}$; Kleiber 1975), and so can dive for longer, than its smaller neighbor.

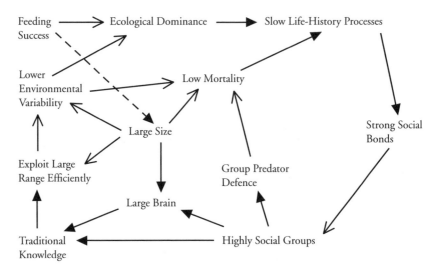

Figure 8.7 How large body size and brain size promote the feedback loop between sociality and slow life history processes. Arrows with solid heads indicate primarily evolutionary processes; other arrows indicate primarily ecological processes. The dashed arrow indicates the additional factor promoting this feedback in elephants and sperm whales: the enhanced feeding success produced by their specialized noses allows large size.

Brains are also especially useful in complex social circumstances (Byrne and Whiten 1988; Barton 1996). So our large, social, *K*-selected animal is likely both to be able to afford a large brain and to be able to benefit from one. This big brain may be secondarily useful in helping it to make sensible foraging decisions (Byrne 1999; figure 8.7).

Thus, as summarized above and in figures 8.5–8.7, we can see how slowly reproducing, social, large-bodied and large-brained animals may have evolved, and perhaps how a convergence between two such species from very different habitats emerged. A question arises: Why, if this feedback loop is so strong, has it not operated more than twice? In fact, it has: other species, such as humans, chimpanzees (*Pan troglodytes*), bottlenose dolphins, and killer whales, share many characteristics with sperm whales and elephants, but not with the tight convergence indicated in table 8.3. Large size, which permits a large brain, also has drawbacks, especially in the amount of food that must be consumed to maintain such a large body. Humans, chimpanzees, and bottlenose dolphins are not much larger than several less socially and cognitively advanced, but phylogenetically related, members of the same guilds (such as baboons for chimpanzees and humans, and perhaps phocoenid porpoises for bottlenose dolphins). In contrast, their wonderful noses allow sperm whales and elephants to exploit abundant food sources more effectively, and so to grow substantially larger, than other members of their guilds (such as the beaked whales or rhinoceroses, respectively). This large size, I suggest, gives the sociality/K-selection loop (see fig. 8.7) an extra twist for these two species, resulting in the colossal convergence.

This sketch covers many of the convergences between sperm whales and elephants (see section 8.5), but not those between males. Why should males leave their mother's unit, move to a different area, delay serious breeding efforts until their late twenties, and rove between groups of females? I will discuss these issues in section 8.8, following a more detailed consideration of social evolution in females.

8.7 Social Evolution of Female Sperm Whales: Facing the Predators

Like other attributes of a population, social structure can be viewed as a potential adaptation to an animal's environment. Prey and predators are primary aspects of that environment, and much social behavior appears to function by either increasing an animal's ability to obtain food resources or decreasing its vulnerability to predators (Alexander 1974). Competition with conspecifics is also considered an important driving force in social evolution, particularly in the appearance of more complex social structures, such as those based on long-term bonds (Wrangham 1980). Elements of social structures, such as the grouping behavior of females, may affect the value of other social elements, such as male-female bonds. But at the heart of social evolution are the issues of gathering food and avoiding predation (in other words, dealing with the environment), especially, in mammals, for females.

I have discussed the functions of the three most prominent social structures among female sperm whales—groups, social units, and clusters—in sections 6.7 and 6.8. I concluded that all three structures are likely to be principally concerned with reducing predation risk, but in different ways. To summarize, the groups provide greater vigilance against predators and better communal defense, the permanent units serve as a setting for the more complex behavior used to protect calves and provide nomadic females with a minimum group size at all times, and the clusters may reduce predation risk in the short term in the face of sudden attack. In contrast, sperm whale food is probably usually not defensible (for a possible exception, see Nolan et al.'s [2000] account of male sperm whales feeding beside longliners, summarized in section 5.9), and, generally, females will be able to provide rather little assistance to one another in finding or catching it. A possible exception is in the social direction of movement during periods of scarcity, especially by older animals (see section 6.7), which could be a vital function of the social unit.

Predation is likely to be a significant concern for almost all oceanic marine mammals except, perhaps, the adult male sperm whale (see section 6.10) and the killer whale itself. Why, then, don't sperm whale–like social systems proliferate? To some extent, they probably do—female pilot whales seem to live in similar societies, which include groups, permanent units, and clusters with numbers of animals similar to those of the sperms (Ottensmeyer 2001). I would not be surprised if other large odontocetes, such as false killer whales (*Pseudorca crassidens*), also have the same basic social elements. The oceanic dolphins, including the pantropical spotted dolphin (*Stenella attenuata*), spinner dolphin (*S. longirostris*), and striped dolphin (*Stenella coeruleoalba*), are found in groups that vary enormously in size (tens to thousands), substructure, and sex ratio (Scott and Perryman 1991; Perrin et al. 1994). These groups seem to include cluster-like elements (fig. 8.8), although sometimes with many tens of animals moving in a coordinated manner. They may also include unit-like permanent structures, although this is unclear. As in the case of sperm whales, predation risk is considered an important function for these groups of smaller oceanic odontocetes (Norris and Schilt 1988; Connor 2000).

Some female oceanic marine mammals clearly do not possess societies

Figure 8.8 Approximately two-thirds of a school of 455 common dolphins, *Delphinus delphis.* (Photograph by Michael Scott, courtesy of the Inter-American Tropical Tuna Commission.)

anything like that of the sperm whale, and we have already met two of them in section 8.4: the northern bottlenose whale and the elephant seal. The northern bottlenose whale—at least the Nova Scotian population for which we have information on social organization (Gowans et al. 2001)—is, I would argue, not a truly oceanic animal. It does not move over large swaths of ocean, instead remaining in localized canyons, and so can depend on the reliable presence of nearby conspecifics for assistance against predators (see section 8.4). The elephant seals do travel widely, but, perhaps in common with other oceanic pinnipeds, seem to use stealth rather than numbers to protect themselves. While at sea, elephant seals minimize their time at the surface, and even seem to dive deep to rest (Asaga et al. 1994). That this strategy may not be as efficient as the social strategies of the oceanic cetaceans is indicated by the much higher mortalities and shorter life spans of the elephant seals. This difference can be explained in two ways. It may pay elephant seals, and other seals, with their much shorter reproductive lives, to forgo the safety in numbers in order to benefit from less feeding competition. Alternatively, or perhaps additionally, the seals may not possess some of the attributes that make grouping effective for the odontocetes, including speed in the case of the smaller cetaceans and size in the sperms.

That predation may have selected for group living in sperm whales and other cetaceans is not a controversial idea. But the idea that it led to stable social units and strong bonds between sometimes unrelated animals stands somewhat in contrast to the general view for primates, in which intraspecific contest competition for resources is considered the key to the development of long-term bonds among unrelated females (Wrangham 1980). The structure of the ocean may play some role in this contrast (see section 2.9) by ruling out or lessening the effectiveness of many of the nonsocial antipredator strategies of primates and other terrestrial animals (such as hiding from avian predators or climbing trees in response to terrestrial ones). The female sperm whales are left with one another as their principal refuge (see section 5.9).

So, I envisage an ancestral social system in female sperm whales that was more like that of the inshore odontocetes or northern bottlenose whales or pinnipeds: fairly solitary foragers or loose networks of animals sharing home ranges, forming clusters of different sizes. But as the pre-sperms moved out into the open ocean, with its widely dispersed and highly variable resources, extensive movements became important. For animals almost continually on the move through a large ocean, two problems could be alleviated by forming permanent groups: the problem of unexpected predators, and the problem of where to go. A permanent set of companions gave a female better vigilance against predators, better ability to fight them off, and a babysitting system to help protect her calves. Her companions might also possess useful knowledge about where to go when conditions deteriorated.

But what about the largest identified social structure, the acoustic clan? What is the benefit to a female of sharing a dialect with a substantial proportion of the several thousand female sperm whales with ranges overlapping hers, but not all? Barrett-Lennard (2000, 67–68) has found that resident killer whales off Vancouver Island tend to mate with animals using different dialects, so the acoustic clans may function to inhibit inbreeding. However, with the dispersal of male sperm whales from their maternal units and the much larger sizes of the sperm whale clans, this seems an unlikely function for sperm whale clans. I find it hard to identify any discrete potential "function" for these structures. Cultural structures, such as vocal dialects, can often be essentially functionally neutral (Slater 1986), and this might be true of sperm whale clans. However, they may also be markers of important cultural divides in female sperm whale society. A female probably learns her coda dialect and characteristic movement patterns from her mother and other members of her unit. She may also learn much else about day-to-day and minute-to-minute life. This information is probably channeled by conformity, the tendency "to do the done thing." As units grow large and split, descendant units will possess similar cultures and dialects, and if they continue to associate, may actively maintain conformist dialects, which become badges of units that behave in similar ways and may preferentially associate. In this scenario, female sperm whales are members of clans that are a result of the evolution of culture in general (see section 8.11), together with the evolution of coda dialects in particular.

8.8 The Lives of Males

Males, especially mammalian males, are usually subject to a very different set of factors that influence their fitness, and channel their evolution, than females. Superficially, their most important resources are females, their most important behavior is mating, and their most important constraints are female choice and competition with other males. However, success in the mating arena depends on a male's achievements in other parts of his life: avoiding predators, so that he is alive to mate; feeding well, so that he is healthy and perhaps large; and, maybe, forming functional social relationships with other males (e.g., Connor et al. 1992) or females (e.g., Connor et al. 2000a), so as to improve his opportunities in a political society.

Clutton-Brock (1989) identified four factors that determine male breeding behavior:

1. Whether males can, by helping females raise offspring, increase their reproductive rate
2. How far females range
3. How large and stable female groups are
4. The density and distribution of female groups

As far as we can tell, no male cetacean provides paternal care (Connor et al. 2000a), so neither monogamy nor polyandry is a likely mating system for these animals (Clutton-Brock 1989). Female sperm whales range very widely (see section 3.4), ruling out territorial defense as a viable option for the breeding male.

So, faced with mobile females in groups of varying size and stability, what is a male to do? He would seem to have just a few basic options:

• Stay with his maternal unit, relying on groups and aggregations to provide mating opportunities—the "Mom's boys" scenario found in resident killer whales (Bigg et al. 1990) and long-finned pilot whales (Amos et al. 1993), but not among terrestrial mammals
• Join and stay with another group of females, defending them against other males—"female defense polygyny," as found in hamadryas baboons (*Papio hamadryas*) and gelada baboons (*Theropithecus gelada*) (Clutton-Brock 1989)
• Rove between groups of females, as elephants do

Furthermore, under the last two scenarios, he could either work by himself or in alliance with one or more other males.

The male sperm whale roves by himself between groups of females, a strategy that is not well described for any other cetacean or, I think, marine animal in which females travel in permanent units. However, solitary roving is probably common in species in which females are dispersed and their groupings are not permanent, including balaenopterids such as the fin (*Balaenoptera physalus*) and blue (*B. musculus*) whales (Connor et al. 2000a). These observations raise three questions: Why don't male sperm whales permanently accompany groups of unrelated females, why don't they stay with their mother's social unit, and why don't they form alliances?

I modeled the relative advantages of roving among groups and residence within groups for nonterritorial males (Whitehead 1990b), as have others (Sandell and Liberg 1992; Magnusson and Kasuya 1997). The results of these efforts are reasonably consistent, suggesting that, generally, a roving male encounters more receptive females than a resident one when the average travel time between groups of females is less than a female's period of receptivity. We have not measured the travel times of breeding male sperms, but they are much more efficient than we are at finding the females, and we take about 20 hr (see fig. 2.11). Likewise, we do not know for how long females are receptive, but it is probably more than a day (Whitehead 1993). Thus, the roving behavior of the male sperms seems functional and expected when compared with the alternative of permanently accompanying unrelated units.

A very different, and less quantifiable, set of considerations comes into play when the male's alternative to roving is staying with his mother. Resident killer whales and long-finned pilot whales have female social structures similar to those of the sperms, and these males stay with their mothers, so this is not an unreasonable alternative; in fact, it may be ancestral to the current breeding behavior of male sperms. A male that remains in his mother's social unit will benefit from his maternal relatives' assistance in finding food and warding off predators, as well as perhaps in capturing prey and during mating contests (Connor et al. 2000a). He may also be able to increase his own inclusive fitness by assisting his relatives. Set against these benefits are the costs of sharing food and a reduction in the rate of meeting unrelated females due to his ties to his mother, who will have other priorities in addition to his mating opportunities. By going out on his own, he eliminates the benefits he gives and receives as a member of a tight society of relatives, but gains freedom to forage in areas where there are few competitors and to optimize his roving for females.

I suggest that the increased sexual dimorphism of male sperm whales ($3 \times$ ♀ mass), compared with male killer whales ($1.45 \times$ ♀ mass; Dahlheim and Heyning 1999) and male pilot whales ($1.9 \times$ ♀ mass; using information in Bernard and Reilly 1999), tips the balance in favor of roving. Very large mature male sperms can benefit little from the assistance of their much smaller female relatives when competing with one another or trying to impress unrelated females. But the energetic needs of their huge bodies are such that they need to get away from the areas so intensely exploited by the females. Hence they disperse from their maternal units and move to areas unexploited by females, who are presumably designed by natural selection to be the optimal size for mesopelagic foraging at low latitudes (see section 8.9).

There are still more puzzles in the lives of male sperm whales, including the late age—about 27 years—at which they start returning to warmer waters and begin effective breeding. Why so long? Surely they must miss out on good potential mating opportunities by such a delay. But, by delaying their entry into the breeding lists, they avoid the reduced feeding rates of males on the breeding grounds (see table 2.6). They can grow unchecked, getting large fast, so that when they do start breeding they can compete successfully with other males. I have modeled the relative benefits of these options, assuming that breeding delays growth and that larger males usually get preferential access to mates, because of either male-male competition or female choice (Whitehead 1994). The results of the model suggest that males should delay competitive breeding—moving to lower latitudes, in the case of sperm whales—until they can expect to encounter an average of fewer than about two to four larger males attending each receptive female. These results seem roughly consistent with what we see: an average of perhaps two to four different mature males attending a group of females during a period of a few days (Whitehead 1994). Males who entered the breeding lists while smaller than their competitors would, according to my model, have few opportunities to become a father and lose out on valuable growing time. So they wait, and they grow.

So far, the models seem to provide a convincing explanation for the reproductive strategies of male sperm whales. But why don't males form alliances? Would it pay male sperm whales to form alliances in which they would outcompete single animals, as well as alliances with fewer members than theirs, for the attentions of females, but would have to share any resulting mating opportunities with other alliance members? I have addressed this question using a model (H. Whitehead and R. C. Connor, unpublished data), which predicts that alliance formation will be favored whenever an average of more than one breeding male attends a breeding female. This is almost certainly the case with sperm whales, but we see no sign of alliances. Either the males are not behaving adaptively, or more probably, some aspect of the model is wrong. There are several possibilities, but perhaps the most likely false assumption is the most basic: that alliances can outcompete single males. If female choice is important in determining mating success, as some of our results suggest (see section 6.11), and if females choose the male they like irrespective of whether he is an alliance member or not, then there may be no benefit to the males in forming alliances.

I will end this consideration of the lives of males with a few words about their most obvious attribute, size. Male sperm whales are enormous, one-and-a-half times the length and three times the mass of a mature female, forming the most extreme case of sexual dimorphism in mammals, except for a few pinniped species (Ralls et al. 1980). Sexual selection is almost certainly the driving force behind this dimorphism, with larger males outcompeting smaller males or being preferentially selected as partners by females. But why has it gone to such extremes? Among mammals in general, larger species are more sexually dimorphic (e.g., Jarman 1983). Thus, when sexual selection started to increase male sperm whale body size, we might expect it to have operated particularly successfully on one of the largest of mammals. Additionally, there are fewer constraints on large body size in neutrally buoyant seawater than on land, so there is less of a damper on the power of sexual selection. Thus, the marine mammals include the most extreme cases of sexual selection among the mammals: the elephant seals, some other pinnipeds, and the sperm whale.

However, the sperm whale, unlike the highly sexually dimorphic pinnipeds, does not mate on land, where males can establish territories. Attack and defense work differently in the ocean (see section 2.9). Can male sperm whales effectively use their vast sizes to block access to females by smaller males? Perhaps, but it will be much harder than for the terrestrially breeding elephant seals. Among pinnipeds, sexual dimorphism is pronounced only in species that mate on land (Berta and Sumich 1999, 336). In every species in which male pinnipeds are substantially larger than females, breeding males control access to a substantial number of females, their "harem," and the larger the male, the larger the harem (Lindenfors et al. 2002). In the case of the highly dimorphic, but aquatically mating, sperm whale, my suspicion, once again, is that female choice is at work. The males may owe their hugeness less to the competitive advantage that such a body gives than to a female whim: "bigger is better."

8.9   Sexual Segregation *

The sperm whale provides the living world's most dramatic case of segregation of males and females when they are not breeding (see fig. 2.2), but it is not the only species in which the sexes live apart for much of the time. A number of scientists have considered the proximate and functional causes of sexual segregation, especially in ungulates. A recent broad approach is that of Ruckstuhl and Neuhaus (2000), who examined five proposed explanations for sexual segregation and made predictions for both sexually dimorphic and monomorphic species. Here, I will list their hypotheses and, for each, discuss whether the sperm whale fits the predictions made for dimorphic species.

Activity budget hypothesis: Differences between the sexes in their optimal activity budgets, including issues such as swimming and diving capabilities and costs, make it less than optimal for the sexes to

---

* This section was inspired by an unpublished essay by Robert Michaud entitled "Sexual segregation in odontocetes."

synchronize their activities. So, when not breeding, they live apart. While male and female sperm whales probably do have different activity budgets, and so would do better to live separately, this hypothesis does not predict separation by a distance of even tens of kilometers, let alone several thousand, and so is incomplete as an explanation for sexual segregation in sperm whales.

Predation risk hypothesis: If the sexes have different susceptibilities to predation, then the more vulnerable sex—usually females, and especially females with calves—should seek safer waters. When applied to sperm whales, this hypothesis states that sexual segregation is due to the females avoiding colder waters where there are more predators, an argument made by Corkeron and Connor (1999) to explain the seasonal migrations of mysticete whales into the tropics, where, they suggest, there are fewer killer whales.

Forage selection hypothesis: Because the optimal food requirements of the sexes differ, they choose different habitats. This is the traditional explanation for sexual segregation in sperm whales, the argument being that because of their size, males do relatively better at higher latitudes because the squid there are bigger and deeper (e.g., Best 1979). Although this factor may have some role in the latitudinal segregation of the species, males at high latitudes do not necessarily dive deeply (see section 3.2), frequently eat small squid, and may eat fish rather than squid (see section 2.3).

Scramble competition hypothesis: This hypothesis, my favorite, states that males are excluded from the highest-quality habitat by scramble competition with the better-adapted females. This is the only one of Ruckstuhl and Neuhaus's hypotheses that provides an explanation for the influx of nonbreeding male sperm whales into the waters off the Galápagos Islands as female sperm whales left those waters in the 1990s (see box 4.2).

Social preference hypothesis: If the sexes are aggressive toward one another, or prefer the company of their own sex, then sexual segregation should result. In sperm whales there is no indication of aggression between the sexes, and, as in the case of the activity budget hypothesis, the social preference hypothesis does not predict the wide latitudinal segregation of the sexes.

Thus, I believe that the sexual segregation of sperm whales is best explained by males being outcompeted by females in scramble competition for mesopelagic squid, and so spending their nonbreeding time in high-latitude, female-free waters. However, this conclusion leaves open the question of why the females are found only in warmer waters. Possible explanations include superior feeding opportunities and decreased predation risk at low latitudes, as just described. The major predation threat would likely be to calves, but it might not just be the danger of predators that keeps females at low latitudes. Calves could be at risk from the harsher weather nearer the poles, or have poor thermal tolerances. This last hypothesis is supported by the fairly close agreement, among different oceans, between the normal limit of female distribution and the 15°C sea surface temperature isocline (Rice 1989). However, cetaceans much smaller than sperm whale calves, such as harbor porpoises (*Phocoena phocoena*), live successfully in waters much cooler than 15°C.

8.10 Social Evolution in the Open Ocean

Social Life in a Dangerous Ocean

Like virtually all other animals, the creatures of the deep ocean face two primary challenges: to eat and to avoid being eaten. While finding something to eat is not always easy in a fluid, unstructured, homogeneous medium, often the balance seems tipped toward the predator, especially if it is larger. The predator can travel easily, can sense (especially acoustically) at long ranges, and has many avenues of attack (see section 2.9). In contrast, the prey has few, if any, refuges—no caves, burrows, or trees—and is open to attack from all directions. There are only a few antipredator strategies available to oceanic animals:

1. Accept the likelihood of an early death, and reproduce early and, perhaps, often. Most of the smaller oceanic animals, and even some of the larger ones, such as codfish (*Gadus* spp.) and many cephalopods (Mangold 1987), could have been characterized as highly "*r*-selected" before this term fell from grace.

2. Become sufficiently dangerous to discourage predators, in the manner of sharks and killer whales themselves.

3. Live in a body that makes a difficult or unpleasant meal for a predator. Examples of this strategy include marine turtles and the ocean sunfish (*Mola mola*), a very large and sluggish animal of the surface waters that would seem to be a "sitting duck" for predators, but may survive because of its extremely tough skin.

4. Have the ability to outpace predators, or, at least, to reduce the probability of capture by sudden flight (Godin 1997). I suspect that sleek and

speedy (Jefferson et al. 1994) right whale dolphins (*Lissodelphis* spp.) can outpace killer whales, and flying fish (Exocoetidae) escape their oceanic predators by removing themselves from the viscous ocean.

5. Use depth as a refuge. At depth, there is safety from visual predators due to darkness and from air-breathing predators (mammals, birds, and turtles) tied to the surface. Depth is also a partial refuge from oxygen-filtering predators (fish and cephalopods), as the deep waters are low in oxygen and so may restrict the movements of these animals. A large number of animals are found only at depth, and there are others that use the deeps during predictably vulnerable times, such as the creatures that descend during daylight and the elephant seal, which dives to rest.

6. Adopt cryptic coloration, countershading, or transparency. Transparency is particularly favored by small oceanic animals (Smith 1997).

All of these antipredator strategies involve substantial investment in particular anatomical features (large reproductive organs, size and jaws for sharks, tough skin for ocean sunfish, wings for flying fish) or other costs. For instance, animals in the deep do not have direct access to the photic zone near the surface, where most oceanic life begins. But there is another type of antipredator strategy that does not have these immutable costs:

7. Form groups to improve vigilance against predators, to confuse predators, to dilute the effects of predators, to hide behind conspecifics, or to mount a communal defense (Alexander 1974; Keenleyside 1979, 50–61; Godin 1997; Connor 2000).

This social strategy of combating predation is found widely among marine organisms (e.g., Norris and Schilt 1988; Folt and Burns 1999), perhaps especially among those of the near-surface pelagic zone, where predators tend to be particularly active, as there is plenty of oxygen in the water and nearby atmosphere. Fish, squid, dolphins, whales, and even seabirds form groups at or near the surface, probably principally to reduce the predation risks of individuals (fig. 8.9). These groups vary enormously in a variety of ways. Group sizes range from the humpback whales (*Megaptera novaeangliae*) that rest in pairs off Newfoundland (Whitehead 1983) to swarms containing billions of actively aggregating krill. To make sense of this variation in group size, we need to consider costs (see later in this section), benefits (the risk of predation for krill may start to decline only when schools contain huge numbers of animals), and availability of conspecifics (there are not millions of humpback whales to form groups with). Marine schools also vary considerably in physical structure, from loose aggregations of seabirds floating at the surface to highly polarized schools of fast-moving fish. This variation is probably related to the different mechanisms by which predation risk

Figure 8.9 Many marine organisms form groups as an antipredator strategy. (Photograph by Robert Scheibling, courtesy of Robert Scheibling, Dalhousie University.)

is reduced: polarity and synchronized movements are nearly essential for confusing predators, but to benefit from communal vigilance, all an animal needs is alert neighbors.

A third source of variation among marine antipredator groups is particularly relevant in a book about sperm whales: differences in social structure. Many, and almost certainly most, marine groupings are ephemeral. Animals group with those around them, but make no active effort to maintain long-term bonds. This observation fits with some of the hypothesized routes by which these groupings reduce predation risk to individuals. Neither long-term bonding nor genetic relatedness among group members is likely to increase the value of the group to an individual in terms of improving vigilance, diluting the effects of predators, or hiding behind conspecifics.

However, predators might be more effectively confused by groups of long-term companions and/or relatives who look alike and move with greater synchrony. And bonded animals may also mount communal defenses more effectively. Additionally, long-term relationships could be important in guaranteeing the security of group membership, especially when mean group sizes are small. If there are substantial costs to forming large groups (e.g., food sharing; see below) and to being in very small groups (predation risks), then a dynamic fission-fusion society may pose a risk to an individual of being left alone, particularly if animals travel over large areas. A solution is to travel in a stable unit, so that a minimum group size is guaranteed. Thus, when confusion, communal defense, or maintaining a minimum group size are important, predation risk may favor long-term stable groups.

In the case of the sperm whale, two of these factors, communal defense (see section 5.9) and the need to maintain a minimum group size, especially when a female has a dependent calf that must be left at the surface while she dives (see section 6.9), may have selected for the stable units that are so fundamental to their societies. But, for both sperm whales and other oceanic animals, there are other consequences to forming groups.

Social Life in an Unpredictable Ocean

As Alexander (1974) expressed so clearly, grouping has costs as well as benefits. The most important of these costs is having to share food with other group members. But here oceanic animals have an advantage: their food usually comes in large patches, partly as a result of the biological variation induced by the physical variability of the ocean (Steele 1985; see section 2.1), and partly because of grouping of their prey. The prey may group effectively to reduce predation on themselves, but this means that once the predators find them, there is often plenty for all.

However, if the prey's grouping strategy is effective, finding and capturing food may not be simple. Food may be widely spread and elusive, or dangerous, once found. These problems can also be countered socially. Cooperative capture can be an effective strategy against cooperative defense, as seen so vividly in the successful attack of killer whales on sperm whales off California (Pitman et al. 2001; see box 5.3). This strategy is also seen in the cooperative herding of fish by cetacean predators (e.g., Würsig and Würsig 1980; Similä and Ugarte 1993), although such cooperative hunting by oceanic fish appears to be rare (Keenleyside 1979, 41). Prey encounter rates are also increased by sociality when groups spread out to forage over wide ocean areas and then come together again when food is found (Norris and Dohl 1980; Würsig 1986); the success of this strategy has been shown quantitatively by Baird and Dill (1995) in the case of transient killer whales feeding on seals. On a much wider scale, and more speculatively, long-lived nektonic animals may benefit from the social direction of movement over large scales (see section 6.7), or perhaps socially directed changes in prey selection, and so be better able to cope with extreme large-scale shifts in oceanic resource availability, such as those caused by "super El Niños."

Like the antipredator benefits, some of the feeding benefits of grouping can be achieved by casual acquaintances, while others derive from longer-term groupings. In particular, cooperative searching may not require long-term relationships, unless food comes in such small quantities that there are costs to alerting others to its presence. In contrast, cooperative prey capture and social direction of movement over large scales often requires, or is enhanced by, long-term relationships, and perhaps genetic relatedness among group members.

Defense of resources against conspecifics is likely to be impracticable for sperm whales (see section 6.7), and the same probably holds for most oceanic predators. But there may be exceptions. There are a few accounts of groups of killer whales displacing other, usually smaller, groups at resource patches, which include fish schools, patches of pinnipeds, and individual carcasses (Baird 2000).

The Male in the Ocean

For most males, the most valuable resources are receptive females. For a male in the open ocean, these resources are often found in highly mobile groups. What is a male to do? In some species, such as humpback whales (Clapham 2000), many cephalopods (Amaratunga 1987), and many fishes (Barnes and Hughes 1988, 211–13), animals aggregate to breed in particular areas. Here males can display, and perhaps even set up mobile territories, to attract females and repel other males. If there is substrate, the territories may be geographic, and the males may be able to provide useful parental care. But in other species, including sperm whales and some schooling fishes (Keenleyside 1979, 175), mating takes place "on the run," as the females move around actively in search of food or are passively driven by currents. In the case of ephemeral groups of females, the males probably do best by staying in the heart of the action, within the largest groups. When female groups are more permanent, there are more possibilities: accompanying unrelated females, staying with relatives and relying on group mergers to get access to unrelated mates, and roving between groups, either singly or with other males. The potential costs and benefits of these different strategies follow along the lines discussed for sperm whales in section 8.8 and more generally for cetaceans by Connor et al. (2000a).

At smaller scales, once he is grouped with receptive females, a number of factors may influence a male's chances of fathering offspring. These factors include male-male competition, perhaps especially in maneuvering close to females within mobile schools, female choice, and sperm competition. In the great majority of oceanic species we have no idea how these factors may play out, beyond the hints given by anatomical structures, such as relatively large testes indicating the presence of sperm competition (e.g., Brownell and Ralls 1986).

Social Life on Land and at Sea

In some cases, such as that of the sperm whale and the elephant (see section 8.5), social evolution has taken parallel courses on land and in the ocean (see Connor et al. 1998). However, there are some important differences between social structures in these two habitats, both quantitative and qualitative.

The most obvious quantitative difference is in group size: ant colonies are large, but krill swarms can be vast (sometimes spanning over 1 km; Folt and Burns 1999). Hundreds or thousands of ungulates may occasionally be seen together on the African savannah, but dolphin groups of such sizes are not unusual over large parts of the deep ocean (see fig. 8.8). Larger group sizes are made possible in the ocean by ease of movement and large prey schools. They are selected for by the difficulties of combating predation by nonsocial means.

A number of social attributes are present on land but rarely, or never, in the open ocean. The most fundamental of these is territoriality, which is a major determinant of social structure among terrestrial mammals (e.g., Clutton-Brock 1989), but is not practical in the open ocean (see section 2.9). Social behavior and social structures linked to territoriality, including monogamy and other mating systems in which there is paternal care, are also absent. Defense of transient resources is rare, but occasionally present (see above for killer whales). So, following the theoretical arguments for primates (Wrangham 1980), I would expect less internal structuring of groups, including dominance and preferred affiliations, in pelagic animals than is found in some terrestrial species. Another potentially important factor in terrestrial social evolution is infanticide (see Van Schaik and Kappeler 1997). Although good evidence for infanticide has been found among coastal cetaceans (Patterson et al. 1998), I think it is likely to be rare or absent in the open ocean, as the conditions that favor it—contest competition for resources within groups of females, or males being able to mate predictably with the mothers of the infants they kill—are probably absent (see section 2.9). Cannibalism is another matter, and is certainly a part of the feeding ecology of some oceanic animals, such as squid (O'Dor 1998).

Conversely, there are social structures in the ocean that are not found on land, most famously natal group philopatry, in which both sons and daughters remain grouped with their mother for life. This pattern is well described for resident killer whales (Ford et al. 2000, 23–24) and is strongly indicated by both genetic and photoidentification studies of pilot whales (Amos et al. 1993; Ottensmeyer 2001). I suspect it may be quite widespread among the medium-sized and large odontocetes. It is a feasible strategy in the ocean because the natural mobility and lack of territoriality of groups gives males many opportunities to encounter other groups containing potential mates even while remaining close to their mothers (Connor et al. 2000a).

Oceanic animals that can solve the predation problem, either socially or otherwise, may lead exceedingly long lives, permitting low reproductive rates. Many of the larger oceanic animals, particularly cetaceans, but also turtles, some fishes, and seabirds, have very long lives. For instance, many large cetaceans routinely lived into their sixties and seventies before whaling, and some, such as the bowhead (*Balaena mysticetus*) (George et al. 1999) and perhaps the sperm whale (see section 4.3), much longer (Whitehead and Mann 2000). Among other mammals, only humans and elephants on land, and dugongs (*Dugong dugon*) in the coastal zone, live much over 45 years (Whitehead and Mann 2000). Long, safe lives provide a setting in which elaborate social structures, and cultures, may evolve.

8.11 Cultural Evolution in the Ocean

Sperm whale populations appear to be strongly structured along cultural lines (see section 7.2), and their cultural differences seem to have functionally important aspects (see section 7.3). It seems likely (to me, at least) that culture is an important determinant of many aspects of sperm whale behavior, although we have yet to produce hard evidence of this. Cultural structuring of populations has been found in all of the best-studied cetaceans (bottlenose dolphins, killer whales, sperm whales, and humpback whales) as well as some other species (Rendell and Whitehead 2001b). In killer whales, it is clear that culture affects important aspects of behavior, including diet, foraging strategies, and social conventions (Boran and Heimlich 1999; Rendell and Whitehead 2001b), and there are strong indications that this is also true for bottlenose dolphins (Connor 2001). As these two species are by far the best-studied cetaceans, it seems entirely plausible that culture is an important determinant of behavior throughout much of the Cetacea, and that identity within cultural groups may form an important part of how these animals see themselves (assuming that they have a self-image).

This perspective is rejected by true believers in Morgan's (1894) canon: "In no case may we interpret an action as the outcome of the exercise of a higher psychical faculty, if it can be interpreted as the outcome of the exercise of one which stands lower on the psychological scale." For instance, Janik (2001) explains the group-specific coda dialects of sperm whales, and their correlation with mtDNA (see section 7.2), as follows: "Mitochondrial DNA encodes mitochondrial proteins. Differences in their effectiveness would affect the energetic budget of animals and is therefore likely to affect a wide range of behavioral patterns [including coda production]." This is an extraordinarily tortuous explanation for why members of one group go "click-click-click-click-click" and members of another (which may or may not have a different distribution of mtDNA haplotypes) prefer "click-click-click-click-PAUSE-click," when an alternative explanation is that the animals learn their coda patterns from their mothers (Weilgart and Whitehead 1997). But, because it invokes a "lower" mechanism (genetic determinism), Janik prefers it to our original explanation, which requires vocal learning, a mechanism found in humans (as well as songbirds, bottlenose dolphins, and quite a range of other animals; see Catchpole and Slater 1995; Janik and Slater 1997). De Waal (2000, 69) calls such a priori rejection of a characteristic in nonhumans because it exists in humans "anthropodenial." Together with other scientists who have spent time watching animals under natural conditions (de Waal 2000, 68–71; Boesch 2001; Rendell and Whitehead 2001a; Whitehead 2003), I reject this sterile approach to the study of animal behavior. I consider it very likely that culture is widespread among cetaceans, often important in the lives of the animals, and takes some forms rarely found in nonhuman terrestrial animals.

Why? Why has learning from others seemingly become so important among the whales and dolphins? Cultural theorists have found that culture is generally more useful to individuals, and so is more likely to evolve, in variable (although not extremely variable) environments (Boyd and Richerson 1985, 125–28; Laland et al. 1996). Furthermore, stable, vertical (parent to offspring) cultural transmission is favored when the temporal scales of variation are large. Consider a "super El Niño," which hits every century or so and raises sea surface temperatures by about 10°C from California to Chile and across much of the eastern Pacific, greatly affecting the marine environment for many months (Arntz 1986). Normal El Niños (with a roughly 5°C temperature rise) substantially reduce the feeding success of sperm whales off the Galápagos Islands (see section 2.6) and may affect their reproductive rates (Whitehead 1997b). In the "super El Niño" of 1982–

1983, the sperm whales off Peru changed their distribution, moving away from the worst-affected areas (Ramirez and Urquizo 1985). So, it looks as though movement is the principal strategy sperm whales use to survive these events. But where to go? Information possessed by members of a social unit—most likely older members—on favorable feeding areas in previous large El Niños could be extremely valuable, making a substantial difference to the fitness of the unit members. The same argument, that information possessed by the elderly can be enormously important in an environment with large-scale variation over long time scales, has been made by Diamond (1998, 120–23) for humans. But this information might be especially significant in the ocean, which shows great variation over long temporal scales (see section 2.1; fig. 2.1), and whose connectedness makes long-distance movement a viable method for overcoming poor conditions. Even if the appropriate knowledge for the new environmental conditions is not available, however, culture can still help. As animals explore, some will arrive at useful solutions more quickly than others, and with social learning and biased transmission (Boyd and Richerson 1985, 135), these solutions can spread. Thus individuals within the group learn successful strategies for the new environmental conditions more quickly than if they had to depend completely on their own individual learning. This process could lead to cultural group selection, which can occur much more easily than its genetic counterpart (Boyd and Richerson 1985, 240).

Other aspects of the oceanic environment may have increased the propensity for culture to be useful and for cultural facilities to evolve (Rendell and Whitehead 2001a). Barrett-Lennard et al. (2001) suggest that the medium-scale variation of the ocean—with most food coming in large, but well-spaced, patches—may have encouraged culture, as these patches allow, and promote, the formation of stable groups (see section 8.7) as well as communal techniques to efficiently exploit them. They also propose that the acoustic communication channel, used by cetaceans, humans, and songbirds, favors culture. Vocal imitation is probably simpler mechanistically than other forms of imitation (Janik and Slater 1997). Similarly, highly synchronous movement is probably simpler in an aquatic medium, and this may have promoted imitation (Whiten 2001).

Thus the ocean would seem to be an excellent medium for the evolution of vocal, social, and foraging cultures—although not material cultures, with no hands and few suitable materials (Rendell and Whitehead 2001a; Whiten 2001). The killer whale, which arguably possesses the most sophisticated of known vertical cultures other than those of humans, is an animal of this habitat. Other cetaceans with similar social structures and horizontal ranging patterns, such as sperm whales, pilot whales, and oceanic dolphins, may also possess comparably sophisticated vertically transmitted cultures. These vertical cultures have the potential to affect genetic evolution (Laland 1992; Whitehead 1998).

An open-ocean habitat does not automatically lead to strong vertical cultures; other attributes are also important. These attributes include long lives, overlapping generations within stable groups, prolonged parental care, and substantial cognitive abilities (Roper 1986; Rendell and Whitehead 2001b). One or more of these factors are lacking in most oceanic invertebrates, sea turtles, pinnipeds, seabirds, and pelagic fishes.

The ocean also supports important horizontal cultures, which are usually transmitted between members of the same generation. Most notable is the song of the humpback whale, with its gradual evolution (Payne 1999) and occasional revolution (Noad et al. 2000). But there are other candidates, both in the vocal repertoires, migrations, and feeding behaviors of other cetaceans (Rendell and Whitehead 2001a) and in the learned responses to predators of schooling fish (Smith 1997). In contrast, the hypothesis that social facilitation—in this case, the following of experienced animals—is important in the selection of breeding beaches by sea turtles has not been supported by genetic analyses (Bowen and Karl 1997). Horizontally transmitted cultures, while unlikely to affect genetic evolution, may have substantial effects on ecological processes. A switch in food sources, which is likely to have had a cultural component, by a few killer whales from pinnipeds to sea otters (*Enhydra lutris*) during the 1990s seems to have precipitated a chain of events that completely changed the structure of large parts of the Alaskan nearshore ecosystem (Estes et al. 1998).

8.12 Summary

1. The spermaceti organ represents an enormous evolutionary development, probably driven by the selective advantage of long-range echolocation for feeding on mesopelagic squid. It seems to have been secondarily co-opted into functioning as a producer of communicative signals. Some of these signals are used in mating, and so sexual selection has also had a role in shaping the spermaceti organ.

2. The sperm whale's brain is extreme in its absolute size, but relative to body size it is unremarkable among mammals. There are indications that absolute size may better indicate the cognitive abilities of a brain, but that relative size is a better measure of its costs. Thus the sperm whale may possess

substantial cognitive abilities, but pay a relatively small price for them. The architecture of the sperm whale brain suggests strengths in acoustic processing and intelligence. However, there is virtually no direct evidence bearing on the intelligence of the sperm whale, although its complex social system is consonant with the Machiavellian intelligence hypothesis explaining the evolution of advanced cognitive abilities.

3. The sperm whale's social system is nothing like those of the other deep-diving mammals whose social behavior has been studied, but it is remarkably congruent with that of the elephant. Elephants and sperm whales are similar, and similarly extreme, in a range of other biological characteristics.

4. The "colossal convergence" between elephants and sperm whales may be fundamentally due to the extraordinarily modified noses of the two species—the trunk and the spermaceti organ—which gave them a competitive advantage over other members of their ecological guilds. As a result, their fitness might have been largely regulated by intraspecific scramble competition, leading to slower life history processes and setting in motion a feedback loop between "*K*-selected" characteristics and sociality. This process was probably enhanced by the large sizes of the animals, which were made possible by their useful noses, and by the large brains that their large bodies could easily carry.

5. The social structures of female sperm whales, and of many other oceanic mammals, are likely to have been driven by the importance of reducing predation pressure, especially on calves. Animals almost constantly on the move benefit from being members of permanent units, which can reliably provide nearby companions to help protect against predators. Communal knowledge of ephemeral resources over large ranges may also be important.

6. The roving of male sperm whales between groups of females, and the delay of their entry into breeding competition until their late twenties, seem to fit models that compare the relative benefits of different mating strategies. However, similar models predict that male sperm whales should form alliances. That they do not do so could result from female choice being an important determinant of mating success. Female choice is also a potential cause of the large sexual dimorphism of the aquatically mating sperm whale.

7. The extreme sexual segregation of the sperm whale probably results from females preferring warmer waters, perhaps because of the physiological constraints on, or predation risks to, their calves, coupled with males being outcompeted by the females in these waters.

8. Animals of the open ocean have few antipredator strategies available to them. But grouping to reduce predation risks is often possible, and may have fewer costs than for terrestrial species. This tendency to aggregate may in

turn promote aggregation among predators, leading to large groupings being commonplace among open-ocean animals at a variety of trophic levels. Territoriality and resource defense, important drivers of social evolution on land, are largely absent in the open ocean.

9. Attributes of the open ocean, such as its large-scale variation, its connectedness, and its suitability for vocal communication, may have promoted culture in cetacean societies. However, most non-cetaceans in this habitat seem to lack one or more of the prerequisites for the development of strong vertically transmitted cultures.

9 Sperm Whales and the Future

9.1 Ghosts of Past Whaling

Why didn't I foresee this? How could it be otherwise? Whole family pods have been devastated by the loss of their matriarchs. The pods split up and wander aimlessly, without purpose, until they're killed too—because they have lost their survival skills, as well as all our cachalot wisdom and tolerance. Even if we do escape extinction and survive in small numbers, what kind of whales will we have become? And why?
—Baird 1999, 195

The culture of the sperm whale may be sophisticated compared with that of most other animals, but it was no match for that of humans in the eighteenth century, when whalers developed a suite of techniques for catching sperm whales and rendering them into valuable oil (see section 1.6). The open-boat pelagic whaling that took place all around the world during the eighteenth and nineteenth centuries was one of the world's most important industries during the earlier part of the Industrial Revolution. By 1880, when sperm oil had largely been replaced by petroleum and the whalers had generally ceased targeting sperms, sperm whale populations had been substantially reduced (see section 4.4). The animals were for the most part left alone until, after the destruction of the stocks of large baleen whales, modern whalers turned back to sperms in the 1950s.

Despite volumes of records from both hunts, it is unclear, even in the simple terms of proportional population depletion, what the effects of either hunt were on sperm whales. My simplistic model (see section 4.4; fig. 4.7) suggests that current population sizes are about 19–64% of those in 1712, when commercial sperm whaling began. This model gives only a very approximate estimate, but in fact its uncertainty is understated. The model omits a number of potentially important factors. Excluded are some elements of the sperm whale's population biology, such as its spatial distributions, the population's age and sex structure, and the effects of whaling on social structure and thus on population biology.

There are two routes by which whaling may have, and probably has, affected sperm whale populations via their social structures. The first is through the sex ratio. Modern whaling concentrated primarily on males, partially because they are larger and more valuable, but also because of the view that, given the supposed "harem" system of the sperm whale, only one breeding male was needed per group of females, plus a few in reserve (assumed to be 0.3 males per group of females; International Whaling Commission 1982). Thus, it was believed that male numbers could be reduced, as long as there was still an average of 1.3 large males per group of females, without affecting the number of animals being recruited to the population. The assumption of a "harem" mating system was wrong, but the roving system that the sperms actually adopt seems, on the surface, even more resilient to male depletion, as one roving male might be able to inseminate females in several groups (Whitehead 1987).

Whaling, and especially whaling for males, was particularly heavy in the southeastern Pacific, where operations off Chile and Peru were not limited by even the rather ineffective regulations of the International Whaling Commission, as neither country was a member. Off Peru, the proportion of large (>13.5 m) males in the catch fell from 35% in 1958–1961 to 11% in 1975–1977 to 2% in 1979–1981, and in the final year of whaling (1981), only one of the 225 animals killed was a large male (Clarke et al. 1980; Ramirez 1989). It appears that virtually all the large breeding males using Peruvian waters—and probably a much wider area, given their movements (see section 3.5)—had been killed by 1981. As might be expected, the pregnancy rate fell, from 0.28 in 1959–1961 to 0.23 in 1975–1977 (Clarke et al. 1980), although this change was not statistically significant (Chapman 1980). However, that it was biologically significant was indicated by our observations in the area between 1985 and 1995. We estimated the proportion of first-year calves off the Galápagos Islands (1,000 km from the Peruvian whaling grounds) to be 0.037, and off mainland Ecuador (contiguous with the whaling grounds) to be 0.019, in contrast with a rate of about 0.1 expected from population models and observed in other parts of the world (Whitehead et al. 1997). The proportion of mature males in the population was also low—about 4% in what seems to be the principal breeding season—much less than the 15% in the catches of Yankee whalers in the same area between 1830 and 1850 and the 16% predicted by population models (Whitehead 1993). Correlation does not prove causation, but it certainly looks as though the reduction of large mature males can be linked to the low calving rate in the eastern tropical Pacific (an anomalous aspect of this seemingly straightforward conclusion is discussed in section 6.11).

In savannah elephants (*Loxodonta africana*), whose biology is similar in so many ways to that of sperm whales (see section 8.5), exploitation affected population biology via another social route. Social units whose older females had been killed had lower reproductive rates, probably at least partially because of the loss of their matriarchs' knowledge (McComb et al. 2001). If, as is very plausible, there are similar features in sperm whale society, then this could be another way by which exploitation affects the animals not killed.

Because of the slow pace of sperm whale life, these effects linger. Off the Galápagos Islands in 1995, 14 years after the end of Peruvian whaling and 1,000 km distant, we saw no first-year calves in 18 days spent tracking groups of females and immatures. A population cannot persist without calves. The encouraging rise of the population trajectories in figure 4.7 for the last two decades does not consider these social effects, and so may be unrealistically optimistic.

If whaling and other threats are kept in check, however, in not too many years these trends may begin to reverse. With time, there should be large males to mate with and, if sperm whale societies resemble those of elephants in this regard, older females to lead social groups and act as cultural fulcrums.

9.2   The New Threats to Sperm Whales

In the meantime, sperm whale populations are vulnerable to a new range of threats that we are introducing into the ocean. The most obvious of these threats are those that directly kill animals, but other, more insidious threats could also have severe long-term effects on sperm whale populations.

The Japanese Whaling Industry

Harpoons are still the clearest threat to sperm whales, and they are reappearing. In 2000, after a 12-year interruption, the Japanese resumed killing sperm whales in the North Pacific. They circumvented the International Whaling Commission's moratorium on commercial whaling by declaring that the whales were being killed "for scientific purposes." Scientific research was done on the carcasses,* but they were then sold for meat and other products. The Japanese killed five sperm whales in 2000 and eight in 2001. These

---

* See http://www.highnorth.no/hunt-2000/japan/research2000.h

numbers will have virtually no effect on North Pacific sperm whale stocks, but the Japanese whaling industry makes no secret of its wish for this to be only the beginning—selling whale carcasses on the Japanese market has the potential to be extremely profitable.* As history has shown so clearly, whaling is very hard to control, and a poorly controlled hunt would have the potential to quickly wreak havoc on what remains of the sperm whale stocks.

Collisions with Ships

Sperm whales are sometimes killed or injured when they are struck by ships (Reeves and Whitehead 1997; André and Potter 2000; Laist et al. 2001). Laist et al. (2001), following a comprehensive review, listed sperm whales as one of the five species that are "commonly hit in some areas," but their mortality from this source does not approach that of fin whales (*Balaenoptera physalus*), the most commonly struck species. Laist et al. (2001) found eleven cases in which sperm whales were the known or likely species struck, not all of which were fatal. Even though most collisions are not reported, these numbers indicate that ship strikes are not currently an important element in sperm whale population biology. However, the problem is likely to worsen as ships become more numerous, larger, and especially, faster (Laist et al. 2001).

Marine Debris

Another direct cause of mortality is the ingestion of plastic debris (Viale et al. 1992), which perhaps somehow resembles squid. Laist (1997), in his "Comprehensive list of species with entanglement and ingestion records," lists the sperm whale as having records of ingestion of anthropogenic objects that are "more than infrequent." Although we know little of the scale of the effects of marine debris on sperm whale populations, I suspect that, as in the case of collisions with ships, by itself marine debris does not currently substantially affect sperm whale populations.

Interactions with Fisheries

Entrapment in fishing gear is recognized as a major source of mortality for smaller cetaceans (Read 1996). But fishing gear can also kill much larger animals, including sperm whales. Tens of sperm whales died in the Mediterranean Sea after becoming entangled in gill nets that were illegally set for

---

* A problem for the whalers is the high level of pollutants in sperm whale meat (Nielsen et al. 2000), making it a potentially undesirable food source.

swordfish (*Xiphias gladius*) (Di Natale and di Sciara 1994), and mortality after entrapment in gill and drift nets has also been reported from other parts of the world, including the waters off Ecuador (Barlow et al. 1994; Haase and Félix 1994). Once again, while entrapment in fishing gear does kill sperm whales, it seems likely that most populations are not currently threatened by this factor acting alone.

Fishing poses another class of threats to sperm whales, however. Off southern Chile, South Georgia, and the Falkland Islands, sperm whales are frequently seen in the vicinity of ships hauling longlines set for Patagonian toothfish (*Dissostichus eleginoides*) (Ashford et al. 1996; Nolan et al. 2000). At least off Chile, they seem to be occasionally taking fish from the lines (R. Hucke-Gaete, personal communication), but this is not always clear, as fish remains are frequently not found, and if they are, only fish lips remain attached to the hooks. Sometimes whales become entangled in the lines and die, while on other occasions the animal frees itself from the line, leaving badly damaged gear (R. Hucke-Gaete, personal communication). However, perhaps more importantly, there is a perception among some fishermen that the sperms are taking a significant part of their catch, and I have heard unconfirmed reports that in some areas of conflict some fishermen will try to ram or shoot sperm whales. A similar situation is occurring off Alaska, where sperm whales take halibut (*Hippoglossus stenolepis*) and sablefish (*Anoplopoma fimbria*) from longlines and suffer hostile repercussions from angry fishermen (National Marine Fisheries Service 1998). It seems likely that these types of interactions will increase in coming years as fisheries develop and sperm whales learn how to take advantage of them. The interactions just described involve male sperm whales, and as they are much more likely to feed on finfish than females (see section 2.3), they will probably continue to be the most affected sex.

Ultimately, fisheries pose an even more serious threat. Currently there is only a small overlap between the sperm whale diet and human fisheries: mainly the squid *Dosidicus* and a few large deep-water fishes of higher latitudes. But humans are insatiable. We may well develop the desire and technology to catch sperm whale food, and we have a remarkable record of overexploiting marine resources. What will happen to the sperm whale then?

Noise

Ship strikes, debris, and fishing gear are known to kill sperm whales. But some insidious threats whose effects are less dramatic may be even more dangerous. For example, some studies suggest that sperm whales are particularly sensitive to noise pollution, changing their behavior and distributions in response to unnatural low-frequency sounds, which have immense ranges in the ocean. However, the picture is far from clear.

One of the loudest anthropogenic sounds is that made by seismic vessels surveying beneath the ocean for petroleum deposits. Bowles et al. (1994) noted sperm whales seeming to respond to seismic pulses by changing their vocalizations at ranges of several hundred kilometers in the Southern Ocean, and Mate et al. (1994a) noticed that, as seismic surveying began in an area of the Gulf of Mexico, sperms left the affected region. However, other studies, including one of our own (McCall Howard 1999), found little change in sperm whale distributions or behavior during seismic surveys (Moscrop and Swift 1999).

Another type of loud anthropogenic sound, the sonar signals of whaling vessels, tended to scatter groups of sperm whales (Tønnessen and Johnsen 1982, 252). Watkins et al. (1985) reported that the intense military sonar used during the 1983 U.S. invasion of Grenada made the sperm whales in the area skittish for some time afterward. In contrast, few effects were found when a variety of sounds were played back to sperm whales off the Canary Islands (André et al. 1997b), or when male sperm whales were observed in the presence of an explosive detonation (Madsen and Møhl 2000).

Perhaps we can resolve these inconsistencies among observations on the effects of noise by considering some results of the Marine Mammal Research Program of the Acoustic Tomography of Ocean Climate (ATOC) project. ATOC deployed a loud low-frequency sound source off the California coast between 1995 and 1997 to study physical oceanography over large scales, but it was operated in an experimental manner so that the reactions of marine mammals could be observed. Sperm whales tended to stay away from the source when it was on (Calambokidis et al. 1998). When I further analyzed these data, it appeared that this effect was present only when whale densities were low (H. Whitehead, unpublished data). This finding can be interpreted as animals avoiding the noise when there was no particular attraction to the area, but staying and putting up with the noise when valuable resources were present, a pattern that corresponds to humans' reactions to noise (see also André et al. 1997b).

This finding also means that sperm whales may sometimes be exposed to loud noise, but do little to reduce its effects, thus risking anatomical damage. Consistent with this scenario, two sperm whales killed by a collision with a ship off the Canary Islands, an area with substantial shipping, were found to have suffered ear damage (André et al. 1997a). For an animal that probably depends crucially on sound for sensing its environment and communicating, ear damage may be devastating, reducing feeding or mating opportunities and perhaps contributing to ship strikes.

The oceans have become steadily noisier over the past century and a half, and the trend is sure to continue (Gordon and Moscrop 1996). We are using more ships, larger ships, and faster ships. Seismic activity is increasing in many parts of the world as the search for oil and gas is stepped up. We are also adding new types of sounds, including oceanographic sources such as ATOC and the extremely loud Low Frequency Active Sonar that is being deployed by the U.S. Navy and other navies. Beaked whales are killed by loud military sonars (Balcomb and Claridge 2001; National Marine Fisheries Service and United States Navy 2001), and there is good reason to think that sperm whales, because of their deep-diving habits and dependence on sound, may also be particularly vulnerable to such noises.

Chemical Pollution

If we wished, we *could* reduce oceanic noise almost instantaneously, but that is not the case with chemical pollution. Most of what we have put into the ocean will stay there for a long time, and some of the chemicals currently in terrestrial or freshwater environments will inevitably make their way into the ocean, even into the depths (Tanabe et al. 1994). The range of pollutants that may affect cetaceans is wide, but prominent concerns include heavy metals (including mercury, lead, and cadmium), organochlorines (such as DDTs, PCBs, and dioxins), and polycyclic aromatic hydrocarbons (O'Shea 1999). Among oceanic animals, marine mammals are particularly susceptible to chemical pollution because they feed at high trophic levels—pollutants bioaccumulate up the food chain—and because they store the chemicals in their blubber and pass them to their offspring in their milk. There are two principal areas of concern: that an affected animal may have a compromised immune system, and so may be more susceptible to naturally occurring diseases, and that toxic chemicals may damage the reproductive system. The first effect is demonstrated by the correlation between high pollutant levels and disease in the beluga whales (*Delphinapterus leucas*) of the St. Lawrence Estuary (Martineau et al. 1994), and the second by the experimental work of Reijnders (1986) with harbor seals (*Phoca vitulina*).

What of sperm whales? As odontocetes feeding fairly high on the food chain, we might expect them to accumulate high pollutant levels, but this effect is tempered by their usually substantial distance from shore, where most pollution originates. A wide range of heavy metals and organochlorines have been found in sperm whale tissues (O'Shea 1999). Organochlorine levels in sperm whales are normally intermediate between those of the highly affected coastal odontocetes and the cleaner mysticetes, which generally feed at lower trophic levels (e.g., Aguilar 1983; Law et al. 1996). However, heavy metal (e.g., mercury, cadmium) concentrations in sperm whales can at least sometimes be extremely high, reaching concentrations that would be extremely toxic in many other mammals (e.g., Nielsen et al. 2000). No sperm whale mortality has been clearly linked to chemical pollution, although there have been unexplained mass mortality events in the relatively highly polluted waters off northwestern Europe: forty-one dead animals were found off northern Europe in 1988–1989 (Christensen 1990), and twenty-one in the North Sea in 1994–1995 (Law et al. 1996). There is no good evidence that chemical pollution had any part in these deaths, but it does not seem unreasonable that we humans, who so affect the ocean in so many ways, are in some way implicated.

Global Warming

Evidence is growing that Earth's climate is warming in response to anthropogenic activities (Intergovernmental Panel on Climate Change 2001), and that this warming is affecting organisms, populations, and ecosystems (Walther et al. 2002). What about sperm whales? Climatic variation clearly affects the feeding success of sperm whales (see section 2.6), and we have some evidence from the Galápagos Islands that, following a lag, it affects calving rates (Whitehead 1997b). Thus, systematic changes in oceanic climate, which are one consequence of global warming (Walther et al. 2002), have the potential to affect the sperms.

Environmental changes are generally most threatening to species that have small ranges, strict environmental requirements, and limited mobility. The sperm whale, with its huge global range (see section 2.2), broad diet (see section 2.3), and high mobility (see section 3.4) would seem to be well adapted to deal with the effects of global warming. However, two aspects of the sperm whale's biology give some cause for concern. First, the very low potential rate of increase of sperm whale populations (see section 4.3) means that rather small drops in fecundity or survival could affect population viability. Second, there is growing evidence that sperm whales deal with their environment using culturally transmitted knowledge (see section 7.3), which may become out of date or irrelevant as oceanic climate changes. As for so many other species, the effects of global warming on sperm whales are essentially unpredictable.

Management of Threats

Many of the factors just discussed—"scientific whaling," shipping, debris, entanglement in fishing gear, noise, chemical pollution—are known to kill or affect individual sperm whales, but are not thought, by themselves, to be a substantial threat to sperm whale populations. But what about their cumulative effects on a population with a particularly low potential rate of increase? If we keep abusing the ocean with increasing noise pollution, chemical pollution, high-speed ships, and anthropogenic climate change, sperm whales will die at increasing rates. At some time, perhaps not too far in the future, there will come a point at which mortality overtakes the low reproductive rate of the population. The sperm whale population shows little sign of being divided into discrete geographic populations at scales of less than an ocean basin (see section 4.1). Animals move across and between oceans, just as ships, noise, and chemicals do. The ocean is one organ. We cannot preserve just one part.

Given our inability to manage the relatively straightforward threat posed by whaling, or even to appreciate the effects of it, there would seem to be very little chance of being able to predict the population-level consequences of the much less quantifiable threats posed by noise, chemical pollution, and climate change before it is too late (Whitehead et al. 2000b). We must not be seduced by the false promise that science can understand and reverse all the adverse consequences of our overexploitation of Earth, and scientific research should not be used, as it so often is, in the place of hard management decisions (Reeves and Reijnders 2002). To save the sperm whale, and other oceanic life, we must conserve on a grand scale, and we must do this in the absence of conclusive scientific data on what levels of threats are dangerous.

9.3   Studying Sperm Whales in the Future

Although we need to be aware of the substantial limitations of current scientific methodology, scientific research is important in trying to understand the threats we pose to sperm whales and in trying to mitigate them. However, I think there is a more fundamental role for science: getting to know the whale itself. In this book I have described how sperm whales relate to their environment and how they move through the ocean, I have estimated how many there are, and I have uncovered the basics of their social systems. But there is no mention of how the sperm whale sees its world. Does the sperm whale view its prey, those strange mesopelagic squid (see figs. 2.12 and

2.13), like a hunting lion, a grazing cow, or in some completely different way? Does it see its lifelong companions as attractive stimuli, or as beings with personalities and desires? I have ideas on these subjects (see section 9.4), but no convictions, and nothing even approaching scientific data, despite nearly 20 years of a working life, and several years at sea, largely focused on these animals. From this standpoint, it is extraordinary to read the works of those who have spent comparable times studying the great apes and get a feel for their insight into the minds of their subjects (e.g., Boesch and Boesch-Achermann 2000, 225–57). Will we ever approach such a view of sperm whales?

Because of the difficulties involved in observing oceanic animals, and especially deep-diving ones, sperm whale scientists depend on technology much more than, say, primatologists do. By the standards of cetacean science, most of the work described in this book uses quite primitive field technology: ocean-going sailing vessels, 35 mm cameras, standard hydrophones, and tape recorders. Its technical sophistication lies mainly in the analyses, especially the genetic studies (see appendix, section A.5) and analyses of social structure (Whitehead 1997a). However, a range of rapidly developing techniques are giving us much deeper insights into what the animals are doing under water (Read 1998; Whitehead et al. 2000a). For instance, hydrophone arrays can tell us the spatial arrangement of vocalizing animals under water, and tags attached to animals can collect data on movement, the oceanic environment, vocalizations, physiology, and potentially, the focal animal's social environment (through visual, acoustic, and perhaps sonar records).

There have been some successful deployments of both hydrophone arrays around sperm whales and tags attached to them (e.g., Watkins and Schevill 1977a; Watkins et al. 1993, 1999; Møhl et al. 2000; Madsen et al. 2002), but these have been very few and of short duration. To date, deployment of these tools has proved difficult and uncertain. This will undoubtedly change. Small, noninvasive tags are being routinely attached to killer whales and some other small cetaceans, and the data are being retrieved consistently (Baird 1998). Hydrophone arrays are steadily becoming simpler to deploy (Sayigh et al. 1993; Miller and Tyack 1998) and providing more detailed information (e.g., Miller and Tyack 1998). And there are other innovative approaches being used to obtain a better view of the behavior of cetaceans, such as lighter-than-air balloons (Nowacek et al. 2001).

Suppose we could tell the identities and spatial arrangements of participants in a coda exchange through an acoustic array, or watch, from a sperm whale's perspective, the physical interactions within socializing groups, individually identifying the social partners of a focal animal, through a visually recording tag. Suppose we could do these things regularly. Then, I think, we could begin to approach the insights that primatologists have into their subjects. These are not unrealistic expectations for the next decade of sperm whale research.

With these prospects, and with further developments in the standard genetic, photoidentification, and acoustic recording techniques used today (see the appendix), the future looks bright. However, we need to be thoughtful. Technology can take on a life of its own, driving science, and driving it in unpredictable and occasionally unproductive directions. Both for the validity of the science and the welfare of the animals, we must guard against our devices affecting the animals. And we need to keep referencing our studies to the biology of the animals, ground-truthing the output of technology with direct observation whenever possible.

Our knowledge of sperm whale societies has many gaps, including: *
- interactions among individual sperm whales
- the details of calf care
- the social context and function of vocalizations made by sperm whales
- the significance of culture in determining behavior
- spatial and temporal variation in the distribution of sperm whale prey
- the foraging behavior of individual whales at depth
- the identity and importance of non-mammalian competitors
- the long-term movements of mature males between and within feeding and breeding grounds
- mate selection

Our ability to investigate even the most inaccessible parts of the universe is accelerating rapidly. I suspect that in a remarkably short time, using the methods outlined in the appendix or suggested above, as well as others we can now only dream of or have not yet imagined, many of the current mysteries of the sperm whale will be solved. But this research will undoubtedly throw up another suite of questions. Where will it lead? My suspicion is that we will uncover increasing levels of social, cultural, and cognitive complexity.

9.4   The Nature and Future of the Sperm Whale

In the late 1990s two remarkable novels were published: *White as the Waves,* a retelling of *Moby-Dick* from the perspective of the whale (Baird 1999), and *The White Bone,* about the destruction of elephant society as seen by ele-

---

* This list of gaps is largely based upon a list suggested by one of the anonymous reviewers of this book's proposal.

phants (Gowdy 1998). Both novels use what is known of the biology and social lives of their subject species to build pictures of elaborate societies, cultures, and cognitive abilities. Their females are concerned with religion and environment as well as the survival of calves; their males inhabit a rich social and ecological fabric of which mating is only a small part. A reductionist might class these portraits with *Winnie-the-Pooh* as fantasies on the lives of animals. But for me they ring true, and may well come closer to the natures of these animals than the coarse numerical abstractions that come from my own scientific observations.

These books are built on what we have found out about sperm whale society and similar, but more detailed, work by elephant scientists (e.g., Lee 1991b). I think the communication should be reciprocal. We need to take these constructions, note the large parts that are consistent with what we now know, and use them as hypotheses to guide our work. Sperm whale culture may be restricted to coda types and movement patterns. But it could also include whole suites of techniques for making a living from an unpredictable ocean and relating to other sperms. It might encompass abstract concepts, perhaps even religion. Sperm whale society may be little more complex than the representation of the cluster diagram in figure 6.10, but there may be very much more (fig. 9.1). And we won't find it if we don't look.

We also won't find it if the habitat is destroyed and the whales are gone. Has the sperm whale a future over more than a few generations? That will depend on whether humans can act collectively to limit our exploitation of the ocean. Some signs, such as the development of marine protected areas and the Convention on the Law of the Sea, are hopeful. Others, such as the resumption of sperm whaling by Japan and the rapidly increasing levels of ocean noise, are not. The long-term future of life in the ocean hangs in the balance.

9.5   Summary

1. Whaling affected sperm whale social structure as well as population numbers, and so seems to have lowered population growth rates. These effects have persisted long after the exploitation ceased.

2. Sperm whales are killed by collisions with ships, ingesting marine debris, and entrapment in fishing gear, and confrontations with fisheries are becoming prominent in some parts of the world. The inconsistent information on the reactions of sperm whales to noise may result from responses that are contingent on the resource value of the noisy area. Both noise pollution

Figure 9.1 A social unit of female sperm whales: a set of animals that maintain proximity to one another and have homogeneous association indices, or perhaps much more? (Copyright Emese Kazár; drawing based on photographs of Flip Nicklin, such as fig. 5.17.)

and chemical pollution are potential threats to sperm whales, but their significance is not well understood. Many of these threats seem to be increasing in severity. Conservation actions must be taken in the absence of clear scientific data.

3. Technological advances should soon give us a much clearer image of the fine-scale behavior and social life of the sperm whale.

4. Working hypotheses for the next phase of sperm whale research should include the possibility that these animals possess elaborate and multi-layered social relationships, societies, and cultures.

Appendix: How We Study the Sperm Whale

A.1 Examining Carcasses

For most of the past 200 years, the humans most concerned with the sperm whale have been whalers. Their activities gave them intimate access to the sperm whale's body. Although the interests of most whalers did not stray far from the most efficient way to turn a large carcass into valuable products, there were some who were intrigued by the animal itself. This was especially the case with the most scientifically educated members of the whaleship's crew, the surgeons. In the early nineteenth century, men like Frederick Bennett and Thomas Beale made good use of the ample opportunities that they had to examine sperm whale carcasses (Beale 1839; Bennett 1840).

The following century, along with explosive harpoons and steam-powered catcher vessels, brought a new kind of observer to the sperm whale hunt: the professional scientist. Unlike their amateur predecessors, these men (they were almost all men) had a mandate: to collect information that would improve the efficiency of the whaling industry. Their charge included technical matters such as developing the best techniques for processing carcasses, but it also concerned the management of whaling. The scientists believed that if they had a good understanding of the life history and population biology of the whale, then whaling could be regulated on a sustainable basis. By about 1980, this was generally recognized as a delusion: the life history characteristics of the whale makes sustainable management difficult in theory, and the pressures of commerce and politics made it impossible in practice. The 1986 moratorium on commercial whaling was largely a consequence of this realization.

But, by then, the scientists had developed a suite of methods for studying the life histories and population biologies of whales from carcasses, and had amassed a large set of data on all the commercially important species, including the sperm. Some of the data were not especially reliable (see Best 1989). But a tremendous amount had been learned.

These whaling industry scientists studied the distributions and movements of the whales using catch-per-unit-effort statistics as well as metal "Discovery" tags bearing serial numbers, which were shot into whales and later recovered when the animals were killed (Brown 1978). They measured, and occasionally weighed, the carcasses. They described morphology, and they collected ectoparasites as keys to migrations. They pulled and sectioned teeth because distinctive layers are laid down seasonally, so that by counting these layers, the animal can be aged (e.g., Bow and Purday 1966). But much of their most informative data came from internal examinations of thousands of carcasses.

To look at feeding, they examined stomach contents, especially the lower rostra (beaks) of squid in the case of sperm whales, and they measured blubber thickness as a key to long-term feeding success (e.g., Clarke 1980; Clarke et al. 1988; Best 1999). Reproductive biology was a particular focus because reproductive parameters were vital elements of population models. For males, the scientists measured volumes of seminal fluid and testis weights, while for females, the uterus was examined for fetuses, the mammary glands for milk, and the ovaries for the "corpora" that give a summary of reproductive history (corpora lutea, which indicate ovulation, and corpora albicantia, which record past ovulations) (e.g., Best et al. 1984).

Between 1945 and 1988, scientists and technicians armed with these methods worked in the shore stations and on the factory vessels of a number of countries (but perhaps especially Australia, Chile, Japan, New Zealand, Peru, South Africa, the Soviet Union, and the United States), examined hundreds of thousands of sperm whale carcasses, made their analyses, and drew their conclusions. Some of these conclusions, such as those on stock structure, were contentious at the time and have been largely superseded by more modern research (e.g., the molecular genetic studies of Lyrholm et al. 1999). In contrast, studies of reproduction and feeding using carcasses have been particularly useful in drawing a picture of the behavior and ecology of the sperm whale.

A.2   Studying Living Animals

During the course of the studies on carcasses, inferences about behavior were made. For instance, the different ectoparasite fauna found on large males and on females that were caught from the same group suggested, correctly, to Best (1979) that the relationship between males and females was transitory. However, the carcass research could not address behavior directly.

This was the goal of the *Tulip* project, which Jonathan Gordon and I began in 1982 (see section 1.6). During the course of 3 years off Sri Lanka, Jonathan developed a suite of methods for studying the behavior of the living sperm whale (Whitehead and Gordon 1986; Gordon 1987a, 1991b). Since then, these techniques have been refined and useful new tools have been added (e.g., Dawson et al. 1995; Goold 1996).

Tulip was a 10 m ocean-going auxiliary sloop with a rubber-mounted diesel engine (see fig. 1.10). Rather fortuitously, she turned out to be a suitable platform for long-term behavioral studies of female and immature sperm whales. However, subsequently, both Jonathan and I have used rather larger boats, the 14 m *Song of the Whale* and the 13 m *Balaena,* respectively (fig. A.1). Compared with 10 m *Tulip,* the 13–14 m size range increases our comfort, cruising range, and research possibilities—for instance, permitting 3-week trips to sea instead of the 2-week trips that we managed with *Tulip.* The costs and complexity of operation of these boats were not drastically greater than those for *Tulip,* and I believe their size to be nearly ideal for pelagic work on sperm whale behavior. However, the same kinds of studies have been done from larger vessels (e.g., Kahn 1991), although expenses are much greater. There is no absolute necessity to use sailing vessels, as our boats usually maneuver around the whales under power. However, sailing vessels can stay at sea for longer than powered vessels of the same size, are quieter and more stable, and have a built-in high vantage point. Male sperm whales at high latitudes have been studied from a much wider range of vessels, including high-speed rigid-hull inflatables (Jaquet et al. 2000) and whaling vessels converted to whale-watching (Ciano and Huele 2000), although in these studies there has been no attempt to follow animals for more than a few hours.

The boats we use to study sperm whales are not unusual examples of vessels of their classes, but they usually have a number of additions, such as a crow's nest (see fig. A.1). However, because the sperm whale is a particularly vocal animal, their most important specialized equipment is the hydrophones used to listen to, and record, underwater sounds. Omnidirectional hydrophones are used to record the clicking patterns of the sperms, and if recordings are made on more than two hydrophones, they can provide useful additional information on the location of the whales.

This information is important for several reasons, but the most basic is that it allows us to approach and track the whales. Even a simple array of two hydrophones streamed on the same cable allows the bearing of the whales relative to the boat to be calculated in real time (for instance, using the "Rainbow Click" software; Gillespie 1997), although there is left-

Figure A.1 Balaena, a 13 m ocean-going cutter ("Valiant 40" class) used extensively for sperm whale research around the South Pacific. Special features include the "crow's nest," which provides a high vantage point. (Photograph by Flip Nicklin, courtesy of Minden Pictures.)

Figure A.2 A directional hydrophone, mounted in a bomb-shaped housing behind the research vessel, is used for obtaining a bearing on sperm whale clicks and thus tracking sperm whale groups.

right ambiguity. Other directional hydrophone systems can be used to locate and track sperm whales. For instance, we use a single hydrophone positioned along the horizontal axis of a cone covered with acoustically reflective material, which preferentially listens along the axis of the cone (fig. A.2). This is mounted within a bomb-shaped, acoustically transparent, water-filled housing in such a way that the hydrophone and cone can rotate about a vertical axis. Thus, by turning a wheel at deck level, we can listen in any direction and determine the approximate bearing of a clicking whale. Such systems can be used under way, and even with the engine running, as long as the clicks are reasonably loud. As sperm whale clicks can be heard at about 7 km through a hydrophone, we have a powerful tool for approaching and tracking the animals as long as they are vocalizing.

Sperm whales make clicks during most of the time they are under water (see section 5.4), so a sperm whale at depth is usually locatable and trackable. Their times at the surface are of two principal types, the roughly 8 min intervals between dives and the much longer periods of socializing/resting. The former are too short to disrupt acoustic tracking, while socializing/resting sperm whales usually move slowly and may be making codas, creaks, or other vocalizations (see section 5.5). Additionally, socializing/resting usually occurs during the daylight hours (see section 5.3), when animals can be tracked visually. This combination of characteristics makes it possible to track sperm whales using passive acoustics and vision for substantial periods. With single large males, tracks of 12 hr or so are possible, after which prolonged silent periods may cause the animals to be lost (e.g., Mullins et al. 1988). Although females and immatures in a group tend to synchronize their acoustic and visually observable behavior, the presence of multiple acoustic sources and viewable animals allows tracking to be more consistent, and tracks of several days are not uncommon (e.g., box 6.3). A drawback of this passive acoustic/visual tracking method is that, in areas with reasonably high densities of sperm whales, the identities of the tracked males or groups of females may be inadvertently changed, especially during the night, so that a different animal or group is being tracked in the morning, although contact with clicking sperms was never lost.

Tracking whales obviously tells us much about their movement behavior (see section 3.4), but it also provides an excellent opportunity to collect, either systematically or haphazardly, a range of other useful data. These data include recordings of the whales' vocalizations and notes or images of their visually observable behavior.

There are a variety of ways in which the behavior of animals can be recorded (Altmann 1974). For many purposes the preferred technique is "focal animal sampling," in which an individual animal is chosen randomly or systematically and its behavior is recorded for a predefined length of time (Mann 1999, 2000). Behavior may be noted systematically at regular intervals (e.g., "behavioral state") or as it occurs (e.g., breaches or changes in group composition). From these data, activity budgets can be calculated and other summary analyses carried out. However, these analyses are valid only if the animal can be followed reliably for the predetermined length of time, and if the selection of animals, and the researcher's ability to follow them, is unbiased by their behavioral state. Unfortunately, these conditions rarely hold for sperm whales, and we have been able to use focal animal sampling in only a few special circumstances: following solitary males on the Scotian Shelf (Mullins et al. 1988), and looking at the dynamics of cluster membership during the 8 min or so at the surface between dives (see section 6.8).

Instead of focal animal sampling, we use focal group sampling, in which a group is followed and the behavioral state (foraging or socializing/resting) and other data (such as mean cluster size and presence of large males or small calves) are recorded each hour. During our studies off the Galápagos Islands in 1985 and 1987, we used the much more intensive method of "scan sampling," in which the locations (relative to the boat), compositions (presence of large males or calves), headings, speeds, and behavior of all visible clusters were recorded every 5 min during daylight. Concurrently, we made 4 min recordings of the underwater vocalizations of the whales every hour. In later years, as our focus shifted to other areas of research and our acoustic studies focused on the coda, behavioral records were simplified and recordings were made haphazardly when codas were heard or might be expected.

During daylight, the animals can be approached—gently, as sperm whales are timid animals—and a variety of other data collected, including sonar or depth sounder traces of the whales' dives (e.g., Papastavrou et al. 1989) and physical material from the whales. Sperm whale defecations, which are clearly visible as brown patches in the water, can be noted as a measure of feeding success (see section 2.6) and gathered using long-handled dip nets. Diet can be examined by looking at squid beaks from the feces (Smith and Whitehead 2000; see section 2.3), and, potentially, other prey remnants, or their DNA, can also be examined.

The feces also contain DNA from the whales themselves (e.g., Parsons 2001), but until now, genetic studies of living sperm whales have used samples from biopsy darts or abrasive projectiles that are shot at the whales using crossbows or guns, or from sloughed skin (Whitehead et al. 1990). After some experimentation, we prefer sloughed skin (fig. A.3) as a source of DNA. It is collected by dip net from the wake of a swimming whale, so gathering it is not invasive and does not directly disturb the animals. Additionally, the position of the boat when gathering sloughed skin—directly behind the whale—is the same as that used to take identification photographs, so sloughed skin samples are more often linked to photoidentifications than those from biopsy darts. However, biopsy darts collect higher-quality DNA than is available in sloughed skin, giving this method an advantage, especially in the analysis of nuclear DNA. The material from biopsy darts contains blubber as well as

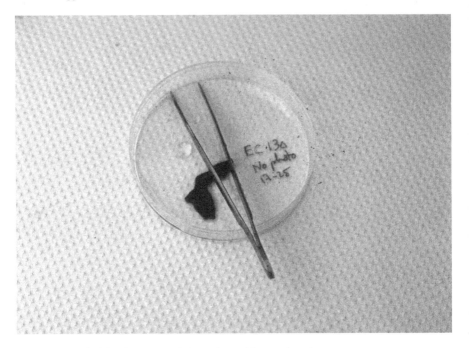

Figure A.3 Sloughed skin from sperm whales can be used for genetic analyses.

skin, so, unlike sloughed skin, it can be used for toxicological analyses (O'Shea 1999) as well as dietary studies using methods such as fatty acid signatures and stable isotopes (e.g., Hooker et al. 2001).

A.3   Photoidentification and Photographic Measurement: The Key to Individuals

His broad fins are bored, and scalloped out like a lost sheep's ear!
—Melville 185, 301

Despite the value of squid beaks, DNA extracted from sloughed skin, and sonar tracks, by far the most useful data we have been able to collect from living sperm whales are photographs, especially photographs allowing the identification of individuals. With individuals identified, long-distance movements can be traced, population sizes estimated, and social structure unraveled. Although other parts of sperm whales are individually distinctive, it is the tail, or flukes, that is most easily photographed and distinguished.

Sperm whales raise their flukes above the surface nearly every time they make a deep dive, about once every 45–50 min (see table 5.6), as well as when lobtailing (see fig. 5.23) and occasionally in the course of other activities. The great majority of individual identifications are made during fluke-ups, as the whales start their dives. The usual practice is to follow the whale about 50 m astern while it is breathing at the surface between dives, then photograph the flukes as they are lifted out of the water, trying to capture them at their most vertical. The cameras used are usually 35 mm single-lens reflex with telephoto (often

300 mm) lenses and black-and-white or color film, although, in the near future, digital cameras will become the norm.

The photoidentification process usually follows the procedure described by Arnbom (1987). After the film is developed, photographs are matched against each other and against pre-existing catalogs, using the pattern of marks and scars along the trailing edge of the flukes (fig. A.4). Useful categories of marks are nicks, distinct nicks, scallops, waves, holes, toothmarks, and missing portions. Each photograph is given a grade, usually following Arnbom's (1987) "Q-value" scale, indicating the quality of the image in terms of focus, contrast, the size and orientation of the flukes in the image, and parts obscured, ranging from $Q = 1$ for a barely usable image to $Q = 5$ for an excellent one. The Q-value of an image is independent of how well marked the animal is. With good-quality images, $Q > 4$, virtually all sperm whales are individually identifiable (Dufault and Whitehead 1995a; Childerhouse and Dawson 1996).

I have developed a computer program that assists the matching process (Whitehead 1990a), and have recently updated it to work with digital images. It pulls likely matches from a digital catalog and image bank. However, in common with other computer programs used for photoidentification, it cannot definitively show that a particular fluke marking pattern is not in the catalog. Thus, fluke matching, especially against a large catalog, is still skilled and time-consuming work. However, sperm whale photoidentification is rather easier than that for most other cetaceans, as the flukes are large and generally well marked, and the pattern of interest occurs almost entirely along one dimension (the fluke edge). These characteristics make both visual and computer-assisted matching simpler than that, for example, for fin whales *(Balaenoptera physalus),* which are identified using two-dimensional coloration patterns (Agler et al. 1990).

The photoidentification process is subject to two kinds of matching errors: matching images from different animals (false positives), and failing to match images from the same animal (false negatives). False positives are rarely much of a problem if only good-quality images are used and the matching is done carefully and well checked. False negatives are a greater concern. They can result from trying to match poor-quality images and/or poorly marked flukes, or from changes in the markings on the flukes between identifications. Two studies have looked at change in the fluke markings of sperm whales (Dufault and Whitehead 1995a; Childerhouse and Dawson 1996). These studies found that marks are gained more often than they are lost, and that change rates are sufficiently low that animals should usually be recognized if there is not too long a gap (< 5 yr) between the photographs. Although neither study could quantify the rate at which marks change dramatically enough over short time periods to make the animals completely unrecognizable, this rate seems to be small. Despite these encouraging findings, it is likely that a low level of false negatives remains in our sperm whale fluke matching data, and this should be borne in mind when using our results. For instance, permanent changes to fluke markings, making animals unrecognizable, will be confounded with mortality and emigration in population models (e.g., Whitehead et al. 1997).

Photographs provide other useful information in addition to individual identity. The marks and scars used to identify each individual are probably incurred during interactions between the whale and its environment, including, in the case of toothmarks, potential predators. Thus, the distribution of marks on an individual's flukes says something about its experiences. So similarities in fluke markings on whales have been used to infer common experiences, especially experiences with predators (Dufault and Whitehead 1995b, 1998).

Photographs of other body parts have also proved useful. On the top of the dorsal fin, there is often a whitish or cream-colored callus, a secondary sexual character (see fig. 1.5; Kasuya and Ohsumi 1966; Clarke and Paliza 1994). These calluses are found most often in mature females (Kasuya and Ohsumi 1966; Clarke and Paliza 1994); thus, they can thus be

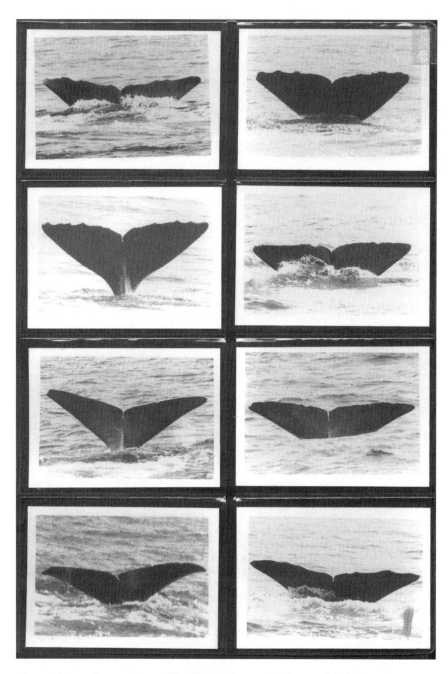

Figure A.4 A page from a catalogue of identification photographs of sperm whale flukes.

Figure A.5 A measuring photograph of sperm whales, showing animals roughly parallel to the horizon.

used to indicate the presence of mature females (e.g., Gordon 1987b; Waters and White-head 1990b). However, as calluses are also occasionally found on immature males, and some sexually mature females do not have calluses, the dorsal fin callus is not a reliable indicator of sex in any particular case. Instead, its incidence is perhaps best used to suggest the overall age-sex composition of groups.

Photographs can also be used to measure the size of sperm whales. The technique developed by Gordon (1990) during the *Tulip* project is suitable for the sailing vessels used in most studies of females and immatures at low latitudes. A photograph (e.g., fig. A.5) is taken showing the whale and the horizon from a position up the vessel's mast. Based on knowledge of the height of the camera above the water surface (usually 7–10 m), the focal length of the lens (usually about 50 mm), and the radius of Earth, the range to the whale can be calculated from the distance between the whale and the horizon on the image, and, with this figure, the true distance between any two visible points on the whale can be estimated. This method produces useful length estimates (with CVs ∼0.05) in conditions with low swell, a clear horizon, and the whale nearly perpendicular to the horizon (Gordon

1990). An alternative technique, developed by Dawson et al. (1995), uses stereo-images from two cameras mounted on a pole to estimate range and thus to scale the image of the whale. It is more suitable for small outboard-powered vessels and produces length estimates with precision (CVs ~0.04) similar to that of Gordon's method. In the future, photographic measuring techniques developed for other species (e.g., Spitz et al. 2000) may also prove useful with sperm whales.

A.4   Quantifying Sounds

Like photographs, recordings of sounds have a range of uses for the sperm whale scientist, in addition to their own intrinsic interest. They serve as beacons for the animals, and so provide information on their location. They can also indicate behavior, the number of animals present, the presence of mature males (from hearing slow clicks; see section 5.2), length, and individual identity. The sounds of the sperm whale, principally the usual click trains, are very suitable for many of these secondary uses. They are audible at long ranges of at least several kilometers; they are quite predictable, being made most of the time when the whales are at depth; and their sharp onset structure is helpful in calculating differences in arrival time at different hydrophones in an array, and so location with respect to the array (e.g., Madsen et al., in press).

The sperm whale sounds are recorded on reel-to-reel, audiocassette, or video or digital-audio tape, or directly onto a computer hard drive, although some processing can be done in real time, especially using the versatile "Rainbow Click" program designed for analyzing sperm whale sounds (Gillespie 1997). Clicks can be examined and counted using pressure-time "waveform" displays produced by an oscilloscope, while frequency-time "spectrograph" displays may indicate features that might be useful for individual identification (see below; fig. 5.1). Rainbow Click gives a useful display that combines temporal, frequency, and locational information. Using such displays, click series can be sorted into patterns, such as usual clicks, creaks, and codas (see section 5.2).

Clicks can be counted automatically and used to estimate the number of animals vocalizing and thus located within acoustic range of the hydrophones (Whitehead and Weilgart 1990). There is also useful information within each click, most notably the inter-pulse interval. Most sperm whale clicks are made up of pulses, and the interval between the pulses reflects the time it takes sound to travel twice the length of the spermaceti organ, and so indicates the length of the whale (Norris and Harvey 1972; Møhl et al. 1981; Gordon 1991a; Goold 1996; Møhl 2001). Acoustic measurement of sperm whales using this technique is thus practicable (e.g., Adler-Fenchel 1980). The best current way to do this seems to be to use cepstrums (the cepstrum is the frequency spectrum of the frequency spectrum of the click), as proposed by Goold (1996), but it is important to note that the inter-pulse interval may change as the whale dives and the spermaceti organ is subjected to increasing pressure (Goold 1996; Madsen et al. 2002).

The acoustic identification of individuals has long been a goal of sperm whale scientists (Watkins 1980). The most sophisticated effort to date seems to be that of Dougherty (1999), who reports nearly perfect discrimination among ten click sequences from different whales when information, such as frequency spectra and inter-pulse intervals, from several clicks is combined, but he does not investigate whether his routines can match the same individual on different occasions. This is a crucial issue, as the characteristics of a click are highly influenced by the depths of the whale and the hydrophone and by the water masses through which the click travels, as well as by the orientation of the hydrophone relative to the whale (Møhl et al. 2000).

A.5   Unraveling Genes

The techniques of molecular genetics are advancing rapidly, and they are allowing extraordinary insights into animal societies from the tiniest of samples. Currently, three types of molecular sequences from sperm whale samples have been examined: the *SRY* and *ZFX* genes, which indicate sex (Richard et al. 1994); the maternally inherited mitochondrial genome, especially the "D-loop" or "control region" (Dillon and Wright 1993); and highly variable nuclear "microsatellites" (Richard et al. 1996b).

The two sex-indicator markers are used in a fairly straightforward way to determine the gender of an animal. Fortunately, they complement each other: *SRY* definitively identifies males, and *ZFX* identifies females. Data from mtDNA sequences and microsatellites can be run through a wide variety of analytical methods, with a range of possible biological objectives. Some of the possible techniques and outputs are illustrated by recent work on sperm whales (e.g., Dillon 1996; Richard et al. 1996a; Lyrholm and Gyllensten 1998; Bond 1999; Lyrholm et al. 1999; Mesnick 2001). These studies had two principal objectives: to investigate the role of kinship in sperm whale social structure (see section 6.6) and to examine the geographic structure of sperm whale populations (see section 4.1). However, these studies had other useful, and sometimes unexpected, by-products, including an estimate of the age at which males disperse from their maternal unit (Richard et al. 1996a) and the discovery of very low diversity in the sperm whale's mitochondrial genome (Lyrholm et al. 1996).

However, this is just the very beginning. The current genetic techniques are becoming much more powerful as sequencing becomes faster and cheaper, and a range of new molecular markers with novel properties have been developed (Sunnucks 2000).

References

Adler-Fenchel, H. S. 1980. Acoustically derived estimates of the size distribution of a sample of sperm whales (*Physeter catodon*) in the western North Atlantic. *Can. J. Fish. Aquat. Sci.* 37:2358–61.

Agler, B. A., J. A. Beard, R. S. Bowman, H. D. Corbett, S. W. Frohock, M. P. Hawvermale, S. K. Katona, S. S. Sadove, and I. E. Seipt. 1990. Fin whale (*Balaenoptera physalus*) photographic identification: Methodology and preliminary results from the western North Atlantic. *Rep. Int. Whal. Commn.* (Special Issue) 12:349–56.

Aguayo, A. 1963. Observaciones sobre la maduréz sexual del cachalot macho (*Physeter catodon*) capturado en aguas chilenas. *Montemar* 11:99–125.

Aguilar, A. 1983. Organochlorine pollution in sperm whales, *Physeter macrocephalus,* from the temperate waters of the eastern North Atlantic. *Mar. Pollut. Bull.* 9:349–52.

Alexander, R. D. 1974. The evolution of social behavior. *Annu. Rev. Ecol. Syst.* 5:325–83.

Allen, K. R. 1980. The influence of schooling behavior on CPUE as an index of abundance. *Rep. Int. Whal. Commn.* (Special Issue) 2:141–46.

Altmann, J. 1974. Observational study of behavior: Sampling methods. *Behaviour* 49:227–67.

Amaratunga, T. 1987. Population biology. In *Cephalopod life cycles,* vol. 2, edited by P. Boyle, 239–52. London: Academic Press.

Amos, B. 1993. Use of molecular probes to analyse pilot whale pod structure: Two novel analytical approaches. *Symp. Zool. Soc. Lond.* 66:33–48.

———. 1996. Levels of variability in cetacean populations have probably changed little as a result of human activities. *Rep. Int. Whal. Commn.* 46:657–58.

Amos, B., C. Schlötterer, and D. Tautz. 1993. Social structure of pilot whales revealed by analytical DNA profiling. *Science* 260:670–72.

Amos, W. 1999. Culture and genetic evolution in whales. *Science* 284:2055a.

André, M. 1997. Distribution and conservation of the sperm whale (*Physeter macrocephalus*) in the Canary Islands. Ph.D. dissertation, University of Las Palmas de Gran Canaria, Spain.

André, M., and C. Kamminga. 2000. Rhythmic dimension in the echolocation click trains of sperm whales: A possible function of identification and communication. *J. Mar. Biol. Assoc. UK* 80:163–69.

André, M., C. Kamminga, and D. Ketten. 1997a. Are low frequency sounds a marine hearing hazard: A case study in the Canary Islands. *Proc. I.O.A.* 19:77–84.

André, M., and J. R. Potter. 2000. Fast-ferry acoustic and direct physical impact on cetaceans: Evidence, trends and potential mitigation. In *Proceedings of the fifth European conference on underwater acoustics, ECUA 2000,* edited by M. E. Zakharia, P. Chevret, and P. Dubail, 491–96. Lyon, France: ECUA.

André, M., M. Terada, and Y. Watanabe. 1997b. Sperm whale (*Physeter macrocephalus*) behavioural response after playback of artificial sounds. *Rep. Int. Whal. Commn.* 47:499–504.

Arch, J. K. 2000. Group function of northern bottlenose whales (*Hyperoodon ampullatus*) in the Gully: Evidence from group structure and disturbance reactions. M.Sc. thesis, Dalhousie University, Halifax, Nova Scotia.

Arnbom, T. 1987. Individual identification of sperm whales. *Rep. Int. Whal. Commn.* 37:201–4.

Arnbom, T., V. Papastavrou, L. S. Weilgart, and H. Whitehead. 1987. Sperm whales react to an attack by killer whales. *J. Mammal.* 68:450–53.

Arnbom, T., and H. Whitehead. 1989. Observations on the composition and behaviour of groups of female sperm whales near the Galápagos Islands. *Can. J. Zool.* 67:1–7.

Arnold, G. P. 1979. Squid: A review of their biology and fisheries. *Lab. Leafl. MAFF Direct. Fish. Res., Lowestoft* 48:1–37.

Arntz, W. E. 1986. The two faces of El Niño 1982–83. *Meeresforschung* 31:1–46.

Asaga, T., Y. Naito, B. J. Le Boeuf, and H. Sakurai. 1994. Functional analysis of dive types of female northern elephant seals. In *Elephant seals: Population ecology, behavior, and physiology,* edited by B. J. Le Boeuf and R. M. Laws, 310–27. Berkeley: University of California Press.

Ashford, J. R., P. S. Rubilar, and A. R. Martin. 1996. Interactions between cetaceans and longline fishery operations around South Georgia. *Mar. Mammal Sci.* 12:452–56.

Au, W. W. L. 1993. *The sonar of dolphins.* New York: Springer Verlag.

Backus, R. H., and W. E. Schevill. 1966. *Physeter* clicks. In *Whales, dolphins and porpoises,* edited by K. S. Norris, 510–27. Berkeley: University of California Press.

Baird, A. 1999. *White as the waves: A novel of Moby Dick.* St. John's, Newfoundland: Tuckamore Books.

Baird, R. W. 1998. Studying diving behavior of whales and dolphins using suction-cup attached tags. *Whalewatcher* 32:3–7.

———. 2000. The killer whale: Foraging specializations and group hunting. In *Cetacean societies,* edited by J. Mann, R. C. Connor, P. Tyack, and H. Whitehead, 127–53. Chicago: University of Chicago Press.

Baird, R. W., and L. M. Dill. 1995. Occurrence and behavior of transient killer whales: Seasonal and pod-specific variability, foraging behavior and prey handling. *Can. J. Zool.* 73:1300–1311.

Baker, C. S., A. Perry, J. L. Bannister, M. T. Weinrich, R. B. Abernethy, J. Calambokidis, J. Lien, R. H. Lambertsen, J. U. Ramirez, O. Vasquez, P. J. Clapham, A. Alling, S. J. O'Brien, and S. R. Palumbi. 1993. Abundant mitochondrial DNA variation and world-wide population structure in humpback whales. *Proc. Natl. Acad. Sci. USA* 90:8239–43.

Balcomb, K. C., and D. E. Claridge. 2001. A mass stranding of cetaceans caused by naval sonar in the Bahamas. *Bahamas J. Sci.* 5:2–12.

Barlow, J. 1995. The abundance of cetaceans in California waters. Part I: Ship surveys in summer and fall of 1991. *Fish. Bull., US* 93:1–14.

Barlow, J., R. W. Baird, J. E. Heyning, K. Wynne, A. M. Manville, L. F. Lowry, D. Hanan, J. Sease, and V. N. Burkanov. 1994. A review of cetacean and pinniped mortality in coastal fisheries along the west coast of the USA and Canada and the east coast of the Russian Federation. *Rep. Int. Whal. Commn.* (Special Issue) 15:405–26.

Barlow, J., and B. L. Taylor. 1998. Preliminary abundance of sperm whales in the northeastern temperate Pacific estimated from combined visual and acoustic surveys. Paper SC/50/CAWS20 presented to the Scientific Committee of the International Whaling Commission.

Barnes, R. H. 1991. Indigenous whaling and porpoise hunting in Indonesia. In *Cetaceans and cetacean research in the Indian Ocean Sanctuary,* edited by S. Leatherwood and G. P. Donovan, 99–106. Nairobi: UNEP Marine Mammal Technical Report.

Barnes, R. S. K., and R. N. Hughes. 1988. *An introduction to marine ecology.* 2d ed. Oxford: Blackwell.

Barrett-Lennard, L. 2000. Population structure and mating patterns of killer whales (*Orcinus orca*) as revealed by DNA analysis. Ph.D. dissertation, University of British Columbia, Vancouver.

Barrett-Lennard, L. G., H. Yurk, and J. K. B. Ford. 2001. A sound approach to the study of culture. *Behav. Brain Sci.* 24:325–26.

Barton, R. A. 1996. Neocortex size and behavioural ecology in primates. *Proc. R. Soc. Lond., B* 263:173–77.

Beale, T. 1839. *The natural history of the sperm whale.* London: John van Voorst.

Beckoff, M., and J. A. Byers. 1998. *Animal play: Evolutionary, comparative, and ecological approaches.* New York: Cambridge University Press.

Begon, M., and M. Mortimer. 1986. *Population ecology.* 2d ed. Sunderland, MA: Sinauer Associates.

Bejder, L., D. Fletcher, and S. Bräger. 1998. A method for testing association patterns of social animals. *Anim. Behav.* 56:719–25.

Benjaminsen, T. 1972. On the biology of the bottlenose whale, *Hyperoodon ampullatus* (Forster). *Norw. J. Zool.* 20:233–41.

Benjaminsen, T., and I. Christensen. 1979. The natural history of the bottlenose whale, *Hyperoodon ampullatus* (Forster). In *Behavior of marine animals,* vol. 3, edited by H. E. Winn and B. L. Olla, 143–64. New York: Plenum.

Bennett, F. D. 1840. *Narrative of a whaling voyage around the globe from the year 1833 to 1836.* London: Richard Bentley.

Benson, A. J., and A. W. Trites. 2002. Ecological effects of regime shifts in the Bering Sea and eastern North Pacific Ocean. *Fish Fisheries* 3:95–113.

Beran, M. J., K. R. Gibson, and D. M. Rumbaugh. 1999. Predicted hominid performance on the transfer index: Body size and cranial capacity as predictors of transfer ability. In *The descent of mind,* edited by M. Corballis and S. E. G. Lea, 87–97. Oxford: Oxford University Press.

Bergman, C. M., J. A. Schaefer, and S. N. Luttich. 2000. Caribou movement as a correlated random walk. *Oecologia* 123:364–74.

Bernard, H. J., and S. B. Reilly. 1999. Pilot whales *Globicephala* Lesson, 1828. In *Handbook of marine mammals,* vol. 6, edited by S. H. Ridgway and R. Harrison, 245–79. San Diego: Academic Press.

Berta, A., and J. L. Sumich. 1999. *Marine mammals: Evolutionary biology.* San Diego: Academic Press.

Berzin, A. A. 1972. *The sperm whale.* Jerusalem: Israel Program for Scientific Translations.

———. 1978. Whale distribution in tropical eastern Pacific waters. *Rep. Int. Whal. Commn.* 28:173–77.

Best, P. B. 1969. The sperm whale (*Physeter catodon*) off the west coast of South Africa. 4. Distribution and movements. *Invest. Rep. Div. Sea Fish. S. Afr.* 78:1–12.

———. 1979. Social organization in sperm whales, *Physeter macrocephalus.* In *Behavior of marine animals,* vol. 3, edited by H. E. Winn and B. L. Olla, 227–89. New York: Plenum.

———. 1983. Sperm whale stock assessments and the relevance of historical whaling records. *Rep. Int. Whal. Commn.* (Special Issue) 5:41–55.

———. 1989. Some comments on the BIWS catch record data base. *Rep. Int. Whal. Commn.* 39:363–69.

———. 1999. Food and feeding of sperm whales *Physeter macrocephalus* off the west coast of South Africa. *S. Afr. J. Mar. Sci.* 21:393–413.

Best, P. B., P. A. S. Canham, and N. Macleod. 1984. Patterns of reproduction in sperm whales, *Physeter macrocephalus. Rep. Int. Whal. Commn.* (Special Issue) 6:51–79.

Bester, M. N., and I. S. Wilkinson. 1994. Population ecology of southern elephant seals at Marion Island. In *Elephant seals: Population ecology, behavior, and physiology,* edited by B. J. Le Boeuf and R. M. Laws, 85–97. Berkeley: University of California Press.

Bigg, M. A., P. F. Olesiuk, G. M. Ellis, J. K. B. Ford, and K. C. Balcomb. 1990. Social organization and genealogy of resident killer whales (*Orcinus orca*) in the coastal waters of British Columbia and Washington State. *Rep. Int. Whal. Commn.* (Special Issue) 12:383–405.

Biggs, D. C., R. R. Leben, and J. G. Ortega-Ortiz. 2000. Ship and satellite studies of mesoscale circulation and sperm whale habitats in the Northeast Gulf of Mexico during GulfCet II. *Gulf Mex. Sci.* 18:15–22.

Boesch, C. 2001. Sacrileges are welcome in science! Opening a discussion about culture in animals. *Behav. Brain Sci.* 24:327–28.

Boesch, C., and H. Boesch-Achermann. 2000. *The chimpanzees of the Taï Forest: Behavioural ecology and evolution.* Oxford: Oxford University Press.

Bond, J. 1999. Genetic analysis of the sperm whale (*Physeter macrocephalus*) using microsatellites. Ph.D. dissertation, University of Cambridge, Cambridge, U.K.

Boran, J. R., and S. L. Heimlich. 1999. Social learning in cetaceans: Hunting, hearing and hierarchies. *Symp. Zool. Soc. Lond.* 73:282–307.

Bow, J. M., and C. Purday. 1966. A method of preparing sperm whale teeth for age determination. *Nature* 210:437–38.

Bowen, B. W., and S. A. Karl. 1997. Population genetics, phylogeography, and molecular evolution. In *The biology of sea turtles,* edited by P. L. Lutz and J. A. Musick, 29–50. Boca Raton, FL: CRC Press.

Bowles, A. E., M. Smultea, B. Würsig, D. P. DeMaster, and D. Palka. 1994. Relative abundance and behavior of marine mammals exposed to transmissions from the Heard Island Feasibility Test. *J. Acoust. Soc. Am.* 96:2469–84.

Boyd, I. L. 2001. Culling predators to protect fisheries: A case of accumulating uncertainties. *Trends Ecol. Evol.* 16:281–82.

Boyd, R., and P. J. Richerson. 1985. *Culture and the evolutionary process.* Chicago: University of Chicago Press.

———. 1996. Why culture is common, but cultural evolution is rare. *Proc. Brit. Acad.* 88:77–93.

Boyer, W. D. 1946. Letter to the editor. *Nat. Hist.* 55:96.

Brabyn, M., and R. V. C. Frew. 1994. New Zealand herd stranding sites do not relate to geomagnetic topography. *Mar. Mammal Sci.* 10:195–207.

Brabyn, M. W., and I. G. McLean. 1992. Oceanography and coastal topography of herd-stranding sites for whales in New Zealand. *J. Mammal.* 73:469–76.

Bradbury, J. W., and S. L. Vehrencamp. 1998. *Principles of animal communication.* Sunderland, MA: Sinauer Associates.

Branch, T. A., and D. S. Butterworth. 2001. Estimates of abundance south of 60°S for cetacean species sighted frequently on the 1978/79 to 1997/98 IWC/IDCR-SOWER sighting surveys. *J. Cet. Res. Manage.* 3:251–70.

Brennan, B., and P. Rodriguez. 1994. Report of two orca attacks on cetaceans in Galápagos. *Notic. Galápagos* 54:28–29.

Brown, S. G. 1978. Whale marking techniques. In *Animal marking,* edited by B. Stonehouse, 71–80. London: Macmillan.

———. 1981. Movements of marked sperm whales in the Southern Hemisphere. *Rep. Int. Whal. Commn.* 31:835–37.

Brownell, R. L., and K. Ralls. 1986. Potential for sperm competition in baleen whales. *Rep. Int. Whal. Commn.* (Special Issue) 8:97–112.

Buckland, S. T., D. R. Anderson, K. P. Burnham, and J. L. Laake. 1993. *Distance sampling: Estimating abundance of biological populations.* New York: Chapman and Hall.

Bullen, F. 1899. *The cruise of the Cachalot.* London: Smith Elder.

Butterworth, D. S., D. L. Borchers, S. Chalis, J. B. De Decker, and F. Kasamatsu. 1995. Estimates of abundance for Southern Hemisphere blue, fin, sei, humpback, sperm, killer and pilot whales from the 1978/79 to 1990/91 IWC/IDCR sighting survey cruises, with extrapolations to the area south of 30° for the first five species based on Japanese scouting vessel data. Paper SC/46/SH24 submitted to the Scientific Committee of the International Whaling Commission. Abstract, *Rep. Int. Whal. Commn.* 45:444.

Byrne, R. W. 1999. Human cognitive evolution. In *The descent of mind,* edited by M. C. Corballis and S. E. G. Lea, 71–87. Oxford: Oxford University Press.

Byrne, R. W., and A. Whiten. 1988. *Machiavellian intelligence.* Oxford: Clarendon.

Calambokidis, J., T. E. Chandler, D. P. Costa, C. W. Clark, and H. Whitehead. 1998. Effects of the ATOC sound source on the distribution of marine mammals observed from aerial surveys off central California. Abstract, World Marine Mammal Science Conference, p. 22.

Caldwell, D. K., M. C. Caldwell, and D. W. Rice. 1966. Behavior of the sperm whale *Physeter catodon* L. In *Whales, dolphins and porpoises,* edited by K. S. Norris, 677–717. Berkeley: University of California Press.

Caldwell, M. C., D. K. Caldwell, and P. L. Tyack. 1990. A review of the signature whistle hypothesis for the Atlantic bottlenose dolphin, *Tursiops truncatus.* In *The bottlenose dolphin: Recent progress in research,* edited by S. Leatherwood and R. R. Reeves, 199–234. San Diego: Academic Press.

Caro, T. M., and M. D. Hauser. 1992. Is there teaching in non-human animals? *Q. Rev. Biol.* 67:151–74.

Carrier, D. R., S. M. Deban, and J. Otterstrom. 2002. The face that sunk the *Essex:* Potential function of the spermaceti organ in aggression. *J. Exp. Biol.* 205:1755–63.

Catchpole, C. K., and P. J. B. Slater. 1995. *Bird song: Biological themes and variations.* Cambridge: Cambridge University Press.

Caughley, G. 1977. *Analysis of vertebrate populations.* New York: John Wiley.

Cavalli-Sforza, L. L., and M. W. Feldman. 1981. *Cultural transmission and evolution: A quantitative approach.* Princeton, NJ: Princeton University Press.

Chapman, D. G. 1980. Report of the Special Meeting on Sperm Whale Assessments, Annex M (SC/SP78/34). Comparison of pregnancy and recently ovulated rate at Paita 1959–64 vs. 1976–78. *Rep. Int. Whal. Commn.* (Special Issue) 2:132.

Chase, O. 1963. *Shipwreck of the whaleship Essex.* New York: Corinth Books.

Childerhouse, S. J., and S. M. Dawson. 1996. Stability of fluke marks used in individual photoidentification of sperm whales at Kaikoura, New Zealand. *Mar. Mammal Sci.* 12:447–51.

Childerhouse, S. J., S. M. Dawson, and E. Slooten. 1995. Abundance and seasonal residence of sperm whales at Kaikoura, New Zealand. *Can. J. Zool.* 73:723–31.

Christal, J. 1998. An analysis of sperm whale social structure: Patterns of association and genetic relatedness. Ph.D. dissertation, Dalhousie University, Halifax, Nova Scotia.

Christal, J., and H. Whitehead. 1997. Aggregations of mature male sperm whales on the Galápagos Islands breeding ground. *Mar. Mammal Sci.* 13:59–69.

———. 2000. A week in the life of a sperm whale group. In *Cetacean societies,* edited by J. Mann, R. C. Connor, P. L. Tyack, and H. Whitehead, 157–59. Chicago: University of Chicago Press.

———. 2001. Social affiliations within sperm whale (*Physeter macrocephalus*) groups. *Ethology* 107:323–40.

Christal, J., H. Whitehead, and E. Letteval. 1998. Sperm whale social units: Variation and change. *Can. J. Zool.* 76:1431–40.

Christensen, I. 1990. A note on recent strandings of sperm whales (*Physeter macrocephalus*) and other cetaceans in Norwegian waters. *Rep. Int. Whal. Commn.* 40:513–15.

Christensen, I., T. Haug, and N. Oien. 1992. Seasonal distribution, exploitation and present abundance of stocks of large baleen whales (Mysticeti) and sperm whales (*Physeter macrocephalus*) in Norwegian and adjacent waters. *ICES J. Mar. Sci.* 49:341–55.

Church, A. C. 1938. *Whale ships and whaling.* New York: Bonanza Books.

Ciano, J. N., and R. Huele. 2000. Photoidentification of sperm whales at Bleik Canyon, Norway. *Mar. Mammal Sci.* 17:175–80.

Clapham, P. J. 2000. The humpback whale: Seasonal breeding and feeding in a baleen whale. In *Cetacean societies,* edited by J. Mann, R. C. Connor, P. L. Tyack, and H. Whitehead, 173–96. Chicago: University of Chicago Pres.

Clapham, P. J., S. B. Young, and R. L. J. Brownell. 1999. Baleen whales: Conservation issues and the status of the most endangered populations. *Mammal Rev.* 29:35–60.

Clarke, M. R. 1962. Significance of cephalopod beaks. *Nature* 193:560–61.

———. 1970. The function of the spermaceti organ of the sperm whale. *Nature* 238:405–6.

———. 1976. Observation on sperm whale diving. *J. Mar. Biol. Assoc. UK* 56:809–10.

———. 1977. Beaks, nets and numbers. *Symp. Zool. Soc. Lond.* 38:89–126.

———. 1978. Buoyancy control as a function of the spermaceti organ in the sperm whale. *J. Mar. Biol. Assoc. UK* 58:27–71.

———. 1980. Cephalopoda in the diet of sperm whales of the Southern Hemisphere and their bearing on sperm whale biology. *Discovery Rep.* 37:1–324.

———. 1987. Cephalopod biomass: Estimation from predation. In *Cephalopod life cycles,* vol. 2, edited by P. R. Boyle, 221–37. London: Academic Press.

Clarke, M. R., and N. MacLeod. 1976. Cephalopod remains from sperm whales caught off Iceland. *J. Mar. Biol. Assoc. UK* 56:733–49.

Clarke, M. R., N. MacLeod, and O. Paliza. 1976. Cephalopod remains from the stomachs of sperm whales caught off Peru and Chile. *J. Zool.* 180:477–93.

Clarke, M. R., H. R. Martins, and P. Pascoe. 1993. The diet of sperm whales (*Physeter macrocephalus* Linnaeus 1758) off the Azores. *Phil. Trans. R. Soc. Lond., B* 339:67–82.

Clarke, M. R., and R. Young. 1998. Description and analysis of cephalopod beaks from stomachs of six species of odontocete cetaceans stranded on Hawaiian shores. *J. Mar. Biol. Assoc. UK* 78:623–41.

Clarke, R. 1956. Sperm whales of the Azores. *Discovery Rep.* 28:237–98.

Clarke, R., A. Aguayo, and O. Paliza. 1980. Pregnancy rates of sperm whales in the southeast Pacific between 1959 and 1962 and a comparison with those from Paita, Peru between 1975 and 1977. *Rep. Int. Whal. Commn.* (Special Issue) 2:151–58.

Clarke, R., and O. Paliza. 1988. Intraspecific fighting in sperm whales. *Rep. Int. Whal. Commn.* 38:235–41.

———. 1994. Sperm whales of the southeast Pacific. Part V. The dorsal fin callus. *Invest. Cetacea* 25:9–91.

———. 2000. The food of sperm whales of the Southeast Pacific. *Mar. Mammal Sci.* 17:427–29.

Clarke, R., O. Paliza, and A. Aguayo. 1988. Sperm whales of the southeast Pacific. Part IV. Fatness, food and feeding. *Invest. Cetacea* 21:53–195.

————. 1994. Sperm whales of the southeast Pacific. Part VI. Growth and breeding in the male. *Invest. Cetacea* 25:93–224.

Clutton-Brock, T. H. 1989. Mammalian mating systems. *Proc. R. Soc. Lond., B* 236: 339–72.

Colnett, J. 1798. *A voyage to the South Atlantic and round Cape Horn into the Pacific Ocean, for the purpose of extending the spermaceti whale fisheries.* London: W. Bennett.

Connor, R. C. 2000. Group living in whales and dolphins. In *Cetacean societies,* edited by J. Mann, R. C. Connor, P. L. Tyack, and H. Whitehead, 199–218. Chicago: University of Chicago Press.

————. 2001. Individual foraging specializations in marine mammals: Culture and ecology. *Behav. Brain Sci.* 24:329–30.

Connor, R. C., J. Mann, P. L. Tyack, and H. Whitehead. 1998. Social evolution in toothed whales. *Trends Ecol. Evol.* 13:228–32.

Connor, R. C., A. J. Read, and R. Wrangham. 2000a. Male reproductive strategies and social bonds. In *Cetacean societies,* edited by J. Mann, R. C. Connor, P. L. Tyack, and H. Whitehead, 247–69. Chicago: University of Chicago Press.

Connor, R. C., R. A. Smolker, and A. F. Richards. 1992. Two levels of alliance formation among male bottlenose dolphins (*Tursiops* sp.). *Proc. Natl. Acad. Sci. USA* 89: 987–90.

Connor, R. C., R. S. Wells, J. Mann, and A. J. Read. 2000b. The bottlenose dolphin: Social relationships in a fission-fusion society. In *Cetacean societies,* edited by J. Mann, R. C. Connor, P. L. Tyack, and H. Whitehead, 91–126. Chicago: University of Chicago Press.

Cooke, J. G. 1985. On the relationship between catch per unit effort and whale abundance. *Rep. Int. Whal. Commn.* 35:511–19.

————. 1986a. On the utility of mark-recapture experiments for the detection of trends in whale stocks. *Rep. Int. Whal. Commn.* 36:451–55.

————. 1986b. A review of some problems relating to the assessment of sperm whale stocks, with reference to the western North Pacific. *Rep. Int. Whal. Commn.* 36: 187–90.

Cooke, J. G., and W. K. de la Mare. 1983. Description of and simulation studies on the length-specific sperm whale assessment technique. *Rep. Int. Whal. Commn.* 33: 741–45.

Corballis, M. C. 1999. Phylogeny from apes to humans. In *The descent of mind,* edited by M. C. Corballis and S. E. G. Lea, 40–70. Oxford: Oxford University Press.

Corkeron, P. J., and R. C. Connor. 1999. Why do baleen whales migrate? *Mar. Mammal Sci.* 15:1228–45.

Costa, D. P., and T. M. Williams. 1999. Marine mammal energetics. In *Biology of marine mammals,* edited by J. E. Reynolds and S. A. Rommel, 176–217. Washington, D.C.: Smithsonian Institution Press.

Cranford, T. W. 1999. The sperm whale's nose: Sexual selection on a grand scale? *Mar. Mammal Sci.* 15:1133–57.

Culik, B., J. Hennicke, and T. Martin. 2000. Humboldt Penguins outmanoevering El Niño. *J. Exp. Biol.* 203:2311–22.

Dahlheim, M. E., and J. E. Heyning. 1999. Killer whale *Orcinus orca* (Linnaeus, 1758). In *Handbook of marine mammals,* vol. 6, edited by S. H. Ridgway and R. Harrison, 281–322. San Diego: Academic Press.

Dalebout, M. L., S. K. Hooker, and I. Christensen. 2001. Genetic diversity and population structure among northern bottlenose whales, *Hyperoodon ampullatus,* in the western North Atlantic. *Can. J. Zool.* 79:478–84.

Darwin, C. 1882. *A naturalist's voyage.* London: John Murray.

da Silva, J., and J. M. Terhune. 1988. Harbour seal grouping as an anti-predator strategy. *Anim. Behav.* 36:1309–16.

Davis, R. W., and G. S. Fargion. 1996. *Distribution and abundance of cetaceans in the north-central and western Gulf of Mexico, final report.* Vol. 2. New Orleans, LA: U.S. Department of the Interior.

Dawson, S. M., C. J. Chessum, P. J. Hunt, and E. Slooten. 1995. An inexpensive, stereographic technique to measure sperm whales from small boats. *Rep. Int. Whal. Commn.* 45:431–36.

Deecke, V. B., J. K. B. Ford, and P. Spong. 2000. Dialect change in resident killer whales: Implications for vocal learning and cultural transmission. *Anim. Behav.* 40:629–38.

de la Mare, W. K., and J. G. Cooke. 1985. Analyses of the sensitivity of the length-specific estimation procedure to some departures from underlying assumptions. *Rep. Int. Whal. Commn.* 35:193–97.

de Waal, F. B. M. 1999. Cultural primatology comes of age. *Nature* 399:635–36.

———. 2000. *The ape and the sushi master.* New York: Basic Books.

Diamond, J. 1998. *Why is sex fun? The evolution of human sexuality.* New York: Basic Books.

Dillon, M. C. 1996. Genetic structure of sperm whale populations assessed by mitochondrial DNA sequence variation. Ph.D. dissertation, Dalhousie University, Halifax, Nova Scotia.

Dillon, M. C., and J. M. Wright. 1993. Nucleotide sequence of the D-loop region of the sperm whale (*Physeter macrocephalus*) mitochondrial genome. *Mol. Biol. Evol.* 10:296–305.

Di Natale, A., and A. Mangano. 1985. Mating and calving of the Sperm Whale in the central Mediterranean Sea. *Aquat. Mamm.* 1:7–9.

Di Natale, A., and G. N. di Sciara. 1994. A review of the passive fishing nets and traps used in the Mediterranean Sea and of their cetacean by-catch. *Rep. Int. Whal. Commn.* (Special Issue) 15:189–202.

Donovan, G. P. 1991. A review of IWC stock boundaries. *Rep. Int. Whal. Commn.* (Special Issue) 13:39–68.

Dougherty, A. M. 1999. Acoustic identification of individual sperm whales (*Physeter macrocephalus*). M.Sc. thesis, University of Washington, Seattle.

Douglas-Hamilton, I., R. F. W. Barnes, H. Shoshani, A. C. Williams, and A. J. T. Johnsingh. 2001. Elephants. In *The new encyclopedia of mammals,* edited by D. Macdonald, 436–45. Oxford: Oxford University Press.

Dufault, S., and H. Whitehead. 1995a. An assessment of changes with time in the marking patterns used for photo-identification of individual sperm whales, *Physeter macrocephalus. Mar. Mammal Sci.* 11:335–43.

———. 1995b. An encounter with recently wounded sperm whales (*Physeter macrocephalus*). *Mar. Mammal Sci.* 11:560–63.

———. 1995c. The geographic stock structure of female and immature sperm whales in the South Pacific. *Rep. Int. Whal. Commn.* 45:401–5.

———. 1998. Regional and group-level differences in fluke markings and notches of sperm whales. *J. Mammal.* 79:514–20.

Dufault, S., H. Whitehead, and M. Dillon. 1999. An examination of the current knowledge on the stock structure of sperm whales (*Physeter macrocephalus*) worldwide. *J. Cet. Res. Manage.* 1:1–10.

Duncan, D. D. 1941. Fighting giants of the Humboldt. *Natl. Geogr.* 79:373–400.

Efron, B., and G. Gong. 1983. A leisurely look at the bootstrap, the jackknife, and cross-validation. *Am. Stat.* 37:36–48.

Ellis, R. 2002. Whaling, traditional. In *Encyclopedia of marine mammals,* edited by W. F. Perrin, B. Würsig, and J. G. M. Thewissen, 1316–28. San Diego: Academic Press.

Ellsworth, S. G. 1990. *The journals of Addison Pratt.* Salt Lake City: University of Utah Press.

Elsner, R. 1999. Living in water: Solutions to physiological problems. In *Biology of marine mammals,* edited by J. E. Reynolds and S. A. Rommel, 73–116. Washington, D.C.: Smithsonian Institution Press.

Emlen, S. T. 1991. Evolution of cooperative breeding in birds and mammals. In *Behavioural ecology: An evolutionary approach,* 3d ed., edited by J. R. Krebs and N. B. Davies, 301–37. Oxford: Blackwell.

Estes, J. A., M. T. Tinker, T. M. Williams, and D. F. Doak. 1998. Killer whale predation on sea otters linking oceanic and nearshore ecosystems. *Science* 282:473–74.

Evans, K., M. Morrice, M. Hindell, and D. Thiele. 2002. Three mass strandings of sperm whales (*Physeter macrocephalus*) in southern Australian waters. *Mar. Mammal Sci.* 18:622–43.

Evans, P. G. H. 1987. *The natural history of whales and dolphins.* London: Facts on File.

Evans, P. G. H., and J. A. Raga. 2002. *Marine mammals: Biology and conservation.* Amsterdam: Kluwer.

FAO. 2000. FISHSTAT plus: Universal software for fishery statistical time series. FAO Fisheries Department, Fishery Information, Data and Statistics Unit.

Folt, C. L., and C. W. Burns. 1999. Biological drivers of zooplankton patchiness. *Trends Ecol. Evol.* 14:300–305.

Ford, J. K. B. 1991. Vocal traditions among resident killer whales (*Orcinus orca*) in coastal waters of British Columbia. *Can. J. Zool.* 69:1454–83.

Ford, J. K. B., G. M. Ellis, and K. C. Balcomb. 2000. *Killer whales.* 2d ed. Vancouver: University of British Columbia Press.

Fowler, C. W. 1984. Density dependence in cetacean populations. *Rep. Int. Whal. Commn.* (Special Issue) 6:373–79.

Frantzis, A. 1998. Does acoustic testing strand whales? *Nature* 392:29.

Fristrup, K. M., and G. R. Harbison. 2002. How do sperm whales catch squids? *Mar. Mammal Sci.* 18:42–54.

Galef, B. G. 1992. The question of animal culture. *Hum. Nat.* 3:157–78.

Gambell, R. 1968. Aerial observations of sperm whale behaviour based on observations, notes and comments by K. J. Pinkerton. *Norsk Hvalfangsttid.* 57:126–38.

———. 1972. Sperm whales off Durban. *Discovery Rep.* 35:199–358.

Gambell, R., C. Lockyer, and G. J. B. Ross. 1973. Observations on the birth of a sperm whale calf. *S. Afr. J. Sci.* 69:147–48.

Garner, M. H., W. H. Garner, and F. R. N. Gurd. 2001. Recognition of primary sequence variations among sperm whale myoglobin components with successive proteolysis procedures. *J. Biol. Chem.* 249:1513–18.

Gaskin, D. E. 1964. Recent observations in New Zealand waters on some aspects of behaviour of the sperm whale (*Physeter macrocephalus*). *Tuatara* 12:106–14.

———. 1967. Luminescence in a squid *Moroteuthis* sp. (probably *ingens* Smith) and a possible feeding mechanism in the sperm whale *Physeter catodon* L. *Tuatara* 15:86–88.

———. 1970. Composition of schools of sperm whales *Physeter catodon* Linn. east of New Zealand. *NZ J. Mar. Freshwater Res.* 4(4):456–71.

———. 1982. *The ecology of whales and dolphins.* London: William Heinemann.

Gaskin, D. E., and M. W. Cawthorne. 1967. Diet and feeding habits of the sperm whale (*Physeter catodon*) in the Cook Strait region of New Zealand. *NZ J. Mar. Freshwater Res.* 1:159–79.

George, J. C., J. Bada, J. Zeh, L. Scott, S. E. Brown, T. O'Hara, and R. Suydam. 1999. Age and growth estimates of bowhead whales (*Balaena mysticetus*) via aspartic acid racemization. *Can. J. Zool.* 77:571–80.

Geraci, J. R., J. Harwood, and V. J. Lounsbury. 1999. Marine mammal die-offs. In *Conservation and management of marine mammals,* edited by J. R. Twiss and R. R. Reeves, 367–95. Washington, D.C.: Smithsonian Institution.

Gibson, K. R. 1999. Social transmission of facts and skills in the human species: Neural mechanisms. In *Mammalian social learning,* edited by H. O. Box and K. R. Gibson, 351–66. Cambridge: Cambridge University Press.

Gillespie, D. 1997. An acoustic survey for sperm whales in the Southern Ocean sanctuary conducted from the RSV *Aurora Australis. Rep. Int. Whal. Commn.* 47:897–907.

Gilmore, R. M. 1959. On the mass strandings of sperm whales. *Pac. Nat.* 1:9–16.

Godin, J.-G. J. 1997. Evading predators. In *Behavioural ecology of teleost fishes,* edited by J.-G. J. Godin, 191–236. Oxford: Oxford University Press.

Goodall, J. 1986. *The chimpanzees of Gombe: Patterns of behavior.* Cambridge, MA: Harvard University Press.

Goold, J. C. 1996. Signal processing techniques for acoustic measurement of sperm whale body lengths. *J. Acoust. Soc. Am.* 100:3431–41.

———. 1999. Behavioural and acoustic observations of sperm whales in Scapa Flow, Orkney Islands. *J. Mar. Biol. Assoc. UK* 79:541–50.

Goold, J. C., and S. E. Jones. 1995. Time and frequency domain characteristics of sperm whale clicks. *J. Acoust. Soc. Am.* 98:1279–91.

Gordon, J. C. D. 1987a. Behaviour and ecology of sperm whales off Sri Lanka. Ph.D. dissertation, University of Cambridge, Cambridge, U.K.

———. 1987b. Sperm whale groups and social behaviour observed off Sri Lanka. *Rep. Int. Whal. Commn.* 37:205–17.

———. 1990. A simple photographic technique for measuring the length of whales from boats at sea. *Rep. Int. Whal. Commn.* 40:581–88.

———. 1991a. Evaluation of a method for determining the length of sperm whales (*Physeter macrocephalus*) from their vocalizations. *J. Zool.* 224:301–14.

———. 1991b. The World Wildlife Fund's Indian Ocean Sperm Whale Project: An example of cetacean research within the Indian Ocean Sanctuary. In *Cetaceans and cetacean research in the Indian Ocean Sanctuary,* edited by S. Leatherwood and G. P. Donovan, 219–39. UNEP Technical Report Number 3, Nairobi, Kenya.

———. 1998. *Sperm whales.* Grantown-on-Spey, Scotland: Colin Baxter.

Gordon, J. C. D., R. Leaper, F. G. Hartley, and O. Chappell. 1992. Effects of whale-watching vessels on the surface and underwater acoustic behaviour of sperm whales off Kaikoura, New Zealand. Science & Research Series no. 52. Department of Conservation, Wellington, New Zealand.

Gordon, J. C. D., and A. Moscrop. 1996. Underwater noise pollution and its significance for whales and dolphins. In *The conservation of whales and dolphins: Science and practice,* edited by M. P. Simmonds and J. Hutchinson, 281–319. Chichester, U.K.: Wiley.

Gordon, J. C. D., A. Moscrop, C. Carlson, S. Ingram, R. Leaper, J. Matthews, and K. Young. 1998. Distribution, movements and residency of sperm whales off the Commonwealth of Dominica, Eastern Caribbean: Implications for the development and regulation of the local whale watching industry. *Rep. Int. Whal. Commn.* 48:551–57.

Gordon, J. C. D., and L. Steiner. 1992. Ventilation and dive patterns in sperm whales, *Physeter macrocephalus,* in the Azores. *Rep. Int. Whal. Commn.* 42:561–65.

Gosho, M. E., D. W. Rice, and J. M. Breiwick. 1984. The sperm whale *Physeter macrocephalus. Mar. Fish. Rev.* 46(4):54–64.

Gowans, S. E. 1999. Social organization and population structure of northern bottlenose whales in the Gully. Ph.D. dissertation, Dalhousie University, Halifax, Nova Scotia.

Gowans, S. E., and L. Rendell. 1999. Head-butting in northern bottlenose whales (*Hyperoodon ampullatus*): A possible function for big heads? *Mar. Mammal Sci.* 15: 1342–50.

Gowans, S. E., H. Whitehead, J. K. Arch, and S. K. Hooker. 2000. Population size and residency patterns of northern bottlenose whales (*Hyperoodon ampullatus*) using the Gully, Nova Scotia. *J. Cet. Res. Manage.* 2:201–10.

Gowans, S. E., H. Whitehead, and S. K. Hooker. 2001. Social organization in northern bottlenose whales (*Hyperoodon ampullatus*): Not driven by deep water foraging? *Anim. Behav.* 62:369–77.

Gowdy, B. 1998. *The white bone.* Toronto: HarperFlamingo.

Green, K., and H. R. Burton. 1993. Comparison of the stomach contents of southern elephant seals, *Mirounga leonina,* at Macquarie and Heard Islands. *Mar. Mammal Sci.* 9:10–22.

Gregr, E. J., L. Nichol, J. K. B. Ford, G. Ellis, and A. W. Trites. 2000. Migration and population structure of northeastern Pacific whales off coastal British Columbia: An analysis of commercial whaling records from 1908–1967. *Mar. Mammal Sci.* 16:699–727.

Gregr, E. J., and A. W. Trites. 2001. Predictions of critical habitat for five whale species in the waters of coastal British Columbia. *Can. J. Fish. Aquat. Sci.* 58:1265–85.

Griffin, R. B. 1999. Sperm whale distributions and community ecology associated with a warm-core ring off Georges Bank. *Mar. Mammal Sci.* 15:33–51.

Gulland, J. A. 1974. Distribution and abundance of whales in relation to basic productivity. In *The whale problem,* edited by W. E. Schevill, 27–51. Cambridge, MA: Harvard University Press.

Gunnlaugsson, T., and J. Sigurjónsson. 1990. NASS-87: Estimation of whale abundance based on observations made onboard Icelandic and Faroese survey vessels. *Rep. Int. Whal. Commn.* 40:571–80.

Gyllensten, U., D. Wharton, A. Josefsson, and A. C. Wilson. 1991. Paternal inheritance of mitochondrial DNA in mice. *Nature* 352:255–57.

Haase, B., and F. Félix. 1994. A note on the incidental mortality of sperm whales (*Physeter macrocephalus*) in Ecuador. *Rep. Int. Whal. Commn.* (Special Issue) 15:481–83.

Hain, J. H. W., M. A. M. Hyman, R. D. Kenney, and H. E. Winn. 1985. The role of cetaceans in the shelf-edge region of the northeastern United States. *Mar. Fish. Rev.* 47:13–17.

Hamilton, W. D. 1964. The genetical evolution of social behaviour. *J. Theor. Biol.* 7:1–52.

Hammond, P. S. 1986. Estimating the size of naturally marked whale populations using capture-recapture techniques. *Rep. Int. Whal. Commn.* (Special Issue) 8:253–82.

Hanlon, R. T., and B.-U. Budelmann. 1987. Why cephalopods are probably not "deaf." *Am. Nat.* 129:312–17.

Heezen, B. C. 1957. Whales entangled in deep sea cables. *Norsk Hvalfangsttid.* 46(12): 665–81.

Herman, L. M., and W. N. Tavolga. 1980. The communication systems of cetaceans. In *Cetacean behavior: Mechanisms and functions,* edited by L. M. Herman, 149–209. New York: Wiley-Interscience.

Hernández-García, V. 1995. The diet of the swordfish *Xiphias gladius* Linnaeus, 1758, in the central east Atlantic, with emphasis on the role of cephalopods. *Fish. Bull., US* 93:403–11.

Heyning, J. E. 1997. Sperm whale phylogeny revisited: Analysis of the morphological evidence. *Mar. Mammal Sci.* 13:596–613.

Hinde, R. A. 1976. Interactions, relationships and social structure. *Man* 11:1–17.

Hindell, M. A., and C. R. McMahon. 2000. Long distance movement of a southern elephant seal (*Mirounga leonina*) from Macquarie Island to Peter 1 Øy. *Mar. Mammal Sci.* 16:504–7.

Hochachka, P. W. 1992. Metabolic biochemistry and the making of a mesopelagic mammal. *Experientia* 48:570–75.

Hoelzel, A. R. 2002. *Marine mammal biology: An evolutionary approach.* Oxford: Blackwell.

Holthuis, L. B. 1987. The scientific name of the sperm whale. *Mar. Mammal Sci.* 3: 87–89.

Hooker, S. K. 1999. Resource and habitat use of northern bottlenose whales in the Gully: Ecology, diving and ranging behaviour. Ph.D. dissertation, Dalhousie University, Halifax, Nova Scotia.

Hooker, S. K., and R. W. Baird. 1999. Deep-diving behaviour of the northern bottlenose whale, *Hyperoodon ampullatus* (Cetacea: Ziphiidae). *Proc. R. Soc. Lond., B* 266: 71–76.

Hooker, S. K., S. J. Iverson, P. Ostrom, and S. C. Smith. 2001. Diet of northern bottlenose whales as inferred from fatty acid and stable isotope analyses of biopsy samples. *Can. J. Zool.* 79:1442–54.

Hooker, S. K., and H. Whitehead. 2002. Click characteristics of northern bottlenose whales (*Hyperoodon ampullatus*). *Mar. Mammal Sci.* 18:69–80.

Hooker, S. K., H. Whitehead, S. Gowans, and R. W. Baird. 2002. Fluctuations in distribution and patterns of individual range use of northern bottlenose whales. *Mar. Ecol. Prog. Ser.* 225:287–97.

Hope, P. L., and H. Whitehead. 1991. Sperm whales off the Galápagos Islands from 1830–50 and comparisons with modern studies. *Rep. Int. Whal. Commn.* 41:273–86.

Hopkins, W. J. 1922. *She blows! And sparm at that!* New York: Houghton Mifflin.

Horn, H. S., and D. I. Rubenstein. 1984. Behavioural adaptations and life history. In *Behavioural ecology: An evolutionary approach,* 2d ed., edited by J. R. Krebs and N. B. Davies, 279–98. Oxford: Blackwell Science Publications.

Horning, M., and F. Trillmich. 1999. Lunar cycles in diel prey migrations exert a stronger effect on the diving of juveniles than adult Galápagos fur seals. *Proc. R. Soc. Lond., B* 266:1127–32.

Horwood, J. W. 1980. Comparative efficiency of catcher boats with and without Asdic in Japanese pelagic whaling fleets. *Rep. Int. Whal. Commn.* (Special Issue) 2:245–49.

Houvenaghel, G. T. 1978. Oceanographic conditions in the Galápagos Archipelago and their relationships with life on the islands. In *Upwelling ecosystems,* edited by R. Boje and M. Tomczak, 181–200. New York: Springer Verlag.

Humphrey, N. K. 1976. The social function of intellect. In *Growing points in ethology,* edited by P. P. G. Bateson and R. A. Hinde, 303–17. Cambridge: Cambridge University Press.

Husson, A. M., and L. B. Holthuis. 1974. *Physeter macrocephalus* Linnaeus, 1758, the valid name for the sperm whale. *Zool. Meded. (Leiden)* 48:205–17.

Intergovernmental Panel on Climate Change. 2001. *Climate change 2001.* Cambridge: Cambridge University Press.

International Fund for Animal Welfare. 1996. Report of the workshop on the special aspects of watching sperm whales. Roseau, Dominica. 36 pp.

International Whaling Commission. 1980a. Report of the Special Meeting on Sperm

Whale Assessments, La Jolla, 27 November to 8 December 1978. *Rep. Int. Whal. Commn.* (Special Issue) 2:107–36.

———. 1980b. *Sperm whales.* Reports of the International Whaling Commission, Special Issue, vol. 2. Cambridge: International Whaling Commission.

———. 1980c. Sub-committee on sperm whale research needs. *Rep. Int. Whal. Commn.* (Special Issue) 2:124.

———. 1982. Report of the sub-committee on sperm whales. *Rep. Int. Whal. Commn.* 32:68–86.

———. 1987. Report of the sub-committee on sperm whales. *Rep. Int. Whal. Commn.* 37:60–67.

———. 1988. Report of the sub-committee on sperm whales. *Rep. Int. Whal. Commn.* 38:67–75.

Ivashin, M. V. 1967. Whale globe-trotter. *Priroda (Moscow)* 8:105–7.

Janik, V. M. 2001. Is social learning unique? *Behav. Brain Sci.* 24:337–38.

Janik, V. M., and P. J. B. Slater. 1997. Vocal learning in mammals. *Advances in the Study of Behavior* 26:59–99.

Jaques, T. G. 1996. Preface. *Bull. Inst. R. Sci. Nat. Belgique, Biol., Suppl.* 67:5–6.

Jaquet, N. 1996a. Distribution and spatial organization of groups of sperm whales in relation to biological and environmental factors in the South Pacific. Ph.D. dissertation, Dalhousie University, Halifax, Nova Scotia.

———. 1996b. How spatial and temporal scales influence understanding of sperm whale distribution: A review. *Mammal Rev.* 26:51–65.

Jaquet, N., S. Dawson, and L. Douglas. 2001. Vocal behavior of male sperm whales: Why do they click? *J. Acoust. Soc. Am.* 109:1154–2259.

Jaquet, N., S. Dawson, and E. Slooten. 2000. Seasonal distribution and diving behaviour of male sperm whales off Kaikoura: Foraging implications. *Can. J. Zool.* 78:407–19.

Jaquet, N., and D. Gendron. 2002. Distribution and relative abundance of sperm whales in relation to key environmental features, squid landings and the distribution of other cetacean species in the Gulf of California, Mexico. *Mar. Biol.* 141:591–601.

Jaquet, N., and H. Whitehead. 1996. Scale-dependent correlation of sperm whale distribution with environmental features and productivity in the South Pacific. *Mar. Ecol. Prog. Ser.* 135:1–9.

———. 1999. Movements, distribution and feeding success of sperm whales in the Pacific Ocean, over scales of days and tens of kilometers. *Aquat. Mamm.* 25:1–13.

Jaquet, N., H. Whitehead, and M. Lewis. 1996. Coherence between 19th century sperm whale distributions and satellite-derived pigments in the tropical Pacific. *Mar. Ecol. Prog. Ser.* 145:1–10.

Jarman, P. J. 1974. The social organization of antelope in relation to their ecology. *Behaviour* 48:215–67.

———. 1983. Mating system and sexual dimorphism in large, terrestrial, mammalian herbivores. *Biol. Rev.* 58:485–520.

Jauniaux, T., L. Brosens, E. Jacquinet, D. Lambrigts, M. Addink, C. Smeenk, and F. Coignoul. 1998. Postmortem investigations on winter stranded sperm whales from the coasts of Belgium and the Netherlands. *J. Wildl. Dis.* 34:99–109.

Jefferson, T. A., M. W. Newcomer, S. Leatherwood, and K. Van Waerebeek. 1994. Right whale dolphins *Lissodelphis borealis* (Peale, 1848) and *Lissodelphis peronii* (Lacépède, 1804). In *Handbook of marine mammals*, vol. 5, edited by S. H. Ridgway and R. Harrison, 355–62. London: Academic Press.

Jefferson, T. A., P. J. Stacey, and R. W. Baird. 1991. A review of killer whale interactions with other marine mammals: Predation to co-existence. *Mammal Rev.* 4:151–80.

Jerison, H. J. 1973. *Evolution of the brain and intelligence.* New York: Academic Press.

Jonsgård, Å. 1968a. Another note on the attacking behaviour of killer whale (*Orcinus orca*). *Norsk Hvalfangsttid.* 57(6):175–76.

———. 1968b. A note on the attacking behaviour of killer whale (*Orcinus orca*). *Norsk Hvalfangsttid.* 57(4):84.

Joseph, J., W. Klawe, and P. Murphy. 1988. *Tuna and billfish: Fish without a country.* La Jolla, CA: Inter-American Tropical Tuna Commission.

Kahn, B. 1991. The population biology and social organization of sperm whales (*Physeter macrocephalus*) off the Seychelles: Indications of recent exploitation. M.Sc. thesis, Dalhousie University, Halifax, Nova Scotia.

Kahn, B., H. Whitehead, and M. Dillon. 1993. Indications of density-dependent effects from comparisons of sperm whale populations. *Mar. Ecol. Prog. Ser.* 93:1–7.

Kasuya, T. 1991. Density-dependent growth in North Pacific sperm whales. *Mar. Mammal Sci.* 7:230–57.

———. 1999. Examination of the reliability of catch statistics in the Japanese coastal sperm whale fishery. *J. Cet. Res. Manage.* 1:109–22.

Kasuya, T., and T. Miyashita. 1988. Distribution of sperm whale stocks in the North Pacific. *Sci. Rep. Whales Res. Inst.* 39:31–75.

Kasuya, T., and S. Ohsumi. 1966. A secondary sexual character of the sperm whale. *Sci. Rep. Whales Res. Inst.* 20:89–94.

Kato, H. 1984. Observation of tooth scars on the head of male sperm whale, as an indication of intra-sexual fightings. *Sci. Rep. Whales Res. Inst.* 35:39–46.

Kato, H., and T. Miyashita. 2000. Current status of the North Pacific sperm whales and its preliminary abundance estimates. Paper SC/50/CAWS2 presented to the Scientific Committee of International Whaling Commission.

Katona, S., and H. Whitehead. 1988. Are Cetacea ecologically important? *Oceanogr. Mar. Biol. Ann. Rev.* 26:553–68.

Kawakami, T. 1980. A review of sperm whale food. *Sci. Rep. Whales Res. Inst.* 32:199–218.

Kazar, E. 2002. Revised phylogeny of the Physeteridae (Mammalia: Cetacea) in the light of *Placoziphius* Van Beneden, 1869 and *Aulophyseter* Kellog, 1927. *Bull. Inst. R. Sci. Nat. Belgique, Sci. Terre* 72:151–70.

Keenleyside, M. H. A. 1979. *Diversity and adaptation in fish behaviour.* Berlin: Springer-Verlag.

Kirpichnikov, A. A. 1950. Present-day distribution of sperm whales in the world ocean according to commercial data. *Byull. Mosk. Ova. Inspyt. Prir. Otd. Biol.* 55:11–25.

Kleiber, M. 1975. *The fire of life.* New York: Krieger.

Kleiman, D. G., and J. R. Malcolm. 1981. The evolution of male parental investment in mammals. In *Parental care in mammals,* edited by D. J. Gubernick and P. H. Klopfer, 347–87. New York: Plenum.

Klein, D. R. 1999. Comparative social learning among arctic herbivores: The caribou, muskox and arctic hare. *Symp. Zool. Soc. Lond.* 72:126–40.

Klinowska, M. 1985a. Cetacean live stranding dates relate to geomagnetic disturbances. *Aquat. Mamm.* 11:109–19.

———. 1985b. Cetacean live stranding sites relate to geomagnetic topography. *Aquat. Mamm.* 11:27–32.

Kojima, T. 1951. On the brain of the sperm whale (*Physeter catodon* L.). *Sci. Rep. Whales Res. Inst.* 6:49–72.

Krebs, C. J. 1989. *Ecological methodology.* New York: Harper and Row.

Laist, D. W. 1997. Impacts of marine debris: Entanglement of marine life in marine debris including a comprehensive list of species with entanglement and ingestion records.

In *Marine debris: Sources, impacts, and solutions,* edited by J. M. Coe and D. B. Rogers, 99–139. New York: Springer.

Laist, D. W., A. R. Knowlton, J. G. Mead, A. S. Collet, and M. Podesta. 2001. Collisions between ships and whales. *Mar. Mammal Sci.* 17:35–75.

Laland, K. N. 1992. A theoretical investigation of the role of social transmission in evolution. *Ethol. Sociobiol.* 13:87–113.

Laland, K. N., P. J. Richerson, and R. Boyd. 1996. Developing a theory of animal social learning. In *Social learning in animals: The roots of culture,* edited by C. M. Heyes and B. G. J. Galef, 129–54. San Diego: Academic Press.

Lalli, C. M., and T. R. Parsons. 1993. *Biological oceanography: An introduction.* Oxford: Pergamon Press.

Lambertsen, R. H. 1997. Natural disease problems of the sperm whale. *Bull. Inst. R. Sci. Nat. Belgique, Biol., Suppl.* 67:105–12.

Laquerist, B. A., K. M. Stafford, and B. R. Mate. 2000. Dive characteristics of satellite-monitored blue whales *Balaenoptera musculus* off the central California coast. *Mar. Mammal Sci.* 16:375–91.

Law, R. J., R. L. Stringer, C. R. Allchin, and B. R. Jones. 1996. Metals and organochlorines in sperm whales (*Physeter macrocephalus*) stranded around the North Sea during the 1994/1995 winter. *Mar. Pollut. Bull.* 32:72–77.

Laws, R. M. 1970. Elephants as agents of habitat and landscape change in East Africa. *Oikos* 21:1–15.

Leaper, R., O. Chappell, and J. Gordon. 1992. The development of practical techniques for surveying sperm whale populations acoustically. *Rep. Int. Whal. Commn.* 42:549–60.

Leaper, R., and M. Scheidat. 1998. An acoustic survey for cetaceans in the Southern Ocean Sanctuary conducted from the German government research vessel *Polarstern. Rep. Int. Whal. Commn.* 48:431–37.

Le Boeuf, B. J., D. P. Costa, A. C. Huntley, and S. D. Feldkamp. 1988. Continuous, deep diving in female northern elephant seals, *Mirounga angustirostris. Can. J. Zool.* 66:446–58.

Le Boeuf, B. J., and R. M. Laws. 1994. Elephant seals: An introduction to the genus. In *Elephant seals: Population ecology, behavior, and physiology,* edited by B. J. Le Boeuf and R. M. Laws, 1–26. Berkeley: University of California Press.

Le Boeuf, B. J., Y. Naito, T. Asaga, D. Crocker, and D. P. Costa. 1992. Swim speed in a female northern elephant seal: Metabolic and foraging implications. *Can. J. Zool.* 70:786–95.

Lee, P. C. 1991a. Reproduction. In *The illustrated encyclopedia of elephants,* edited by S. K. Eltringham, 64–77. London: Salamander Books.

———. 1991b. Social life. In *The illustrated encyclopedia of elephants,* edited by S. K. Eltringham, 48–63. London: Salamander Books.

———. 1994. Social structure and evolution. In *Behaviour and evolution,* edited by P. J. B. Slater and T. R. Halliday, 266–303. Cambridge: Cambridge University Press.

Lee, P. C., and C. J. Moss. 1986. Early maternal investment in male and female African elephant calves. *Behav. Ecol. Sociobiol.* 18:353–61.

Letteval, E., C. Richter, N. Jaquet, E. Slooten, S. Dawson, H. Whitehead, J. Christal, and P. McCall Howard. 2002. Social structure and residency in aggregations of male sperm whales. *Can. J. Zool.* 80:1189–96.

Levin, S. A. 1992. The problem of pattern and scale in ecology. *Ecology* 73:1943–67.

Lewis, C. S. 1955. *The voyage of the* Dawn Treader. London: Harper Collins.

Lilly, J. C. 1978. *Communication between man and dolphin: The possibilities of talking with other species.* New York: Crown Publishers.

Lindenfors, P., B. S. Tullberg, and M. Biuw. 2002. Phylogenetic analyses of sexual selection and sexual size dimorphism in pinnipeds. *Behav. Ecol. Sociobiol.* 52:188–93.

Linnaeus, C. 1758. *Systema naturae.* 10th ed. Stockholm: Laurentii Salvii.

Lockyer, C. 1977. Observations on diving behaviour of the sperm whale. In *A voyage of discovery,* edited by M. Angel, 591–609. Oxford: Pergamon.

———. 1981. Estimates of growth and energy budget for the sperm whale, *Physeter catodon. FAO Fish. Ser.* 5:489–504.

Lucas, Z. N., and S. K. Hooker. 2000. Cetacean strandings on Sable Island, Nova Scotia, 1970–1998. *Can. Field Nat.* 114:45–61.

Lyrholm, T., and U. Gyllensten. 1998. Global matrilineal population structure in sperm whales as indicated by mitochondrial DNA sequences. *Proc. R. Soc. Lond., B* 265:1679–84.

Lyrholm, T., O. Leimar, and U. Gyllensten. 1996. Low diversity and biased substitution patterns in the mitochondrial DNA control region of sperm whales: Implications for estimates of time since common ancestry. *Mol. Biol. Evol.* 13:1318–26.

Lyrholm, T., O. Leimar, B. Johanneson, and U. Gyllensten. 1999. Sex-biased dispersal in sperm whales: Contrasting mitochondrial and nuclear genetic structure of global populations. *Proc. R. Soc. Lond., B* 266:347–54.

MacCall, A. D. 1990. *Dynamic geography of marine fish populations.* Seattle: Washington Sea Grant, University of Washington Press.

Madsen, P. T. 2002. Sperm whale sound production—in the acoustic realm of the biggest nose on record. In *Sperm whale sound production,* Ph.D. dissertation, University of Aarhus, Denmark.

Madsen, P. T., and B. Møhl. 2000. Sperm whales (*Physeter catodon L.* 1758) do not react to sounds from detonators. *J. Acoust. Soc. Am.* 107:668–71.

Madsen, P. T., R. Payne, N. U. Kristiansen, M. Wahlberg, I. Kerr, and B. Møhl. 2002. Sperm whale sound production studied with ultrasonic time/depth-recording tags. *J. Exp. Biol.* 205.

Madsen, P. T., M. Wahlberg, and B. Møhl. In press. Male sperm whale (*Physeter macrocephalus*) acoustics in a high latitude habitat: Implications for echolocation and communication. *Behav. Ecol. Sociobiol.*

Magnusson, K. G., and T. Kasuya. 1997. Mating strategies in whale populations: Searching strategy vs. harem strategy. *Ecol. Modell.* 102:225–42.

Mangold, K. 1987. Reproduction. In *Cephalopod life cycles,* vol. 2, edited by P. Boyle, 157–200. London: Academic Press.

Mann, J. 1999. Behavioral sampling methods for cetaceans: A review and critique. *Mar. Mammal Sci.* 15:102–22.

———. 2000. Unraveling the dynamics of social life: Long-term studies and observational methods. In *Cetacean societies,* edited by J. Mann, R. C. Connor, P. L. Tyack, and H. Whitehead, 45–64. Chicago: University of Chicago Press.

Mann, J., R. C. Connor, P. L. Tyack, and H. Whitehead. 2000. *Cetacean societies: Field studies of dolphins and whales.* Chicago: University of Chicago Press.

Mann, J., and B. B. Smuts. 1998. Natal attraction: Allomaternal care and mother-infant separations in wild bottlenose dolphins. *Anim. Behav.* 55:1097–1113.

Mano, S. 1986. The behavior of sperm whales in schools observed from an operating whaler. *Bulletin of the Faculty of Fisheries, Nagasaki University* 60:1–35.

Marsh, H., and T. Kasuya. 1986. Evidence for reproductive senescence in female cetaceans. *Rep. Int. Whal. Commn.* (Special Issue) 8:57–74.

Marshall, G. J. 1998. CRITTERCAM: An animal-borne imaging and data logging system. *Mar. Technol. Soc. J.* 32:11–17.

Martin, A. P., and S. R. Palumbi. 1993. Body size, metabolic rate, generation time, and the molecular clock. *Proc. Natl. Acad. Sci. USA* 90:4087–91.

Martineau, D., S. Deguise, C. Girard, A. Lagacé, and P. Béland. 1994. Pathology and toxicology of beluga whales from the St. Lawrence estuary, Québec, Canada. *Sci. Total Environ.* 154:201–15.

Mate, B. R. 1989. Watching habits and habitats from Earth satellites. *Oceanus* 32:14–18.

Mate, B. R., and J. T. Harvey. 1984. Ocean movements of radio-tagged gray whales. In *The gray whale* Eschrichtius robustus, edited by M. L. Jones, S. L. Schwartz, and S. Leatherwood, 577–89. San Diego: Academic Press.

Mate, B. R., K. A. Rossbach, S. L. Nieukirk, R. S. Wells, A. B. Irvine, M. D. Scott, and A. J. Read. 1995. Satellite-monitored movements and dive behavior of a bottlenose dolphin (*Tursiops truncatus*) in Tampa Bay, Florida. *Mar. Mammal Sci.* 11:452–63.

Mate, B. R., K. M. Stafford, and D. K. Llungblad. 1994a. A change in sperm whale (*Physeter macrocephalus*) distribution correlated to seismic surveys in the Gulf of Mexico. *J. Acoust. Soc. Am.* 965:3268–69.

Mate, B. R., K. M. Stafford, R. Nawojchik, and J. L. Dunn. 1994b. Movements and dive behavior of a satellite-monitored Atlantic white-sided dolphin (*Lagenorhynchus acutus*) in the Gulf of Maine. *Mar. Mammal Sci.* 10:116–21.

Matsushita, T. 1955. Daily rhythmic activity of the sperm whales in the Antarctic. *Bull. Jpn. Soc. Sci. Fish.* 20:770–73.

Matthews, J. N., L. Steiner, and J. Gordon. 2001. Mark-recapture analysis of sperm whale (*Physeter macrocephalus*) photo-ID data from the Azores (1987–1995). *J. Cet. Res. Manage.* 3:219–26.

Matthews, L. H. 1938. The sperm whale, *Physeter catodon*. *Discovery Rep.* 17:93–168.

Maury, M. F. 1851. *Whale chart*. Miscellaneous no. 8514. Washington, D.C.: U.S. Hydrographic Office.

Maynard Smith, J., and J. Haigh. 1974. The hitch-hiking effect of a favourable gene. *Genet. Res.* 23:23–35.

Mayo, C. A., and M. K. Marx. 1990. Surface foraging behavior of the North Atlantic right whale and associated plankton characteristics. *Can. J. Zool.* 68:2214–20.

McCall Howard, M. P. 1999. Sperm whales *Physeter macrocephalus* in the Gully, Nova Scotia: Population, distribution, and response to seismic surveying. B.Sc. honors thesis, Dalhousie University, Halifax, Nova Scotia.

McComb, K., C. Moss, S. M. Durant, L. Baker, and S. Sayialel. 2001. Matriarchs as repositories of social knowledge in African elephants. *Science* 292:491–94.

McConnell, B. J., and M. A. Fedak. 1996. Movements of southern elephant seals. *Can. J. Zool.* 74:1485–96.

McGrew, W. C. 2003. Ten dispatches from the chimpanzee culture wars. In *Animal social complexity: Intelligence, culture, and individualized societies,* edited by F. B. M. de Waal and P. L. Tyack, 419–39. Cambridge, MA: Harvard University Press.

Mchedlidze, G. A. 2002. Sperm whales, evolution. In *Encyclopedia of marine mammals,* edited by W. F. Perrin, B. Würsig, and J. G. M. Thewissen, 1172–74. San Diego: Academic Press.

McKinnell, S. M., R. D. Brodeur, K. Hanawa, A. B. Hollowed, J. J. Polovina, and C.-I. Zhang. 2001. An introduction to the Beyond El Niño conference: Climate variability and marine ecosystem impacts from the tropics to the Arctic. *Prog. Oceanogr.* 49:1–6.

Mead, J. G. 1989a. Beaked whales of the genus *Mesoplodon*. In *Handbook of marine mammals,* vol. 4, edited by S. H. Ridgway and R. Harrison, 349–430. London: Academic Press.

———. 1989b. Bottlenose whales *Hyperoodon ampullatus* (Forster, 1770) and *Hyperoodon planifrons* Flower, 1882. In *Handbook of marine mammals,* vol. 4, edited by S. H. Ridgway and R. Harrison, 321–48. London: Academic Press.

Mellinger, D. K., A. Thode, and A. Martinez. 2002. Passive acoustic monitoring of sperm whales in the Gulf of Mexico, with a model of acoustic detection distance. Proceedings: Twenty-first annual Gulf of Mexico information transfer meeting, January 2002. New Orleans, LA: U.S. Department of the Interior, Minerals Management Service, Gulf of Mexico OCS Region.

Melville, H. 1851. *Moby-Dick; or, The whale.* London: Penguin (1972).

Mesnick, S. L. 2001. Genetic relatedness in sperm whales: Evidence and cultural implications. *Behav. Brain Sci.* 24:346–47.

Mesnick, S. L., K. Evans, B. L. Taylor, J. Hyde, S. Escorza-Treviño, and A. E. Dizon. 2003. Sperm whale social structure: Why it takes a village to raise a child. In *Animal social complexity: Intelligence, culture, and individualized societies,* edited by F. B. M. de Waal and P. L. Tyack, 170–74. Cambridge, MA: Harvard University Press.

Mesnick, S. L., B. L. Taylor, R. G. Le Duc, S. Escorza-Treviño, G. M. O'Corry-Crowe, and A. E. Dizon. 1999. Culture and genetic evolution in whales. *Science* 284: 2055a.

Milinkovitch, M. C., O. Guillermo, and A. Meyer. 1993. Revised phylogeny of whales suggested by mitochondrial ribosomal DNA sequences. *Nature* 361:346–48.

Miller, P., and P. L. Tyack. 1998. A small towed beamforming array to identify vocalizing resident killer whales (*Orcinus orca*) concurrent with focal behavioral observations. *Deep-Sea Res.* 45:1389–1405.

Mitchell, C. L., S. Boinski, and C. P. van Schaik. 1991. Competitive regimes and female bonding in two species of squirrel monkeys (*Saimiri oerstedi* and *S. sciureus*). *Behav. Ecol. Sociobiol.* 28:55–60.

Mitchell, E. 1975. Report of the Scientific Committee, Annex U. Preliminary report on Nova Scotian fishery for sperm whales (*Physeter catodon*). *Rep. Int. Whal. Commn.* 25:226–35.

———. 1977. Evidence that the northern bottlenose whale is depleted. *Rep. Int. Whal. Commn.* 27:195–203.

Møhl, B. 2001. Sound transmission in the nose of the sperm whale *Physeter catodon:* A post mortem study. *J. Comp. Physiol. A* 187:335–40.

Møhl, B., E. Larsen, and M. Amundin. 1981. Sperm whale size determination: Outline of an acoustic approach. *FAO Fish. Ser.* 5:327–32.

Møhl, B., M. Wahlberg, P. T. Madsen, L. A. Miller, and A. Surlykke. 2000. Sperm whale clicks: Directionality and source level revisited. *J. Acoust. Soc. Am.* 107:638–48.

Moore, K. E., W. A. Watkins, and P. L. Tyack. 1993. Pattern similarity in shared codas from sperm whales (*Physeter catodon*). *Mar. Mammal Sci.* 9:1–9.

Morgan, C. L. 1894. *An introduction to comparative psychology.* London: Scott.

Moscrop, A., and R. Swift. 1999. Atlantic frontier cetaceans: Recent research on distribution, ecology and impacts. Report to Greenpeace, U.K.

Mullins, J., H. Whitehead, and L. S. Weilgart. 1988. Behaviour and vocalizations of two single sperm whales, *Physeter macrocephalus,* off Nova Scotia. *Can. J. Fish. Aquat. Sci.* 45:1736–43.

Myers, J. P. 1983. Space, time and the pattern of individual associations in a group-living species: Sanderlings have no friends. *Behav. Ecol. Sociobiol.* 12:129–34.

National Marine Fisheries Service. 1995. Sperm whale (*Physeter macrocephalus*): Northern Gulf of Mexico stock. Stock Assessment Report, 127–29.

———. 1998. Sperm whale (*Physeter macrocephalus*): North Pacific stock. Stock Assessment Report, 111–14.

———. 1999. Our living oceans: Report on the status of U.S. living marine resources, 1999. U.S. Department of Commerce NOAA Tech. Memo. NMFS-F/SPO-41.

———. 2000a. Sperm whale (*Physeter macrocephalus*): Hawaiian stock. Stock Assessment Report, 217–20.

———. 2000b. Sperm whale (*Physeter macrocephalus*): North Atlantic stock. Stock Assessment Report, 54–59.

National Marine Fisheries Service and United States Navy. 2001. Joint interim report: Bahamas marine mammal mass stranding event 15–16 March 2000.

Nielsen, J. B., F. Nielsen, P.-J. Joergensen, and P. Grandjean. 2000. Toxic metals and selenium in blood from pilot whales (*Globicephala melas*) and sperm whales (*Physeter catodon*). *Mar. Pollut. Bull.* 40:348–51.

Nikaido, M., F. Matsuno, H. Hamilton, R. L. Brownell, Y. Cao, W. Ding, Z. Zuoyan, A. M. Shedlock, R. E. Fordyce, M. Hasegawa, and N. Okada. 2001. Retroposon analysis of major cetacean lineages: The monophyly of toothed whales and the paraphyly of river dolphins. *Proc. Natl. Acad. Sci. USA* 98:7384–89.

Nishiwaki, M. 1962. Aerial photographs show sperm whales' interesting habits. *Norsk Hvalfangsttid.* 51:395–98.

Noad, M. J., D. H. Cato, M. M. Bryden, M.-N. Jenner, and K. C. S. Jenner. 2000. Cultural revolution in whale songs. *Nature* 408:537.

Nolan, C. P., G. M. Liddle, and J. Elliot. 2000. Interactions between killer whales (*Orcinus orca*) and sperm whales (*Physeter macrocephalus*) with a longline fishing vessel. *Mar. Mammal Sci.* 16:658–64.

Norris, K. S., and T. P. Dohl. 1980. The structure and functions of cetacean schools. In *Cetacean behavior: Mechanisms and functions,* edited by L. M. Herman, 211–61. New York: Wiley-Interscience.

Norris, K. S., and G. W. Harvey. 1972. A theory for the function of the spermaceti organ of the sperm whale (*Physeter catodon* L.). In *Animal orientation and navigation,* edited by S. R. Galler, K. Schmidt-Koenig, G. J. Jacobs, and R. E. Belleville, 397–417. Washington, D.C.: NASA.

Norris, K. S., and B. Møhl. 1983. Can odontocetes debilitate prey with sound? *Am. Nat.* 122:85–104.

Norris, K. S., and C. R. Schilt. 1988. Cooperative societies in three-dimensional space: On the origins of aggregations, flocks and schools, with special reference to dolphins and fish. *Ethol. Sociobiol.* 9:149–79.

Nowacek, D. P., P. L. Tyack, and R. S. Wells. 2001. A platform for continuous behavioral and acoustic observation of free-ranging marine mammals: Overhead video combined with underwater audio. *Mar. Mammal Sci.* 17:191–99.

O'Dor, R. K. 1998. Can understanding squid life-history strategies and recruitment improve management? *S. Afr. J. Mar. Sci.* 20:193–206.

Oelschläger, H. H. A., and B. Kemp. 1998. Ontogenesis of the sperm whale brain. *J. Comp. Neurol.* 339:210–28.

Ohlsohn, E. 1991. Patterns of vocalizations made by sperm whales off Madeira and the Azores in 1990. Unpublished report. 27 pp.

Ohsumi, S. 1971. Some investigations on the school structure of sperm whale. *Sci. Rep. Whales Res. Inst., Tokyo* 23:1–25.

———. 1977. Age-length key of the male sperm whale in the North Pacific and comparison of growth curves. *Rep. Int. Whal. Commn.* 27:295–300.

———. 1980. Population assessment of the sperm whale in the North Pacific. *Rep. Int. Whal. Commn.* (Special Issue) 2:31–42.

Ohsumi, S., and Y. Masaki. 1975. Japanese whale marking in the North Pacific, 1963–1972. *Bull. Far Seas Fish. Res. Lab.* 12:171–219.

Okutani, T, and T. Nemoto. 1964. Squids as the food of sperm whales in the Bering Sea and Alaskan Gulf. *Sci. Rep. Whales Res. Inst.* 18:111–22.

Olesiuk, P., M. A. Bigg, and G. M. Ellis. 1990. Life history and population dynamics of resident killer whales (*Orcinus orca*) in the coastal waters of British Columbia and Washington State. *Rep. Int. Whal. Commn.* (Special Issue) 12:209–43.

Osborne, R. W. 1986. A behavioral budget of Puget Sound killer whales. In *Behavioral biology of killer whales,* edited by B. Kirkewold and J. S. Lockard, 211–49. New York: Alan R. Liss.

O'Shea, T. J. 1999. Environmental contaminants and marine mammals. In *Biology of marine mammals,* edited by J. E. Reynolds and S. A. Rommel, 485–563. Washington, D.C.: Smithsonian Institution Press.

Ottensmeyer, A. C. 2001. Social structure of long-finned pilot whales from photo-identification techniques. M.Sc. thesis, Dalhousie University, Halifax, Nova Scotia.

Pabst, D. A., S. A. Rommel, and W. A. McLellan. 1999. The functional morphology of marine mammals. In *Biology of marine mammals,* edited by J. E. Reynolds and S. A. Rommel, 15–72. Washington, D.C.: Smithsonian Institution Press.

Packer, C., and A. E. Pusey. 1983. Adaptations of female lions to infanticide by incoming males. *Am. Nat.* 121:716–28.

Palacios, D. M., and B. R. Mate. 1996. Attack by false killer whales (*Pseudorca crassidens*) on sperm whales (*Physeter macrocephalus*) in the Galápagos Islands. *Mar. Mammal Sci.* 12:582–87.

Papastavrou, V., S. C. Smith, and H. Whitehead. 1989. Diving behaviour of the sperm whale, *Physeter macrocephalus,* off the Galápagos Islands. *Can. J. Zool.* 67:839–46.

Parsons, K. 2001. Reliable microsatellite genotyping of dolphin DNA from faeces. *Mol. Ecol. Notes* 1:341–44.

Pásztor, L., É. Kisdi, and G. Meszéna. 2000. Jensen's inequality and optimal life history strategies in stochastic environments. *Trends Ecol. Evol.* 15:117–18.

Paterson, R. A. 1986. An analysis of four large accumulations of sperm whales observed in the modern whaling era. *Sci. Rep. Whales Res. Inst.* 37:167–72.

Patterson, I. A. P., R. J. Reid, B. Wilson, K. Grellier, H. M. Ross, and P. M. Thompson. 1998. Evidence for infanticide in bottlenose dolphins: An explanation for violent interactions with harbour porpoises? *Proc. R. Soc. Lond., B* 256:1167–70.

Pauly, D., A. W. Trites, E. Capuli, and V. Christensen. 1998. Diet composition and trophic levels of marine mammals. *ICES J. Mar. Sci.* 55:467–81.

Pavan, G., T. J. Hayward, J. F. Borsani, M. Priano, M. Manghi, C. Fossati, and J. Gordon. 2000. Time patterns of sperm whale codas recorded in the Mediterranean Sea 1985–1996. *J. Acoust. Soc. Am.* 107:3487–95.

Payne, K. 1999. The progressively changing songs of humpback whales: A window on the creative process in a wild animal. In *The origins of music,* edited by N. L. Wallin, B. Merker, and S. Brown, 135–50. Cambridge, MA: MIT Press.

Payne, R. 1983. *Communication and the behavior of whales.* Boulder, CO: Westview Press.

———. 1995. *Among whales.* New York: Simon and Schuster.

Perrin, W. F., C. E. Wilson, and F. I. Archer. 1994. Striped dolphin *Stenella coeruleoalba* (Meyen, 1833). In *Handbook of marine mammals,* vol. 5, edited by S. H. Ridgway and R. Harrison, 129–59. London: Academic Press.

Perrin, W. F., B. Würsig, and J. G. M. Thewissen. 2002. *Encyclopedia of marine mammals.* San Diego: Academic Press.

Pervushin, A. S. 1966. Nabliudeniia za rodami u kasholotov. *Zool. Zh.* 45:1892–1893.

Philbrick, N. 2000. *In the heart of the sea: The tragedy of the whaleship* Essex. New York: Penguin.

Pitman, R. L., L. T. Balance, S. L. Mesnick, and S. Chivers. 2001. Killer whale predation on sperm whales: Observations and implications. *Mar. Mammal Sci.* 17:494–507.

Pitman, R. L., and S. J. Chivers. 1999. Terror in black and white. *Nat. Hist.* 107:26–29.

Poole, J. H. 1989. Announcing intent: The aggressive state of musth in African elephants. *Anim. Behav.* 37:140–52.

Pryor, K. 1986. Non-acoustic communicative behavior of the great whales: Origins, comparisons, and implications for management. *Rep. Int. Whal. Commn.* (Special Issue) 8:89–96.

Pulliam, H. R., and T. Caraco. 1984. Living in groups: Is there an optimal group size? In *Behavioural ecology: An evolutionary approach,* 2d ed., edited by J. R. Krebs and N. B. Davies, 122–47. Oxford: Blackwell Scientific Publications.

Ralls, K., R. L. Brownell, and J. Ballou. 1980. Differential mortality by sex and age in mammals, with specific reference to the sperm whale. *Rep. Int. Whal. Commn.* (Special Issue) 2:233–43.

Ramirez, P. 1988. Comportamiento reproductivo del "cachalote" (*Physeter catodon* L.). *Bol. Lima* 59:29–32.

———. 1989. Captura de cachalote en Paita: 1976–1981. *Bol. Lima* 63:81–88.

Ramirez, P., and W. Urquizo. 1985. Los cetaceos mayores y el fenómeno "El Niño" 1982–83. In *El fenómeno El Niño y su impacto en la fauna marina,* edited by W. Arntz, A. Landa, and J. Tarazona, 201–6. Boletin Instituto del Mar Peru (special issue).

Read, A. J. 1996. Incidental catches of small cetaceans. In *The conservation of whales and dolphins: Science and practice,* edited by M. P. Simmonds and J. Hutchinson, 89–128. Chichester, U.K.: Wiley.

———. 1998. Possible applications of new technology to marine mammal research and management. Report to Marine Mammal Commission, Washington, D.C. 36 pp.

Reeves, R. R., and P. J. H. Reijnders. 2002. Conservation and management. In *Marine mammal biology: An evolutionary approach,* edited by A. R. Hoelzel, 388–415. Oxford: Blackwell.

Reeves, R. R., and H. Whitehead. 1997. Status of the sperm whale (*Physeter macrocephalus*) in Canada. *Can. Field Nat.* 111:293–307.

Reijnders, P. J. H. 1986. Reproductive failure of common seals feeding on fish from polluted waters. *Nature* 324:456–57.

Rendell, L., and H. Whitehead. 2001a. Cetacean culture: Still afloat after the first naval engagement of the culture wars. *Behav. Brain Sci.* 24:360–73.

———. 2001b. Culture in whales and dolphins. *Behav. Brain Sci.* 24:309–82.

———. 2003. Vocal clans in sperm whales (*Physeter macrocephalus*). *Proc. Roy. Soc. London B* 270:225–31.

———. In press. Comparing repertoires of sperm whales codas: A multiple methods approach. *Bioacoustics.*

Reyes, J. C., J. G. Mead, and K. Van Waerebeek. 1991. A new species of beaked whale *Mesoplodon peruvianus* sp. n. (Cetacea: Ziphiidae) from Peru. *Mar. Mammal Sci.* 7:1–24.

Reynolds, J. E., and S. A. Rommel. 1999. *Biology of marine mammals.* Washington, D.C.: Smithsonian Institution Press.

Rice, D. W. 1977. Sperm whales in the equatorial eastern Pacific: Population size and social organization. *Rep. Int. Whal. Commn.* 27:333–36.

———. 1978. Sperm whales. In *Marine mammals of eastern North Pacific and Arctic waters.* Seattle, WA: Pacific Search Press.

———. 1989. Sperm whale *Physeter macrocephalus* Linnaeus, 1758. In *Handbook of marine mammals,* vol. 4, edited by S. H. Ridgway and R. Harrison, 177–233. London: Academic Press.

———. 1998. *Marine mammals of the world: Systematics and distribution.* Special publication, no. 4. Lawrence, KS: The Society for Marine Mammalogy.

Rice, D. W., A. A. Wolman, B. R. Mate, and J. T. Harvey. 1986. A mass stranding of sperm whales in Oregon: Sex and age composition of the school. *Mar. Mammal Sci.* 2:64–69.

Richard, K. R. 1995. A molecular genetic analysis of kinship in free-living groups of sperm whales. Ph.D. dissertation, Dalhousie University, Halifax, Nova Scotia.

Richard, K. R., M. C. Dillon, H. Whitehead, and J. M. Wright. 1996a. Patterns of kinship in groups of free-living sperm whales (*Physeter macrocephalus*) revealed by multiple molecular genetic analyses. *Proc. Natl. Acad. Sci. USA* 93:8792–95.

Richard, K. R., S. W. McCarrey, and J. M. Wright. 1994. DNA sequence from the SRY gene of the sperm whale (*Physeter macrocephalus*) for use in molecular sexing. *Can. J. Zool.* 72:873–77.

Richard, K. R., H. Whitehead, and J. M. Wright. 1996b. Polymorphic microsatellites from sperm whales and their use in the genetic identification of individuals from naturally sloughed pieces of skin. *Mol. Ecol.* 5:313–15.

Richter, C. F. 2002. Sperm whales at Kaikoura and the effects of whale-watching on their surface and vocal behaviour. Ph.D. dissertation, Otago University, Dunedin, New Zealand.

Robson, F. D., and P. J. H. van Bree. 1971. Some remarks on a mass stranding of sperm whales, *Physeter macrocephalus,* near Gisborne, New Zealand, on March 18, 1970. *Z. Saugetierkd.* 36:55–60.

Roper, T. J. 1986. Cultural evolution of feeding behaviour in animals. *Sci. Prog.* 70: 571–83.

Ruckstuhl, K. E., and P. Neuhaus. 2000. Sexual segregation in ungulates: A new approach. *Behaviour* 55:361–77.

Rumbaugh, D. S., E. S. Savage-Rumbaugh, and D. A. Washburn. 1996. Toward a new outlook on primate learning and behavior: Complex learning and emergent processes in comparative perspective. *Jpn. Psychol. Res.* 38:113–25.

Samuels, A., and P. Tyack. 2000. Flukeprints: A history of studying cetacean societies. In *Cetacean societies,* edited by J. Mann, R. C. Connor, P. L. Tyack, and H. Whitehead, 9–44. Chicago: University of Chicago Press.

Sandell, M., and O. Liberg. 1992. Roamers and stayers: A model on male mating tactics and mating systems. *Am. Nat.* 139:177–89.

Sayigh, L. S., P. L. Tyack, and R. S. Wells. 1993. Recording underwater sounds of free-ranging dolphins while underway in a small boat. *Mar. Mammal Sci.* 9:209–13.

Scammon, C. M. 1874. *Marine mammals of the northwestern coast of North America.* San Francisco: John H. Carmany and Co.

Schevill, W. E. 1986. The International Code of Zoological Nomenclature and a paradigm: The name *Physeter catodon* Linnaeus 1758. *Mar. Mammal Sci.* 2:153–57.

———. 1987. [Reply to Holthuis 1987]. *Mar. Mammal Sci.* 3:89–90.

Scott, M. D., and W. L. Perryman. 1991. Using aerial photogrammetry to study dolphin school structure. In *Dolphin societies: Discoveries and puzzles,* edited by K. Pryor and K. S. Norris, 227–41. Berkeley: University of California Press.

Scott, T. M., and S. S. Sadove. 1997. Sperm whale, *Physeter macrocephalus,* sightings in the shallow shelf waters off Long Island, New York. *Mar. Mammal Sci.* 13: 317–21.

Seber, G. A. F. 1982. *The estimation of animal abundance and related parameters.* 2d ed. London: Griffin.

Seger, J., and T. Philippi. 1989. Hedging one's evolutionary bets, revisited. *Trends Ecol. Evol.* 4:41–44.

Sergeant, D. E. 1982. Mass strandings of toothed whales (Odontoceti) as a population phenomenon. *Sci. Rep. Whales Res. Inst.* 34:1–47.

Shane, S. H., R. S. Wells, and B. Würsig. 1986. Ecology, behavior and social organization of the bottlenose dolphin: A review. *Mar. Mammal Sci.* 2:34–63.

Shevchenko, V. I. 1975. Kharakter vzaimootnoshenii kasatok i drugikh kitoobraznykh. *Morsk Mlekopitayuschie Chas'* 2:173–74.

Shoshani, J. 1991. Origins and evolution. In *The illustrated encyclopedia of elephants,* edited by S. K. Eltringham, 12–29. London: Salamander Books.

Shuster, G. W. 1983. The Galápagos Islands: A preliminary study of the effects of sperm whaling on a specific whaling ground. *Rep. Int. Whal. Commn.* (Special Issue) 5: 81–82.

Siemann, L. A. 1994. Mitochondrial DNA sequence variation in North Atlantic long-finned pilot whales, *Globicephala melas.* Ph.D. dissertation, Massachusetts Institute of Technology, Cambridge, MA.

Similä, T., and F. Ugarte. 1993. Surface and underwater observations of cooperatively feeding killer whales in northern Norway. *Can. J. Zool.* 71:1494–99.

Simmonds, M. P. 1997. The meaning of cetacean strandings. *Bull. Inst. R. Sci. Nat. Belg. Biol.* 67 (Suppl.):29–34.

Simmonds, M. P., and J. D. Hutchinson. 1996. *The conservation of whales and dolphins: Science and practice.* Chichester, U.K.: Wiley.

Slater, P. J. B. 1986. The cultural transmission of bird song. *Trends Ecol. Evol.* 1:94–97.

Smeenk, C. 1997. Strandings of sperm whales *Physeter macrocephalus* in the North Sea: History and patterns. *Bull. Inst. R. Sci. Nat. Belgique, Biol., Suppl.* 67:12–28.

Smith, R. J. F. 1997. Avoiding and deterring predators. In *Behavioural ecology of teleost fishes,* edited by J.-G. J. Godin, 163–90. Oxford: Oxford University Press.

Smith, S. C., and H. Whitehead. 1993. Variations in the feeding success and behaviour of Galápagos sperm whales (*Physeter macrocephalus*) as they relate to oceanographic conditions. *Can. J. Zool.* 71:1991–96.

———. 2000. The diet of Galápagos sperm whales *Physeter macrocephalus* as indicated by fecal sample analysis. *Mar. Mammal Sci.* 16:315–25.

———. 2001. Reply to R. Clarke and Paliza's comment: "The food of sperm whales of the Southeast Pacific." *Mar. Mammal Sci.* 17:430–31.

Smith, T. 1980. Catches of male and female sperm whales by 2-degree square by Japanese pelagic whaling fleets in the North Pacific, 1966–77. *Rep. Int. Whal. Commn.* (Special Issue) 2:263–75.

Sokal, R. R., and F. J. Rohlf. 1981. *Biometry.* 2d ed. San Francisco: W. H. Freeman.

Soulé, M. E., and M. E. Gilpin. 1991. The theory of wildlife corridor capability. In *The role of corridors,* edited by D. A. Saunders and R. J. Hobbs, 3–8. Chipping Norton, U.K.: Surrey Beatty.

Southall, K. D., G. W. Oliver, J. W. Lewis, B. J. Le Boeuf, D. H. Levenson, and B. L. Southall. 2002. Visual pigment sensitivity in three deep diving marine mammals. *Mar. Mammal Sci.* 18:275–81.

Spitz, S. S., L. M. Herman, and A. A. Pack. 2000. Measuring sizes of humpback whales (*Megaptera novaeangliae*) by underwater videogrammetry. *Mar. Mammal Sci.* 16:664–76.

Stander, P. E. 1992. Cooperative hunting in lions: The role of the individual. *Behav. Ecol. Sociobiol.* 29:445–54.

Starbuck, A. 1878. History of the American whale fishery from its earliest inception to the year 1876. In *United States Commission on Fish and Fisheries, Report of the Commissioner for 1875–1876.* Washington, D.C.: Government Printing Office. Appendix A.

Stearns, S. C. 1992. *The evolution of life histories.* Oxford: Oxford University Press.

Stearns, S. C., and R. F. Hoekstra. 2000. *Evolution.* Oxford: Oxford University Press.

Steele, J. H. 1985. A comparison of terrestrial and marine ecological systems. *Nature* 313:355–58.

———. 1991. Can ecological theory cross the land-sea boundary? *J. Theor. Biol.* 153:425–36.

Stevick, P. T., B. J. McConnell, and P. S. Hammond. 2002. Patterns of movement. In *Marine mammal biology: An evolutionary approach,* edited by A. R. Hoelzel, 185–216. Oxford: Blackwell.

Stewart, B. S., and R. L. DeLong. 1995. Double migrations of the northern elephant seal, *Mirounga angustirostris. J. Mammal.* 76:196–205.

Sukumar, R. 1991. Ecology. In *The illustrated encyclopedia of elephants,* edited by S. K. Eltringham, 78–101. London: Salamander Books.

Sunnucks, P. 2000. Efficient genetic markers for population biology. *Trends Ecol. Evol.* 15:199–203.

Swingland, I. R., P. M. North, A. Dennis, and M. J. Parker. 1989. Movement patterns and morphometrics in giant tortoises. *J. Anim. Ecol.* 58:971–85.

Sydeman, W. J., and S. G. Allen. 1999. Pinniped population dynamics in central California: Correlations with sea surface temperature and upwelling indices. *Mar. Mammal Sci.* 15:446–61.

Tanabe, S., H. Iwata, and R. Tatsukawa. 1994. Global contamination by persistent organochlorines and their toxicological impact on marine mammals. *Sci. Total Environ.* 154:163–77.

Thode, A., D. Mellinger, S. Stiensessen, A. Martinez, and K. Mullin. 2002. Depth-dependent acoustic features of diving sperm whales (*Physeter macrocephalus*) in the Gulf of Mexico. *J. Acoust. Soc. Am.* 112:308–21.

Thouless, C. R. 1995. Long distance movements of elephants in northern Kenya. *Afr. J. Ecol.* 33:321–34.

Tiedemann, R., and M. Milinkovitch. 1999. Culture and genetic evolution in whales. *Science* 284:2055a.

Tillman, M. F., and J. M. Breiwick. 1983. Estimates of abundance for the western North Pacific sperm whale based upon historical whaling records. *Rep. Int. Whal. Commn.* (Special Issue) 5:257–69.

Tomasello, M. 1994. The question of chimpanzee culture. In *Chimpanzee cultures,* edited by R. W. Wrangham, W. C. McGrew, F. B. M. de Waal, and P. G. Heltne, 301–17. Cambridge, MA: Harvard University Press.

Tønnessen, J. N., and A. O. Johnsen. 1982. *The history of modern whaling.* Berkeley: University of California Press.

Townsend, C. H. 1935. The distribution of certain whales as shown by the logbook records of American whaleships. *Zoologica* 19:1–50.

Tucker, V. A. 1975. The cost of moving about. *Am. Sci.* 63:413–19.

Turchin, P. 1998. *Quantitative analysis of movement.* Sunderland, MA: Sinauer Associates.

Twiss, J. R., and R. R. Reeves. 1999. *Conservation and management of marine mammals.* Washington, D.C.: Smithsonian Institution Press.

Tyack, P. L. 1999. Communication and cognition. In *Biology of marine mammals,* edited by J. E. Reynolds and S. A. Rommel, 287–323. Washington, D.C.: Smithsonian Institution Press.

———. 2000. Functional aspects of cetacean communication. In *Cetacean societies,* edited by J. Mann, R. C. Connor, P. L. Tyack, and H. Whitehead, 270–307. Chicago: University of Chicago Press.

Underwood, R. 1981. Companion preference in an eland herd. *Afr. J. Ecol.* 19:341–54.

Van Schaik, C. P., and D. M. Kappeler. 1997. Infanticide risk and the evolution of male-female association in primates. *Proc. R. Soc. Lond., B* 264:1687–94.

Viale, D., N. Verneau, and Y. Tison. 1992. Stomach obstruction in a sperm whale beached on the Lavezzi islands: Macropollution in the Mediterranean. *J. Rech. Oceanogr.* 16:100–102.

Visser, I. N. 1999. A summary of interactions between orca (*Orcinus orca*) and other cetaceans in New Zealand waters. *NZ Nat. Sci.* 24:101–12.

Wada, S. 1980. On the genetic uniformity in the North Pacific sperm whale. *Rep. Int. Whal. Commn.* (Special Issue) 2:205–11.

Wade, P. R., and T. Gerrodette. 1993. Estimates of cetacean abundance and distribution in the eastern tropical Pacific. *Rep. Int. Whal. Commn.* 43:477–93.

Walther, G.-R., E. Post, P. Convey, A. Menzel, C. Parmesan, T. J. C. Beebee, J.-M. Fromentin, O. Hoegh-Guldberg, and F. Bairlein. 2002. Ecological responses to recent climate change. *Nature* 416:389–95.

Ware, D. M. 1975. Growth, metabolism, and optimum swimming speed of a pelagic fish. *J. Fish. Res. Board Can.* 32:33–41.

Waring, G. T., C. P. Fairfield, C. M. Rusham, and M. Sano. 1993. Sperm whales associated with Gulf Stream features off the north-eastern USA shelf. *Fish. Ocean.* 2:101–5.

Waring, G. T., T. Hamazaki, D. Sheehan, G. Wood, and S. Baker. 2001. Characterization of beaked whale (Ziphiidae) and sperm whale (*Physeter macrocephalus*) summer habitat in shelf-edge and deeper waters off the northeast U.S. *Mar. Mammal Sci.* 17:703–17.

Waters, S., and H. Whitehead. 1990a. Aerial behaviour in sperm whales, *Physeter macrocephalus. Can. J. Zool.* 68:2076–82.

———. 1990b. Population and growth parameters of Galápagos sperm whales estimated from length distributions. *Rep. Int. Whal. Commn.* 40:225–35.

Watkins, W. A. 1980. Acoustics and the behavior of sperm whales. In *Animal sonar systems,* edited by R. Busnel and J. F. Fish, 291–97. New York: Plenum Press.

Watkins, W. A., M. A. Daher, N. A. DiMarzio, A. Samuels, D. Wartzok, K. M. Fristrup, D. P. Gannon, P. W. Howey, and R. R. Maiefski. 1999. Sperm whale surface activity from tracking by radio and satellite tags. *Mar. Mammal Sci.* 15:1158–80.

Watkins, W. A., M. A. Daher, N. A. DiMarzio, A. Samuels, D. Wartzok, K. M. Fristrup, P. W. Howey, and R. R. Maiefski. 2002. Sperm whale dives tracked by radio tag telemetry. *Mar. Mammal Sci.* 18:55–68.

Watkins, W. A., M. A. Daher, K. M. Fristrup, T. J. Howald, and G. N. di Sciara. 1993. Sperm whales tagged with transponders and tracked underwater by sonar. *Mar. Mammal Sci.* 9:55–67.

Watkins, W. A., K. E. Moore, C. W. Clark, and M. E. Dahlheim. 1988. The sounds of sperm whale calves. In *Animal sonar,* edited by P. E. Nachtigall and P. W. B. Moore, 99–107. New York: Plenum Press.

Watkins, W. A., K. E. Moore, and P. Tyack. 1985. Sperm whale acoustic behaviors in the southeast Caribbean. *Cetology* 49:1–15.

Watkins, W. A., and W. E. Schevill. 1975. Sperm whales (*Physeter catodon*) react to pingers. *Deep-Sea Res.* 22:123–29.

———. 1977a. Spatial distribution of *Physeter catodon* (sperm whales) underwater. *Deep-Sea Res.* 24:693–99.

———. 1977b. Sperm whale codas. *J. Acoust. Soc. Am.* 62:1486–90.

Watkins, W. A., J. Sigurjónsson, D. Wartzok, R. Maiefski, P. W. Howey, and M. A. Daher. 1996. Fin whale tracked by satellite off Iceland. *Mar. Mammal Sci.* 12:564–69.

Watkins, W. A., and D. Wartzok. 1985. Sensory biophysics of marine mammals. *Mar. Mammal Sci.* 1:219–60.

Weilgart, L. S. 1985. Observations of a sperm whale birth. *Whales Etc.* I:3–5.

———. 1990. Vocalizations of the sperm whale (*Physeter macrocephalus*) off the Galápagos Islands as related to behavioral and circumstantial variables. Ph.D. dissertation, Dalhousie University, Halifax, Nova Scotia.

Weilgart, L. S., and H. Whitehead. 1986. Observations of a sperm whale (*Physeter catodon*) birth. *J. Mammal.* 67:399–401.

———. 1988. Distinctive vocalizations from mature male sperm whales (*Physeter macrocephalus*). *Can. J. Zool.* 66:1931–37.

———. 1993. Coda vocalizations in sperm whales (*Physeter macrocephalus*) off the Galapagos Islands. *Can. J. Zool.* 71:744–52.

———. 1997. Group-specific dialects and geographical variation in coda repertoire in South Pacific sperm whales. *Behav. Ecol. Sociobiol.* 40:277–85.

Weilgart, L. S., H. Whitehead, and K. Payne. 1996. A colossal convergence. *Am. Sci.* 84:278–87.

Weller, D. W., B. Würsig, S. K. Lynn, and A. J. Schiro. 2000. Preliminary findings on the occurrence and site fidelity of photoidentified sperm whales (*Physeter macrocephalus*) in the northern Gulf of Mexico. *Gulf Mex. Sci.* 18:35–39.

Weller, D. W., B. Würsig, H. Whitehead, J. C. Norris, S. K. Lynn, R. W. Davis, N. Clauss, and P. Brown. 1996. Observations of an interaction between sperm whales and short-finned pilot whales in the Gulf of Mexico. *Mar. Mammal Sci.* 12:588–94.

White, G. C., and K. P. Burnham. 1999. Program MARK: Survival estimation from populations of marked animals. *Bird Study (Supplement)* 46:120–38.

Whitehead, H. 1983. Structure and stability of humpback whale groups off Newfoundland. *Can. J. Zool.* 61:1391–97.

———. 1985a. Humpback whale breaching. *Invest. Cetacea* 17:117–55.

———. 1985b. Why whales leap. *Sci. Am.* 252:84–93.

———. 1986. Call me gentle. *Nat. Hist.* 95(6):4–11.

———. 1987. Social organization of sperm whales off the Galapagos: Implications for management and conservation. *Rep. Int. Whal. Commn.* 37:195–99.

———. 1989a. Formations of foraging sperm whales, *Physeter macrocephalus,* off the Galápagos Islands. *Can. J. Zool.* 67:2131–39.

———. 1989b. *Voyage to the whales.* Toronto: Stoddart.

———. 1990a. Computer assisted individual identification of sperm whale flukes. *Rep. Int. Whal. Commn.* (Special Issue) 12:71–77.

———. 1990b. Rules for roving males. *J. Theor. Biol.* 145:355–68.

———. 1993. The behaviour of mature male sperm whales on the Galapagos breeding grounds. *Can. J. Zool.* 71:689–99.

———. 1994. Delayed competitive breeding in roving males. *J. Theor. Biol.* 166:127–33.

———. 1995a. Investigating structure and temporal scale in social organizations using identified individuals. *Behav. Ecol.* 6:199–208.

———. 1995b. Status of Pacific sperm whale stocks before modern whaling. *Rep. Int. Whal. Commn.* 45:407–12.

———. 1996a. Babysitting, dive synchrony, and indications of alloparental care in sperm whales. *Behav. Ecol. Sociobiol.* 38:237–44.

———. 1996b. Variation in the feeding success of sperm whales: Temporal scale, spatial scale and relationship to migrations. *J. Anim. Ecol.* 65:429–38.

———. 1997a. Analyzing animal social structure. *Anim. Behav.* 53:1053–67.

———. 1997b. Sea surface temperature and the abundance of sperm whale calves off the Galápagos Islands: Implications for the effects of global warming. *Rep. Int. Whal. Commn.* 47:941–44.

———. 1998. Cultural selection and genetic diversity in matrilineal whales. *Science* 282:1708–11.

———. 1999a. Culture and genetic evolution in whales. *Science* 284:2055a.

———. 1999b. Testing association patterns of social animals. *Anim. Behav.* 57:F26–29.

———. 1999c. Variation in the visually observable behavior of groups of Galápagos sperm whales. *Mar. Mammal Sci.* 15:1181–97.

———. 2000. Density-dependent habitat selection and the modeling of sperm whale (*Physeter macrocephalus*) exploitation. *Can. J. Fish. Aquat. Sci.* 57:223–30.

———. 2001a. Analysis of animal movement using opportunistic individual-identifications: Application to sperm whales. *Ecology* 82:1417–32.

———. 2001b. Direct estimation of within-group heterogeneity in photoidentification of sperm whales. *Mar. Mammal Sci.* 17:718–28.

———. 2002. Estimates of the current global population size and historical trajectory for sperm whales. *Mar. Ecol. Prog. Ser.* 242:295–304.

Whitehead, H. 2003. Society and culture in the deep and open ocean: The sperm whale and other cetaceans. In *Animal social complexity: Intelligence, culture, and individualized societies,* edited by F. B. M. de Waal and P. L. Tyack, 444–64. Cambridge, MA: Harvard University Press.

Whitehead, H., and T. Arnbom. 1987. Social organization of sperm whales off the Galápagos Islands, February–April 1985. *Can. J. Zool.* 65:913–19.

Whitehead, H., S. Brennan, and D. Grover. 1992. Distribution and behaviour of male sperm whales on the Scotian Shelf, Canada. *Can. J. Zool.* 70:912–18.

Whitehead, H., J. Christal, and S. Dufault. 1997. Past and distant whaling and the rapid decline of sperm whales off the Galápagos Islands. *Conserv. Biol.* 11:1387–96.

Whitehead, H., J. Christal, and P. L. Tyack. 2000a. Studying cetacean social structure in space and time: Innovative techniques. In *Cetacean societies,* edited by J. Mann, R. C. Connor, P. L. Tyack, and H. Whitehead, 65–87. Chicago: University of Chicago Press.

Whitehead, H., M. Dillon, S. Dufault, L. Weilgart, and J. Wright. 1998. Non-geographically based population structure of South Pacific sperm whales: Dialects, fluke-markings and genetics. *J. Anim. Ecol.* 67:253–62.

Whitehead, H., and J. Gordon. 1986. Methods of obtaining data for assessing and modelling sperm whale populations which do not depend on catches. *Rep. Int. Whal. Commn.* (Special Issue) 8:149–66.

Whitehead, H., J. Gordon, E. A. Mathews, and K. R. Richard. 1990. Obtaining skin samples from living sperm whales. *Mar. Mammal Sci.* 6:316–26.

Whitehead, H., and N. Jaquet. 1996. Are the charts of Maury and Townsend good indicators of sperm whale distribution and seasonality? *Rep. Int. Whal. Commn.* 46:643–47.

Whitehead, H., and B. Kahn. 1992. Temporal and geographical variation in the social structure of female sperm whales. *Can. J. Zool.* 70:2145–49.

Whitehead, H., C. D. MacLeod, and P. Rodhouse. 2003. Differences in niche breadth among some teuthivorous mesopelagic marine mammals. *Mar. Mammal Sci.* In press.

Whitehead, H., and J. Mann. 2000. Female reproductive strategies of cetaceans. In *Cetacean societies,* edited by J. Mann, R. Connor, P. L. Tyack, and H. Whitehead, 219–46. Chicago: University of Chicago Press.

Whitehead, H., V. Papastavrou, and S. C. Smith. 1989. Feeding success of sperm whales and sea-surface temperatures off the Galápagos Islands. *Mar. Ecol. Prog. Ser.* 53:201–3.

Whitehead, H., R. R. Reeves, and P. L. Tyack. 2000b. Science and the conservation, protection, and management of wild cetaceans. In *Cetacean societies,* edited by J. Mann, R. C. Connor, P. L. Tyack, and H. Whitehead, 308–32. Chicago: University of Chicago Press.

Whitehead, H., and S. Waters. 1990. Social organisation and population structure of sperm whales off the Galápagos Islands, Ecuador (1985 and 1987). *Rep. Int. Whal. Commn.* (Special Issue) 12:249–57.

Whitehead, H., S. Waters, and T. Lyrholm. 1991. Social organization in female sperm whales and their offspring: Constant companions and casual acquaintances. *Behav. Ecol. Sociobiol.* 29:385–89.

Whitehead, H., and L. Weilgart. 1990. Click rates from sperm whales. *J. Acoust. Soc. Am.* 87:1798–1806.

———. 1991. Patterns of visually observable behaviour and vocalizations in groups of female sperm whales. *Behaviour* 118:275–96.

———. 2000. The sperm whale: Social females and roving males. In *Cetacean societies,* edited by J. Mann, R. C. Connor, P. Tyack, and H. Whitehead, 154–72. Chicago: University of Chicago Press.

Whiten, A. 2001. Imitation and cultural transmission in apes and cetaceans. *Behav. Brain Sci.* 24:359–60.

Whiten, A., J. Goodall, W. C. McGrew, T. Nishida, V. Reynolds, Y. Sugiyama, C. E. G. Tutin, R. W. Wrangham, and C. Boesch. 1999. Cultures in chimpanzees. *Nature* 399:682–85.

Whiten, A., and R. Ham. 1992. On the nature and evolution of imitation in the animal kingdom: Reappraisal of a century of research. *Advances in the Study of Behavior* 21:239–83.

Widder, E. A., D. F. Bernstein, D. F. Bracher, J. F. Case, K. R. Reisenbichler, J. J. Torres, and B. H. Robison. 1989. Bioluminescence in the Monterrey sub-marine canyon: Image analysis of video recordings from a midwater submersible. *Mar. Biol.* 100:541–51.

Williams, T. M. 1999. The evolution of cost efficient swimming in marine mammals: Limits to energetic optimization. *Phil. Trans. R. Soc. Lond., B* 354:193–201.

Wilson, E. O. 1975. *Sociobiology: The new synthesis.* Cambridge, MA: Belknap Press.

Wolff, J. O. 1997. Population regulation in mammals: An evolutionary perspective. *J. Anim. Ecol.* 66:1–13.

Woodroffe, R., and A. Vincent. 1994. Mother's little helpers: Patterns of male care in mammals. *Trends Ecol. Evol.* 9:294–97.

Worthington, L. V., and W. E. Schevill. 1957. Underwater sounds heard from sperm whales. *Nature* 180:291.

Worthy, G. A. J., and J. P. Hickie. 1986. Relative brain size in marine mammals. *Am. Nat.* 128:445–59.

Wrangham, R. W. 1977. Feeding behaviour of chimpanzees in Gombe National Park, Tanzania. In *Primate ecology,* edited by T. H. Clutton-Brock, 504–38. New York: Academic Press.

————. 1980. An ecological model of female-bonded primate groups. *Behaviour* 75:262–300.

Wrangham, R. W., and D. I. Rubenstein. 1986. Social evolution in birds and mammals. In *Ecological aspects of social evolution,* edited by D. I. Rubenstein and R. W. Wrangham, 452–70. Princeton, NJ: Princeton University Press.

Wray, P., and K. R. Martin. 1983. Historical whaling records from the western Indian Ocean. *Rep. Int. Whal. Commn.* (Special Issue) 5:218–42.

Würsig, B. 1986. Delphinid foraging strategies. In *Dolphin cognition and behavior: A comparative approach,* edited by R. J. Schusterman, J. A. Thomas, and F. G. Wood, 347–59. Hillsdale, NJ: Lawrence Erlbaum Associates.

Würsig, B., E. M. Dorsey, M. A. Fraker, R. S. Payne, and W. J. Richardson. 1985. Behavior of bowhead whales, *Balaena mysticetus,* summering in the Beaufort Sea: A description. *Fish. Bull., US* 83:357–77.

Würsig, B., E. M. Dorsey, W. J. Richardson, and R. S. Wells. 1989. Feeding, aerial and play behaviour of the bowhead whale, *Balaena mysticetus,* summering in the Beaufort Sea. *Aquat. Mamm.* 15:27–37.

Würsig, B., and M. Würsig. 1980. Behavior and ecology of the dusky dolphin, *Lagenorhynchus obscurus,* in the South Atlantic. *Fish. Bull., US* 77:871–90.

Yablokov, A. V. 1994. Validity of whaling data. *Nature* 367:108.

Yukhov, V. L., E. K. Vinogradova, and L. P. Medvedev. 1975. Ob'ekty pitaniya kosatok (*Orcinus orca* L.) v Antarktike i sopredel'nykh vodakh. *Morsk Mlekopitayuschie Chas'* 2:183–85.

Zahavi, A. 1977. The testing of a bond. *Anim. Behav.* 25:246–47.

Zemsky, V. A., A. A. Berzin, Y. A. Mikhaliev, and D. D. Tormosov. 1995. Soviet Antarctic pelagic whaling after WWII: Review of actual catch data. *Rep. Int. Whal. Commn.* 45:131–35.

Zenkovich, B. A. 1960. Sea mammals as observed by the round-the-world expedition of the Academy of Sciences of the USSR in 1957–58. *Norsk Hvalfangsttid.* 51:198–210.

Index